T0332377

Traffic and Performance Engineering for Heterogeneous Networks

RIVER PUBLISHERS SERIES IN INFORMATION SCIENCE AND TECHNOLOGY

Volume 1

Consulting Series Editor

KWANG-CHENG CHEN
National Taiwan University
Taiwan

Traffic and Performance Engineering for Heterogeneous Networks

Editor

Demetres D. Kouvatsos

PERFORM – Networks & Performance Engineering Research Unit, University of Bradford, U.K.

Routledge
Taylor & Francis Group

LONDON AND NEW YORK

Published 2009 by River Publishers
River Publishers
Alsbjergvej 10, 9260 Gistrup, Denmark
www.riverpublishers.com

Distributed exclusively by Routledge
4 Park Square, Milton Park, Abingdon, Oxon OX14 4RN
605 Third Avenue, New York, NY 10017, USA

Traffic and Performance Engineering for Heterogeneous Networks / by
Demetres D. Kouvatsos.

Routledge is an imprint of the Taylor & Francis Group, an informa
business

ISBN 978-87-92329-16-5 (print)

While every effort is made to provide dependable information, the
publisher, authors, and editors cannot be held responsible for any errors
or omissions.

To my parent's memory

Table of Contents

Preface

The diversity of methodologies and applications in the literature of traffic and performance engineering for convergent multiservice heterogeneous networks, supported by internetworking and the evolution of diverse access and switching technologies, attests to the breath and richness of recent research and developments towards the design and dimensioning of the next and future generation Internets. However, there are still challenging traffic and performance engineering issues to be resolved before the establishment of a global and wide-scale integrated broadband network infrastructure for the efficient support of multimedia applications.

Traffic and Performance Engineering for Heterogeneous Networks presents recent advances in networks of diverse technology reflecting state-of-the-art technology and research achievements in traffic engineering including traffic modelling and characterisation as well as performance engineering featuring performance evaluation studies and tools for heterogeneous networks. The book includes 20 extended/revised research papers, which have their roots in the series of the HET-NETs International Working Conferences on 'Performance Modelling and Evaluation of Heterogeneous Networks'. These events were staged under the auspices of the EU Networks of Excellence (NoE) Euro-NGI and Euro-FGI and are associated with the NoE Work-packages WP.SEA.6.1 and WP.SEA.6.3, respectively.

The research papers are classified into seven technical parts dealing with current research themes in 'Traffic Modelling and Engineering', 'Queueing and Interconnection Networks', 'Performance Evaluation Studies', 'TCP Performance Engineering', 'Congestion Control', 'Application Layer Multicast' and 'Numerical and Software Tools'.

In *Part One* 'Traffic Modelling and Engineering', Nogueira et al. propose a hierarchical approach for the characterisation of a Markov Modulated Poisson Process (MMPP) over multiple time scales for the modelling of Internet

Protocol (IP) network traffic, which may exhibit burstiness, self-similarity and long-range dependence with significant impact on network performance. Markovich and Krieger adopt recurrent marked point processes as analytic estimation techniques for the statistical characterisation of Internet traffic measurements, which often poses heavy tail characteristics (e.g., real Web-data). Erman et al. report traffic measurements and analytic modelling studies of session characteristics of Bit Torrent traffic and advocate the use of the hyper-exponential and lognormal distributions for modelling, respectively, session inter-arrivals, session durations and session sizes. Susitaival and Aalto study adaptive load balancing in Open Shortest Path First (OSPF) networks and, based on measured link loads, propose a new adaptive and distributed algorithm for gradual load balance. Kouremenos et al. devise a generic method, which is validated via analytic and simulation tests, to model constrained and unconstrained video conferencing traffic from variable bit rate (VBR) video encoders over IP networks.

In *Part Two* 'Queueing and Interconnection Networks', Kempken et al. describe a newly developed software Toolkit for the analysis of a discrete-time pen queueing network consisting of time-slotted queues with renewal arrival processes. Both the transient and steady state behaviour are studied in terms of the buffer occupancy at the network nodes and the loss probabilities due to overflow events. Wu et al. investigate performance degradation in interconnection networks under bursty traffic with non-uniformly distributed message destinations and develop a new analytic model for the computation of the mean message latency in hypercubic networks under bursty traffic with hot-spot destinations.

In *Part Three* 'Performance Evaluation Studies', Maertens et al. derive new closed form expressions of the probability generating function of the system contents and the cell delay associated with a discrete-time single-server system with two priority queues with infinite capacity, subject to the head-of-the-line with priority jumps (HOL-PJ) scheduling discipline. Liotopoulos carries out a simulation study of an input-buffered service system represented by a queue with 'Retreat FIFO' service discipline and deals with vital performance degradation issues associated with head-of-line (HoL) blocking, especially under 'hot-spot' traffic conditions. Fiuk and Czachórski present a model of a networked control system, based on the UPnP device communication control protocols (which are built on top of SOAP and XML standards and the HTTP protocol) and develop an efficient analytic algorithm for the investigation of system's performance.

In *Part Four* 'TCP Performance Engineering', El Khoury et al. investigate the behaviour of scalable transport control protocol (TCP) sharing a bottleneck link with constant bit rate (CBR) exogenous non-controlled traffic. A fluid approximation model is proposed and associated conditions are determined, under which the window evolution of scalable TCP behaves as an additive increase multiplicative decrease (AIMD) paradigm. Kooij et al. identify the problem of IP link dimensioning and extend the commonly used TCP performance model for download times by including the impact of the bi-directional packet loss and the correlations between loss occurrences. Avramova et al. develop heuristic formulae for the dimensioning of buffers in a Data Connection Ltd (DSL) access network carrying persistent TCP-controlled traffic, subject to Under Drop-tail and Random Early Detection (RED) buffer acceptance disciplines. Brachman and Chrost address the problem of throughput degradation of TCP with Non-Congestion Robustness (NCR) protocol during congestion events in wireless-cumwired networks with a bottleneck link. An effective scheme is devised with a new functionality-enabling packet marking, based on the concepts of explicit congestion notification (ECN) and 'A Novel-Active Queue Management (AN-AQM)' algorithm. Chodorek and Chodorek carry out burst control simulation analysis, as a TCP-tolerance mechanism, which spreads out burst over time whilst the source's bit rate stays unchanged and consider QoS applications involving real-time transport protocols for multimedia traffic.

In *Part Five* 'Congestion Control', Almeida and Belo[†] focus on the absence of native support from the current Internet for explicit-rate algorithms and suggest a novel approach for an efficient congestion control scheme based on binary probabilistic marking and enhanced router algorithms, originally devised for explicit-rate marking. Altman et al. focus on active queue management (AQM) techniques for developing routers in the Internet and devise a singular perturbation-based approach to analyse a Random Early Detection (RED) enabled queue evolving in two different time scales.

In *Part Six* 'Application Layer Multicast', Rios et al. address multimedia conferencing issues in the context of collaborative Internet applications and propose a flexible and scalable Peer-to-Peer (P2P) prototype video conferencing system based on a hybrid mechanism that uses an Application Layer Multicast (ALM) algorithm and a standard Session Initiation Protocol (SIP) messaging.

In *Part Seven* 'Numerical and Software Tools', Thornley et al. suggest a robust and automated formulation tool, which generates systems of transformed Kolmogorov balance equations of finite range, for the efficient solution of finite or infinite capacity queues at equilibrium with positive and negative customers, subject to geometrically batched Poisson arrival processes under Markov modulation. Finally, Garcia and Hackbarth describe the design and development of a web portal, as a common web interface for user access, towards the integration of a large set of software tools supporting network planning and simulation, performance measurements, statistical analysis and test beds. The work is of relevance to design and engineering of the next and future generation wireless Internets under the auspices of the EU Networks of Excellence Euro-NGI and Euro-FGI.

I would like to end this preface by expressing my thanks to the EU Networks of Excellence Euro-NGI and Euro-FGI for sponsoring in part the publication of this research book and to the members of HET-NETs Advisory Boards and Programme Committees as well as the external referees worldwide for their invaluable and timely reviews. My thanks are also due to Professor Ramjee Prasad, Director of the Center for TeleInFrastruktur (CTIF), Aalborg University, Denmark, for his encouragement and valuable advice during the preparation of this book.

Demetres D. Kouvatsos

Participants in the Peer Review Process

Samuli Aalto
Ramon Agusti
Sohair Al-Hakeem
Eitan Altman
Jorge Andres
Vladimir Anisimov
Laura Aspirot
Salam Adli Assi
Tulin Atmaca
Zlatka Avramova
Frank Ball
Simonetta Balsamo
Ivano Bartoli
Alejandro Beccera
Monique Becker
Pablo Belzarena
Andre-Luc Beylot
Andreas Binzenhoefer
Jozsef Biro
Pavel Bocharov
Sem Borst
Nizar Bouabdallah
Richard Boucherie
Christos Bouras
Onno Boxma
Chris Blondia
Alexandre Brandwajn
George Bravos

Oliver Brun
Herwig Bruneel
Alberto Cabellos-Aparicio
Patrik Carlsson
Fernando Casadevall
Vicente Casares-Giner
Hind Castel
Llorenc Cerda
Eduardo Cerqueira
Mohamad Chaitou
Ram Chakka
Meng Chen
Stefan Chevul
Tom Coenen
Doru Constantinescu
Marco Conti
Laurie Cuthbert
Tadeusz Czachorski
Koen De Turck
Danny De Vleeschauwer
Stijn De Vuyst
Alexandre Delye Mazieux
Luc Deneire
Felicita Di Giandomenico
Manuel Dinis
Tien Do
Jose Domenech-Benlloch
Rudra Dutta

Joerg Eberspaecher
Antonio Elizondo
Khaled Elsayed
Peder Emstad
David Erman
Melike Erol
Jose Oscar Fajardo Portillo
Fatima Ferreira
Markus Fiedler
Jean-Michel Fourneau
Rod Fretwell
Wilfried Gangsterer
Peixia Gao
Ana Garcia Armada
David Garcia-Roger
Georgios Gardikis
Vincent Gauthier
Alfonso Gazo
Xavier Gelabert Doran
Leonidas Georgiadis
Bart Gijsen
Jose Gil
Cajigas Gillermo
Stefano Giordano
Jose Gonzales
Ruben Gonzalez Benitez
Annie Gravey
Klaus Hackbarth
Slawomir Hanczewski
Guenter Haring
Peter Harrison
Hassan Hassan
Dan He
Gerard Hebuterne
Bjarne Helvik
Robert Hines
Enrique Hernandez
Helmut Hlavacs
Amine Houyou

Hanen Idoudi
Ilias Iliadis
Dragos Ilie
Paola Iovanna
Andrzej Jajszczyk
Lorand Jakab
Sztrik Janos
Robert Janowski
Terje Jensen
Laszlo Jereb
Mikael Johansson
Hector Julian-Bertomeu
Athanassios Kanatas
Tamas Karasz
Johan Karlsson
Stefan Koehler
Daniel Kofman
Vangellis Kollias
Huifang Kong
Kimon Kontovasilis
Rob Kooij
Goerge Kormentzas
Ivan Kotuliak
Harilaos Koumaras
Demetres Kouvatsos
Tasos Kourtis
Udo Krieger
Koenraad Laevens
Samer Lahoud
Jaakko Lahteenmaki
Juha Leppanen
Amaia Lesta
Hanoch Levy
Wei Li
Yue Li
Fotis Liotopoulos
Renato Lo Cigno
Michael Logothetis
Carlos Lopes

Johann Lopez
Andreas Maeder
Tom Maertens
Thomas Magedanz
Sireen Malik
Lefteris Mamatas
Jose Manuel Gimenez-Guzman
Michel Marot
Alberto Martin
Jim Martin
Simon Martin
Ignacio Martinez Arrue
Jose Martinez-Bauset
Martinecz Matyas
Lewis McKenzie
Madjid Merabti
Bernard Metzler
Geyong Min
Isi Mitrani
Nicholas Mitrou
Is-Haka Mkwawa
Hala Mokhtar
Miklos Boda
Sandor Molnar
Edmundo Monteiro
Ioannis Moscholios
Harry Mouchos
Luis Munoz
Maurizio Naldi
Victor Netes
Pal Nilsson
Simon Oechsner
Sema Oktug
Mohamed Ould-Khaoua
Antonio Pacheco
Michele Pagano
Zsolt Pandi
Panagiotis Papadimitriou
Stylianos Papanastasiou

Nihal Pekergin
Izaskun Pellejero
Roger Peplow
Paulo Pereira
Gonzalo Perera
Jordi Perez-Romero
Rubem Perreira
Guido Petit
Maciej Piechowiak
Michal Pioro
Jonathan Pitts
Vicent Pla
Nineta Polemi
Daniel Popa
Adrian Popescu
Dimitris Primpas
David Remondo-Bueno
David Rincon
Roberto Sabella
Francisco Salguero
Sebastia Sallent
Werner Sandmann
Ana Sanjuan
Lambros Sarakis
Wolfgang Schott
Raffaello Secchi
Maria Simon
Swati Sinha Deb
Charalabos Skianis
Amaro Sousa
Dirk Staehle
Maciej Stasiak
Panagiotis Stathopoulos
Bart Steyaert
Zhili Sun
Kannan Sundaramoorthy
Riikka Susitaival
Janos Sztrik
Yutaka Takahashi

Sotiris Tantos
Leandros Tassiulas
Luca Tavanti
Silvia Terrasa
Geraldine Texier
David Thornley
Florence Touvet
Phuoc Tran-Gia
Chia-Sheng Tsai
Thanasis Tsokanos
Krzysztof Tworus
Rui Valadas
Rob Van der Mei
Vassilios Vassilakis
Vasos Vassiliou
Sandrine Vaton
Tereza Vazao
Speros Velentzas

Dominique Verchere
Pablo Vidales
Nguyen Viet Hung
Manolo Villen-Altamirano
Bart Vinck
Jorma Virtamo
Kostas Vlahodimitropoulos
Joris Walraevens
Xin Gang Wang
Wemke Weij
Sabine Wittevrongel
Mehti Witwit
Michael Woodward
George Xilouris
Mohammad Yaghmaee
Bo Zhou
Stefan Zoels
Piotr Zwierzykowski

PART ONE
TRAFFIC MODELLING AND ENGINEERING

1

A Hierarchical Approach Based on MMPPs for Modeling Self-Similar Traffic over Multiple Time Scales

António Nogueira[1], Paulo Salvador[1], Rui Valadas[1]
and António Pacheco[2]

[1]*Institute of Telecommunications Aveiro, University of Aveiro, Campus de Santiago, 3810-193 Aveiro, Portugal; e-mail: {nogueira,salvador,rv}@ua.pt*
[2]*Department of Mathematics and CEMAT, Instituto Superior Técnico – UTL, Av. Rovisco Pais, 1049-001 Lisboa, Portugal; e-mail: apacheco@math.ist.utl.pt*

Abstract

Efficient traffic engineering of IP networks requires the knowledge of the main characteristics of the supported traffic. Several studies have shown that IP network traffic may exhibit properties of burstiness, self-similarity and/or long-range dependence, with significant impact on network performance. In this work, we propose two Markov Modulated Poisson Processes (MMPP), and their associated parameter fitting procedures, that are able to incorporate these characteristics over multiple time scales. The fitting procedure of the first model matches the complete distribution of the arrival process at each time scale of interest. The second proposed model is constructed using a hierarchical procedure that, starting from a MMPP that matches the distribution of packet counts at the coarsest time scale, successively decomposes each MMPP state into new MMPPs that incorporate a more detailed description of the distribution at finner time scales. The traffic process is then represented by a MMPP equivalent to the constructed hierarchical structure. The accuracy of the fitting procedures is evaluated by comparing the Hurst parameter, the probability mass function at each time scale and the queuing behavior (as as-

D. D. Kouvatsos (ed.), Traffic and Performance Engineering for Heterogeneous Networks, 3–40.

sessed by the loss probability and average waiting time), corresponding to the measured and to synthetic traces generated from the inferred models. Several measured traffic traces exhibiting self-similar behavior are considered: the well-known pOct Bellcore trace, a trace of aggregated IP WAN traffic, and a trace corresponding to a peer-to-peer file sharing application. Our results show that the proposed models and parameter fitting procedures are very effective in matching the main characteristics of the measured traces over the different time scales present in data and their performances are quite similar.

Keywords: Traffic modeling, self-similar, time scale, Markov Modulated Poisson Process.

1.1 Introduction

Traffic characterization and modeling comprise important steps towards understanding and solving performance-related problems in future IP networks. An efficient design and control of IP networks needs to take into account the main characteristics of the supported traffic, and therefore accurate and detailed measurements need to be carried out. Traffic modeling refers to the construction of (usually stochastic) models that capture the most important statistical properties of the measured data. Since the work by Leland et al. [1] several studies have shown that network traffic may exhibit properties of burstiness, self-similarity and/or long-range dependence (LRD) [1–7], which have significant impact on network performance.

Burstiness is a traffic behavior showing noticeable periods with arrivals above the mean (bursts) and self-similarity refers to the replication of statistical characteristics over a wide range of time scales. Models like the fractional Gaussian noise (fGN) and the fractional autoregressive integrated moving average (fARIMA) have been proposed to capture burstiness and self-similarity but there is still a lack of analytical results, e.g., to assess queuing behavior.

In general, self-similarity implies LRD, and vice-versa. The impact of LRD on network performance has been addressed by several authors. In [4, 8–10], for example, the case of a single queue is studied and it is concluded that the buffer occupancy is not affected by autocovariance lags that are beyond the so-called critical time scale (CTS) or correlation horizon (CH), which depends on system parameters such as the buffer capacity. Similar conclusions were observed for the case of tandem queues in [11]. Thus, matching the LRD is only required within the time scales specific to the system under study. One of the consequences of this result is that more traditional traffic models, such as Markov Modulated Poisson Processes (MMPPs), can still be used to model traffic exhibiting LRD. Moreover, the use of MMPPs benefits from the existence of several tools for calculating the queuing behavior and the effective bandwidths.

In this work, we consider discrete-time MMMPs (dMMPPs) instead of continuous-time MMPPs, since they are a more natural model for data corresponding to the number of arrivals (packet counts) in a sampling interval. Note that discrete-time and continuous-time MMPPs are basically interchangeable (through a simple parameter rescaling) as models for arrival processes, whenever the sampling interval used for the discrete-time version is small compared with the average sojourn times in the states of the modulating Markov chain.

In this paper we propose two dMMPP traffic models, and their associated parameter fitting procedures, that are able to incorporate traffic characteristics of different time scales. The first approach proposes a parameter fitting procedure for a superposition of Markov Modulated Poisson Processes (MMPPs), which captures self-similar behavior over a range of time scales. Each MMPP models a specific time scale and the parameter fitting procedure matches, at each time scale, a MMPP to a Probability Mass Function (PMF) that describes the contribution of that time scale to the overall traffic behavior. The number of states of each MMPP is not fixed a priori; it is determined as part of the fitting procedure. This model will be designated by superposition model. In the second model, the construction procedure successively decomposes dMMPP states into new dMMPPs, thus refining the traffic process by incorporating the characteristics offered by finer time scales. We start at the largest time scale by inferring a dMMPP that matches the PMF of this time scale. At the next finer time scale, each dMMPP state is decomposed into a new dMMPP that matches the contribution of this time scale to the PMF of the state it descends from. In this way, a child dMMPP provides a detailed description of its parent state PMF. This refinement process is iterated until a pre-defined number of time scales are integrated. Finally, a dMMPP incorporating this hierarchical structure is derived. Similarly to the previous model, the number of states of each dMMPP is not fixed a priori; it is determined as part of the fitting procedure. This second model will be designated by decomposition model. The accuracy of both fitting procedures is evaluated by applying them to several measured traffic traces that exhibit self-similar behavior: the well-known pOct Bellcore trace, a trace of aggregated IP WAN traffic, and a trace corresponding to a file sharing application (Kazaa). We compare the PMF at each time scale, and the queuing behavior (as assessed by the loss probability and average waiting time), corresponding to the measured and to synthetic traces generated from the inferred models. Our results show that the proposed fitting methods are very effective in matching the PMF at the various time scales and lead to an accurate prediction of the queuing behavior.

When measuring network traffic data, we can either record the individual arrival instants or the number of arrivals in a predefined sampling (time) interval. The former approach brings more detail but the latter one has the advantage of producing a fixed amount of data that is known in advance. This allows the recording of longer traces, which clearly pays off the loss of detail in data recording, if the sampling interval is chosen appropriately.

Several fitting procedures have been proposed in the literature for estimating the parameters of MMPPs from empirical data ([12–21], among others). However, most procedures only apply to 2-MMPPs (e.g. [12,14,15,18]). This model can capture traffic burstiness but the number of states is not enough to reproduce variability over a wide range of time scales. On the other hand, the fitting procedures for MMPPs with an arbitrary number of states mainly concentrate on matching first- and/or second-order statistics, without addressing directly the issue of modeling over multiple time scales [13, 16, 17, 19, 21]. Yoshihara et al. [20] developed a fitting method for self-similar traffic based on the superposition of 2-MMPPs, that matches the variance at each time scale. In this way, the resulting MMPP reproduces the variance-scale curve characteristic of self-similar processes. In this paper, we present two procedures that match the complete distribution of the packet counts at each time scale (and not only the variance) in order to accurately reproduce self-similar behavior. Besides, the concept of time scale is not directly related to second-order statistics; instead it refers to the characterization of the traffic process when aggregated over a number of time intervals. Thus, the proposed fitting methods address the scaling paradigm in a more natural way than the one proposed in [21], achieving even higher accuracy in the fitting of self-similar traffic, although the simple superposition approach usually includes a higher number of parameters in the resulting MMPP.

The paper is organized as follows. Section 1.2 introduces self-similarity and long-range dependence, motivating the need for a traffic model that matches the different time scales of the data. Section 1.3 gives the required background on MMPPs. Section 1.4 describes the proposed models and Section 1.5 presents the various steps of the parameter fitting procedures. Section 1.6 briefly describes the data traces used in the numerical evaluation and in Section 1.7 we discuss the obtained results. Finally, Section 1.8 presents the main conclusions.

1.2 Self-Similarity, Long-Range Dependence, and Time Scales

Consider the continuous-time process $Y(t)$ representing the traffic volume (e.g. in bytes) from time 0 up to time t and let $X(t) = Y(t) - Y(t-1)$ be the corresponding increment process (e.g. in bytes/second). Consider also the sequence $X^{(m)}(k)$ which is obtained by averaging $X(t)$ over non-overlapping blocks of length m, that is

$$X^{(m)}(k) = \frac{1}{m} \sum_{i=1}^{m} X((k-1)m + i), \quad k = 1, 2, \ldots \tag{1.1}$$

The fitting procedure developed in this work will be based on the aggregated processes $X^{(m)}(k)$.

We start by introducing the notion of distributional self-similarity. $Y(t)$ is exactly self-similar when it is equivalent, in the sense of finite-dimensional distributions, to $a^{-H} Y(at)$, for all $t > 0$ and $a > 0$, where H $(0 < H < 1)$ is the Hurst parameter. Clearly, the process $Y(t)$ can not be stationary. However, if $Y(t)$ has stationary increments then again $X(k) = X^{(1)}(k)$ is equivalent, in the sense of finite-dimensional distributions, to $m^{1-H} X^{(m)}(k)$. This illustrates that a traffic model developed for fitting self-similar behavior must preferably enable the matching of the distribution on several time scales.

Long-range dependence is associated with stationary processes. Consider now that $X(k)$ is second-order stationary with variance σ^2 and autocorrelation function $r(k)$. Note that, in this case, $X^{(m)}(k)$ is also second-order stationary. The process $X(k)$ has long-range dependence (LRD) if its autocorrelation function is non-summable, that is, $\sum_n r(n) = \infty$. Intuitively, this means that the process exhibits similar fluctuations over a wide range of time scales. Taking the case of the pOct Bellcore trace, it can be seen in Figure 1.1 that the fluctuations over the $0.01, 0.1$ and 1 s time scales are indeed similar.

Equivalently, one can say that a stationary process is LRD if its spectrum diverges at the origin, that is $f(v) \sim c_f |v|^{-\alpha}$, $v \to 0$. Here, α is a dimensionless scaling exponent, that takes values in $(0, 1)$; c_f takes positive real values and has dimensions of variance. On the other hand, a short range dependent (SRD) process is simply a stationary process which is not LRD. Such a process has $\alpha = 0$ at large scales, corresponding to white noise at scales beyond the so-called characteristic scale or correlation horizon. The Hurst parameter H is related with α by $H = (\alpha + 1)/2$.

There are several estimators of LRD. In this study we use the semi-parametric estimator developed in [22], which is based on wavelets. Here, one looks for alignment in the so-called Logscale Diagram (LD), which is a log-log plot of the variance estimates of the discrete wavelet transform coefficients representing the traffic process, against scale, completed with confidence intervals about these estimates at each scale. It can be thought of as a spectral estimator where large scale corresponds to low frequency. The main properties explored in this estimator are the stationarity and short-term correlations exhibited by the process of discrete wavelet transform coefficients and the power-law dependence in scale of the variance of this process. Figure 1.11 shows an example of applying this estimator to the October Bellcore trace, where scale is represented by j and the logarithm of the variance estimate by y_j. Traffic is said to be LRD if, within the limits of the confidence intervals, the log of the variance estimates fall on a straight line, in a range of scales from some initial value j_1 (4, in this case) up to the largest one present in data (11) and the slope of the straight line, which is an estimate of the scaling exponent α, lies in (0, 1).

There is a close relationship between long-range dependent and self-similar processes. In fact, if $Y(t)$ is self-similar with stationary increments and finite variance then $X(k)$ is long-range dependent, as long as $1/2 < H < 1$. The process $X(k)$ is said to be exactly second-order self-similar ($1/2 < H < 1$) if

$$r(n) = 1/2\left[(n+1)^{2H} - 2n^{2H} + (n-1)^{2H}\right] \tag{1.2}$$

for all $n \geq 1$, or is asymptotically self-similar if

$$r(n) \sim n^{-(2-2H)}L(n) \tag{1.3}$$

as $n \to \infty$, where $L(n)$ is a slowly varying function at infinity. In both cases the autocovariance decays hyperbolically, which indicates LRD. Any asymptotically second-order self-similar process is LRD, and vice versa.

1.3 Markov Modulated Poisson Processes

The discrete-time Markov Modulated Poisson Process (dMMPP) is the discrete-time version of the popular (continuous-time) MMPP and may be regarded as an Markov random walk where the increments in each instant

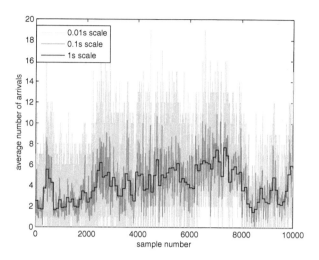

Figure 1.1 LRD processes exhibit fluctuations over a wide range of time scales (Example: trace pOct).

have a Poisson distribution whose parameter is a function of the state of the modulator Markov chain. More precisely, the (homogeneous) Markov chain $(Y, J) = \{(Y_k, J_k), k = 0, 1, \ldots\}$ with state space $I\!N_0 \times S$ is a dMMPP if and only if for $k = 0, 1, \ldots,$

$$P(Y_{k+1} = m, J_{k+1} = j | Y_k = n, J_k = i) = \begin{cases} 0 & m < n \\ p_{ij} \, e^{-\lambda_i} \dfrac{\lambda_i^{m-n}}{(m-n)!} & m \geq n \end{cases}$$

$$(1.4)$$

for all $m, n \in I\!N_0$ and $i, j \in S$, with λ_i, $i \in S$, being nonnegative real constants and $\mathbf{P} = (p_{ij})$ being a stochastic matrix. Note that the distribution of $Y_{k+1} - Y_k$ given $J_k = j$ is Poisson with mean λ_j, so that λ_j represents the mean increment of the process Y when the modulating Markov chain is in state j.

Whenever (1.4) holds, we say that (Y, J) is a dMMPP with set of modulating states S and parameter (matrices) $\mathbf{P}$ and $\mathbf{\Lambda}$, and write

$$(Y, J) \sim \text{dMMPP}_S(\mathbf{P}, \mathbf{\Lambda}) \qquad (1.5)$$

where $\mathbf{\Lambda} = (\lambda_{ij}) = (\lambda_i \delta_{ij})$. The matrix $\mathbf{P}$ is the transition probability matrix of the modulating Markov chain J, whereas $\mathbf{\Lambda}$ is the matrix of Poisson arrival rates. If S has cardinality r, we say that (Y, J) is a dMMPP of order r

(dMMPP$_r$). When, in particular, $S = \{1, 2, \ldots, r\}$ for some $r \in I\!N$, then

$$
\mathbf{P} = \begin{bmatrix} p_{11} & p_{12} & \cdots & p_{1r} \\ p_{21} & p_{22} & \cdots & p_{2r} \\ \cdots & \cdots & \cdots & \cdots \\ p_{r1} & p_{r2} & \cdots & p_{rr} \end{bmatrix} \quad \text{and} \quad \mathbf{\Lambda} = \begin{bmatrix} \lambda_1 & 0 & \cdots & 0 \\ 0 & \lambda_2 & \cdots & 0 \\ \cdots & \cdots & \cdots & \cdots \\ 0 & 0 & \cdots & \lambda_r \end{bmatrix} \quad (1.6)
$$

and we write simply that $(Y, J) \sim \text{dMMPP}_r(\mathbf{P}, \mathbf{\Lambda})$. The stationary distribution of J is denoted by $\pi = [\pi_1 \pi_2, \ldots \pi_r]$.

Consider now the superposition of L independent dMMPPs

$$
(Y^{(l)}, J^{(l)}) \sim r_l\text{-dMMPP}_{r_l}(\mathbf{P}^{(l)}, \mathbf{\Lambda}^{(l)}) \tag{1.7}
$$

where $l = 1, 2, \ldots, L$ and $J^{(1)}, J^{(2)}, \ldots, J^{(L)}$ are ergodic chains in steady-state and the dimension of each dMMPP is not necessarily the same. For $l = 1, 2, \ldots, L$ we denote by $\pi^{(l)} = [\pi_1^{(l)} \pi_2^{(l)} \ldots \pi_{r_l}^{(l)}]$ the stationary distribution of $J^{(l)}$.

The result of the superposition is the process

$$
(Y, J) = \left(\sum_{l=1}^{L} Y^{(l)}, (J^{(1)}, J^{(2)}, \ldots, J^{(L)}) \right) \sim \text{dMMPP}_S(\mathbf{P}, \mathbf{\Lambda})
$$

where

$$
\begin{aligned}
S &= \{1, 2, \ldots, r_1\} \times \ldots \times \{1, 2, \ldots, r_L\} & (1.8) \\
\mathbf{P} &= \mathbf{P}^{(1)} \otimes \mathbf{P}^{(2)} \otimes \ldots \otimes \mathbf{P}^{(L)} & (1.9) \\
\mathbf{\Lambda} &= \mathbf{\Lambda}^{(1)} \oplus \mathbf{\Lambda}^{(2)} \oplus \ldots \oplus \mathbf{\Lambda}^{(L)} & (1.10)
\end{aligned}
$$

where $r_1, r_2, \ldots, r_L$, represent the dimensions of each one of the L dM-MPPs, with $\oplus$ and $\otimes$ denoting the Kronecker sum and the Kronecker product, respectively. Note that the Markov chain J is also in steady-state.

In our approaches L, that represents the number of time scales considered, is fixed *a priori* and the dimensions of the dMMPPs, $r_1, r_2, \ldots, r_L$, are computed as part of the fitting procedures.

1.4 Proposed Multi-Scale Markovian models

The proposed Markovian models are constructed based on the PMF of the arrival process at each time scale, thus enabling them to capture the traffic self-similar behavior over a range of time scales. The number of time scales

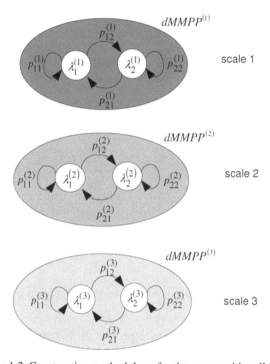

Figure 1.2 Construction methodology for the superposition dMMPP.

to consider, L, is fixed a priori and time scales are numbered in an ascending way, from $l = 1$ (corresponding to the largest scale) to $l = L$ (corresponding to the finest one). The main difference between both construction models lies on the fact that while the superposition model includes a dMMPP for each time scale the decomposition model includes several dMMPPs at each scale, specifically one dMMPP for each state of all the dMMPPs that belong to the immediately higher time scale. In this way, the equivalent dMMPP will be obtained in different ways for both models.

1.4.1 Superposition Model

The first traffic model is based on the superposition of dMMPPs, each one representing a specific time scale. Figure 1.2 illustrates the dMMPP construction methodology for the simple case of having only three time scales and two-state dMMPPs at each time scale. The dMMPP associated to time scale l will be designated by dMMPP$^{(l)}$, and its corresponding number of states by $N_{(l)}$.

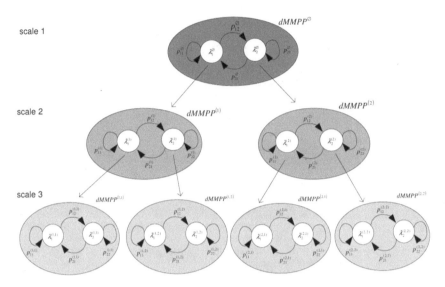

Figure 1.3 Construction methodology for the decomposition dMMPP.

The flow diagram of the inference procedure is represented in Figure 1.4 where, basically, four major steps can be identified:

(i) calculation of the data vectors (corresponding to the average number of arrivals per time interval) at each time scale, by applying and iterative aggregation process that starts at the finest time scale and ends at the largest one;

(ii) calculation of the empirical PMF corresponding to the largest time scale and inference of the corresponding dMMPP;

(iii) for the other time scales (starting from the largest to the finest one), calculation of the empirical PMF, calculation of its deconvolution from the empirical PMF of the preceding time scale and inference of a dMMPP that adjusts the resulting empirical PMF (the shaded part of the flow diagram);

(iv) calculation of matrices $\mathbf{\Lambda}$ and $\mathbf{P}$ of the final dMMPP through the superposition of the different dMMPPs that were inferred for each time scale.

The different steps of the inference procedure will be detailed in Section 1.5.

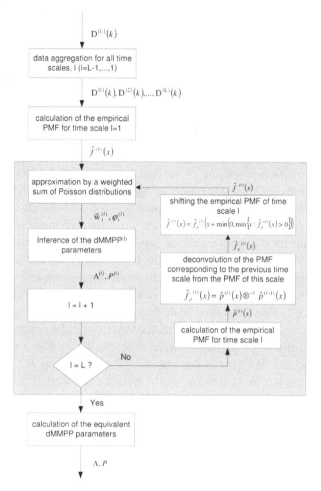

Figure 1.4 Flow diagram of the inference procedure for the superposition model.

1.4.2 Decomposition Model

The second proposed traffic model is constructed based on a decomposition process that successively decomposes the states of a dMMPP corresponding to a certain time scale on new dMMPPs belonging to the immediately finer time scale, refining in this way the traffic process by including the characteristics that are offered by successively finer time scales. The procedure starts at the largest time scale by inferring a dMMPP that adjusts the PMF corresponding to that scale. As part of the parameter inference procedure, each time interval of the data sequence is attributed to a state of the dMMPP; in this way, a new PMF will be associated to each state of the dMMPP. On the next finer time scale, each state of the dMMPP is decomposed into a new dMMPP that adjusts the contribution of that time scale to the PMF of the state from which the dMMPP descends. In this way, a child dMMPP gives a more detailed description of the PMF corresponding to its parent state. This refinement process is iterated until a pre-defined number of time scales has been integrated into the model. Finally, a dMMP that incorporates this hierarchical structure is derived.

The construction procedure of the decomposition model can be described by a tree where, with the exception of the root node, each node of the tree corresponds to a state of a dMMPP and each level of the tree corresponds to a time scale. Figure 1.3 illustrates the construction methodology of this dMMPP, again for the simple case of considering only three time scales and two-sate dMMPPs at each time scale. Each state of a dMMPP will be represented by a vector that indicates the path from the highest level predecessor (that is, the state at the largest scale, $l = 1$, from which it descends) to itself. So, a state that is located at time scale l will be represented by a vector of the type $\mathbf{s} = (s_1, s_2, \ldots, s_l)$, $s_i \in I\!N$. Each dMMPP will be represented by the state that originated it, that is, its parent state. So, we consider that dMMPP$^\mathbf{s}$ represents the dMMPP that is generated by state $\mathbf{s}$ and $\{1, 2, \ldots, N_\mathbf{s}\}$ is the set of its states, where $N_\mathbf{s}$ designates the number of states. The root node of the tree corresponds to a virtual node, designated by $\mathbf{s} = \emptyset$, and is used to represent the dMMPP that is located at the largest time scale, $l = 1$. This dMMPP will be designated by root dMMPP. In this way, the states of the dMMPPs that belong to the tree structure are characterized by vectors

$$\mathbf{s} = (s_1, s_2, \ldots, s_l), \ l \in I\!N \tag{1.11}$$

with $s_{i+1} \in \{1, 2, \ldots, N_{\mathbf{s}_{i]}}\}$, $i = 0, 1, \ldots, l-1$; here, $\mathbf{s}_{j]}$ designates the sub-vector of $\mathbf{s}$ given by $(s_1, s_2, \ldots, s_j)$, with $j < |\mathbf{s}|$ and $\mathbf{s}_{0]} = \emptyset$, where $|\mathbf{s}|$ represents the length of vector $\mathbf{s}$. Note that, with this notation, a vector $\mathbf{s}$ can represent both state $\mathbf{s}$ and the dMMPP generated by $\mathbf{s}$. Besides, the time scale where dMMPP$^\mathbf{s}$ belongs to is given by $|\mathbf{s}| + 1$.

Finally, consider that $E^\mathbf{s}$ represents the set of time intervals associated to state $\mathbf{s}$, that is, to dMMPP$^\mathbf{s}$. Using this notation, the set of time intervals associated to dMMPP$^\emptyset$ will be $E^\emptyset = \{1, 2, \ldots, N\}$, where N corresponds to the number of time intervals existing in the finest time scale. Starting from $E^\emptyset$, the sets $E^\mathbf{s}$ are successively fractioned at each time scale in a hierarchical way. Thus, if states $\mathbf{s}$ and $\mathbf{t}$ are such that $|\mathbf{s}| = |\mathbf{t}| = l$ and $\mathbf{s} \neq \mathbf{t}$, then

$E^s \cap E^t = \emptyset$ and $\bigcup_{s:|s|=l} E^s = E^\emptyset$. Besides, if state $\mathbf{s}$ is a parent of state $\mathbf{t}$, that is, $\mathbf{t} = (\mathbf{s}, j)$, then $E^t \subseteq E^s$ and $\bigcup_{j=1,\dots,N_s} E^{(s,j)} = E^s$.

The inference procedure is represented in flow diagram of Figure 1.5 where, basically, the following major steps can be identified:

(i) calculation of the data vectors (corresponding to the average number of arrivals per time interval) at each time scale, by applying and iterative aggregation process that starts at the finest time scale and ends at the largest one;

(ii) calculation of the empirical PMF corresponding to the largest time scale, $l = 1$, and inference of the corresponding dMMPP;

(iii) for the other time scales (starting from the largest to the finest one), $l = 2, \dots, L - 1$, and for every state of each parent dMMPP, the inference procedure identifies the time intervals that are associated to that state, calculates the corresponding empirical PMF and infer the dMMPP that adjusts the contribution of the time scale to the PMF of that state (the shaded part of the flow diagram);

(iv) finally, calculation of the matrices $\mathbf{\Lambda}$ and $\mathbf{P}$ of the equivalent dMMPP that incorporates the previously described hierarchical structure.

Note that both parameter inference procedures are based on the aggregated processes $X^{(m)}(k)$, that were presented in Section 1.2 during the definition of self-similarity and, in both traffic models, the number of states of the dMMPPs is calculated as part of the inference procedures.

Both inference methods have some common steps. In the next sections, all steps of both inference procedures will be described in detail, not only the common ones but also the steps that are different.

1.5 Steps of the Parameter Inference Procedures

1.5.1 Aggregation Process

This first step is common to both inference procedures. After defining the sample interval at the finest time scale, Δt, the number of time scales, L, and the aggregation level, a, the aggregation process starts by calculating the data sequence corresponding to the number of arrivals at the finest time scale (the scale where the length of the discretization interval and the sampling interval are equal), $D^{(L)}(k), k = 1, 2, \dots, N$. Then, the data sequences corresponding to the remaining time scales, $D^{(l)}(k), l = L-1, \dots, 1$, are calculated and they correspond to the average number of arrivals on intervals of length $\Delta t a^{(L-l)}$.

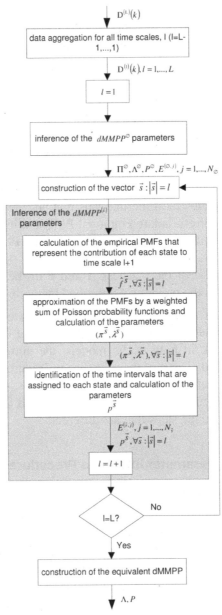

Figure 1.5 Flow diagram of the inference procedure for the decomposition model.

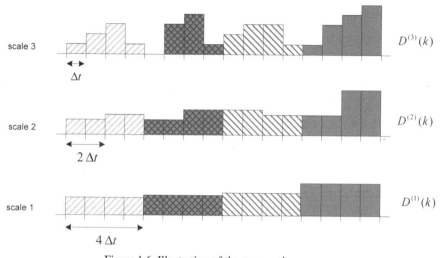

Figure 1.6 Illustration of the aggregation process.

These data sequences are obtained in the following way:

$$D^{(l)}(k) = \begin{cases} \Psi\left(\frac{1}{a}\sum_{i=0}^{a-1} D^{(l+1)}(k + ia^{L-l-1}))\right), & \frac{k-1}{a^{L-l}} \in \mathbb{N}_0 \\ D^{(l)}(k-1), & \frac{k-1}{a^{L-l}} \notin \mathbb{N}_0 \end{cases} \qquad (1.12)$$

where $\Psi(x)$ represents a function that rounds x to its nearest integer. Note that the block length mentioned on equation (1.1), corresponding to the aggregation process, is related with a and l by $m = a^{L-l}$. Note also that all data sequences have the same length N and $D^{(l)}(k)$ is made of sub-sequences composed by a^{L-l} necessarily equal values; these sub-sequences are designated as *l-sequences*. The empirical distribution of $D^{(l)}(k)$ will be designated by $\hat{p}^{(l)}(x)$.

Figure 1.6 illustrates the aggregation process for the particular case of considering only three time scales and an aggregation level of $a = 2$. The top part of the picture corresponds to the finest time scale (scale 3) and represents the number of arrivals per sampling interval. At time scale 2, the figure represents the average number of arrivals per time interval of length $2\Delta t$, while at time scale 1 it represents the average number of arrivals per time interval of length $4\Delta t$, since the aggregation level is equal to 2.

1.5.2 Inference of the Partial dMMPPs

On both proposed traffic models, all partial dMMPPs are inferred in order to adjust an empirical PMF. For the largest time scale, this PMF corresponds to the most aggregated data sequence, $D^{(1)}(k)$. For all other time scales l, $l = 2, \ldots, L$, the associated dMMPPs will model only the traffic components related to each scale, but at this point both traffic models diverge in the way partial dMMPPs are calculated. On the superposition model, there will be only one dMMPP per time scale, inferred according to the flow diagram of Figure 1.4 whose main steps are explained in detail in the next subsections. On the decomposition model, a dMMPP will be inferred for each state that is located on the immediately higher time scale. So, for each dMMPP and for each time scale, the PMF that is adjusted represents the contribution of that scale to the PMF corresponding to its parent state. The parameter inference procedure for each dMMPP of the tree structure includes several steps, that are highlighted on the shaded part of the flow diagram of Figure 1.5 and are also explained in detail in the next subsections.

1.5.2.1 Calculation of the Empirical PMFs

The operations that are involved in this step are basically the same for both proposed traffic models, although the number of PMFs per time scale is different in both cases. For the largest time scale, the considered PMF is simply the empirical one, for both traffic models. For all other time scales, $l = 2, \ldots, L$, the PMFs that are calculated for each model are different.

On the superposition model, each dMMPP is inferred from the PMF that represents its contribution for a given time scale. The traffic components that represent the contribution of time scale $l, l = 2, \ldots, L$, are obtained through the deconvolution between the empirical PMFs corresponding to scale l itself and to the previous time scale, $l - 1$, that is, $\hat{f}_p^{(l)}t(x) = [\hat{p}^{(l)} \otimes^{-1} \hat{p}^{(l-1)}](x)$. However, this operation can lead to non-zero values for the probabilities of obtaining negative arrival rates for the dMMPP$^{(l)}$, that will occur whenever $\min\{x : \hat{p}^{(l-1)}(x) > 0\} < \min\{x : \hat{p}^{(l)}(x) > 0\}$.

On the decomposition model, the construction of a dMMPP located at time scale l and generated from state $\mathbf{s}$ corresponds also to the deconvolution of empirical PMFs, but now calculated over the set of time intervals $E^{\mathbf{s}}$, corresponding to this time scale, $l = |\mathbf{s}| + 1$, and to the previous one, $l - 1 = |\mathbf{s}|$, that is, $\hat{f}_p^{\mathbf{s}}(x) = [\hat{p}^{\mathbf{s},|\mathbf{s}|+1} \otimes^{-1} \hat{p}^{\mathbf{s},|\mathbf{s}|}](x)$, where $\hat{p}^{\mathbf{s},l}(x)$ represents the PMF obtained from the data sequence $D^{(l)}(k), k \in E^{\mathbf{s}}$. Note that, even though the two empirical PMFs are obtained from the same set of time intervals, they are aggregated on different levels. Once again, these operations can result on non-zero values for the probabilities of obtaining negative arrival rates for the dMMPP^s, that will occur whenever $\min\{x : \hat{p}^{s,|s|}(x) > 0\} < \min\{x : \hat{p}^{s,|s|+1}(x) > 0\}$.

In order to correct these meaningless results, from a physical point of view, that can occur on both traffic models, the dMMPP$^\chi$ will be adjusted to

$$\hat{f}^\chi(x) = \hat{f}_p^\chi(x + e^\chi) \tag{1.13}$$

where $e^\chi = \min(0, \min\{x : \hat{f}_p^\chi(x) > 0\})$, which guarantees $\hat{f}^\chi(x) = 0$, $x < 0$, where we are considering that superscript χ replaces (l) on the description of the superposition model and s on the description of the decomposition model. The additional factors that are now being introduced will be removed in the final step of the inference procedure.

1.5.2.2 Inference of the dMMPPs Parameters

This subsection will present the step of both inference procedures corresponding to the estimation of the characterizing parameters of each dMMPP, that is, the matrices corresponding to the transition probabilities and the Poisson arrival rates for each state. In the remaining part of this paper, the same notation that was introduced in the previous section will be maintained: superscript χ will replace (l) or s, depending if we considering the superposition or the decomposition model, respectively.

The first step of the dMMPP$^\chi$ parameter inference procedure is the approximation of $\hat{f}^\chi(x)$ by a weighted sum of Poisson probability functions. This operation is based on an algorithm, introduced in [21], that progressively subtracts a Poisson probability function from function $\hat{f}^\chi(x)$. The most important steps of this algorithm are graphically represented on the flow diagram of Figure 1.7 and will be explained in detail in the next paragraphs.

Let us represent the n-th Poisson probability function, with mean value φ_n^χ, by $g_{\varphi_n^\chi}(x)$ and lets define $h_n^\chi(x)$ as the difference between $\hat{f}^\chi(x)$ and the weighted sum of Poisson probability functions corresponding to the n-th iteration. Initially, one makes $h_1^\chi(x) = \hat{f}^\chi(x)$ and, at each step, the procedure starts by detecting the maximum value of $h_n^\chi(x)$. The corresponding value of x, $\varphi_n^\chi = \arg\max_x h_n^\chi(x)$, will be considered as the n-th Poisson arrival rate of dMMPP$^\chi$. Then, the weights of each Poisson probability function, $\mathbf{w}_n^\chi = [w_{1n}^\chi, w_{2n}^\chi, \ldots, w_{nn}^\chi]$, are calculated through the following set of linear

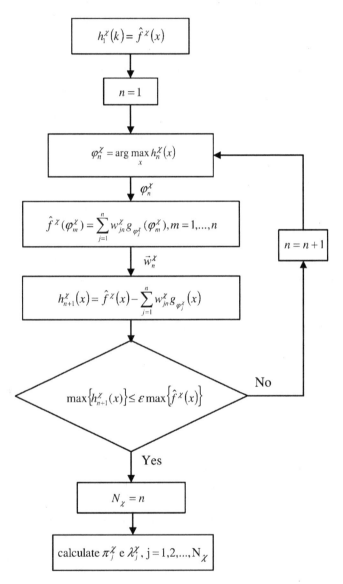

Figure 1.7 Algorithm to calculate the number of states and the Poisson arrival rates of the dMMPP$^\chi$.

equations:

$$\hat{f}^{\chi}(\varphi_m^{\chi}) = \sum_{j=1}^{n} w_{jn}^{\chi} g_{\varphi_j^{\chi}}(\varphi_m^{\chi}) \tag{1.14}$$

for $m = 1, \ldots, n$. This assures that the adjustment between $\hat{f}^{\chi}(x)$ and the weighted sum of Poisson probability functions is exact at points φ_m^{χ}, for $m = 1, 2, \ldots, n$. The final step of each iteration is the calculation of the new difference function

$$h_{n+1}^{\chi}(x) = \hat{f}^{\chi}(x) - \sum_{j=1}^{n} w_{jn}^{\chi} g_{\varphi_j^{\chi}}(x) \tag{1.15}$$

The algorithm will stop when the maximum value of $h_{n+1}^{\chi}(x)$ is less than or equal to a pre-defined percentage of the maximum value of $\hat{f}^{\chi}(x)$. At this time, the number of states of the dMMPP$^{\chi}$, N_{χ}, is made equal to n.

After calculating the value of N_{χ}, the dMMPP$^{\chi}$ parameters, $\{(\pi_j^{\chi}, \lambda_j^{\chi}),$ $j = 1, 2, \ldots, N_{\chi}\}$, are made equal to

$$\pi_j^{\chi} = w_{jN_{\chi}}^{\chi} \quad \text{and} \quad \lambda_j^{\chi} = \varphi_j^{\chi} \tag{1.16}$$

Note that the number of states of each dMMPP depends on the precision level that is used for approximating $\hat{f}^{\chi}(x)$ by the weighted sum of Poisson probability functions.

The next step of the parameter inference procedure associates, for each time scale l, one of the dMMPP$^{\chi}$ states with each time interval of the arrival process. Remember that the data sequences aggregated on time scale l contain a^{L-l} successive equal values, and have been designated as l-sequences. For the decomposition model, the set of time intervals associated to dMMPP^s is given by E^s and the objective of this step is to divide E^s in sub-sets $E^{(s,j)}$, $j = 1, \ldots, N_s$. For the superposition model, the set of time intervals associated to any dMMPP$^{(l)}$, $l = 1, 2, \ldots, L$, is equal to $\{1, 2, \ldots, N\}$, where N corresponds to the number of time intervals existing in the finest time scale, that is, it is equal to set $E^{\emptyset}$ that was defined for the decomposition model. The procedure responsible for assigning states considers only the first time interval of each l-sequence, defined by $i = a^{L-\chi'}(k-1) + 1, k \in \mathbb{N}$, with $i \in E^s$ for the decomposition model and $i \in \{1, 2, \ldots, N\}$ for the superposition model, where symbol χ' is equal to $(|s|+1)$ for the decomposition model and equal to (l) for the superposition model. The state that is assigned to the i-th l-sequence is randomly calculated according to the probability vector

$\theta^{\chi}(i) = \{\theta_1^{\chi}(i), \ldots, \theta_{N_{\chi}}^{\chi}(i)\}$, with

$$\theta_n^{\chi}(i) = \frac{g_{\lambda_n^{\chi}}(D^{\chi'}(i))}{\sum_{j=1}^{N_{\chi}} g_{\lambda_j^{\chi}}(D^{\chi'}(i))} \tag{1.17}$$

for $n = 1, \ldots, N_{\chi}$, where λ_j^{χ} represents the Poisson arrival rate of the dMMPP$^{\chi}$ j-th state, and $g_{\lambda}(y)$ represents a Poisson probability distribution function with mean λ. The elements of this vector represent the probability that state n has originated the number of arrivals $D^{\chi'}(i)$ at time interval i belonging to time scale χ'.

After this step, the dMMPP$^{\chi}$ transition probabilities, p_{od}^{χ}, with $o, d = 1, \ldots, N_{\chi}$, can be inferred by simply counting the number of transitions that occur between each pair of states. If n_{od}^{χ} represents the number of transitions from state o to state d of the dMPPP$^{\chi}$, then

$$p_{od}^{\chi} = \frac{n_{od}^{\chi}}{\sum_{m=1}^{N_{\chi}} n_{om}^{\chi}}, \quad o, d = 1, \ldots, N_{\chi} \tag{1.18}$$

The transition probability and the Poisson arrival rate matrices corresponding to dMMPP$^{\chi}$ are then given by

$$\mathbf{P}^{\chi} = \begin{bmatrix} p_{11}^{\chi} & p_{12}^{\chi} & \cdots & p_{1N_{\chi}}^{\chi} \\ p_{21}^{\chi} & p_{22}^{\chi} & \cdots & p_{2N_{\chi}}^{\chi} \\ \cdots & \cdots & \cdots & \cdots \\ p_{N_{\chi}1}^{\chi} & p_{N_{\chi}2}^{\chi} & \cdots & p_{N_{\chi}N_{\chi}}^{\chi} \end{bmatrix} \quad \text{and} \quad \mathbf{\Lambda}^{\chi} = \begin{bmatrix} \lambda_1^{\chi} & 0 & \cdots & 0 \\ 0 & \lambda_2^{\chi} & \cdots & 0 \\ \cdots & \cdots & \cdots & \cdots \\ 0 & 0 & \cdots & \lambda_{N_{\chi}}^{\chi} \end{bmatrix} + e^{\chi}\mathbf{I} \tag{1.19}$$

The diagonal matrix of the steady state probabilities will be designated by Π^{χ}.

Figure 1.8 schematically illustrates the main steps of the construction process for the superposition model, considering only the first two time scales. As was previously said, the empirical PMF corresponding to the largest time scale (scale 1) is estimated and the dMMPP that best adjusts it is inferred. For the next immediate scale (scale 2), the empirical PMF is estimated and then it is deconvolved from the PMF corresponding to time scale 1. The dMMPP that describes the contribution of time scale 2 for the arrival process is calculated based on the PMF that results from this deconvolution operation.

Figure 1.9 schematically illustrates the main steps of the construction process for the decomposition model, considering only the first two time scales. For the largest time scale (scale 1), the empirical PMF is estimated

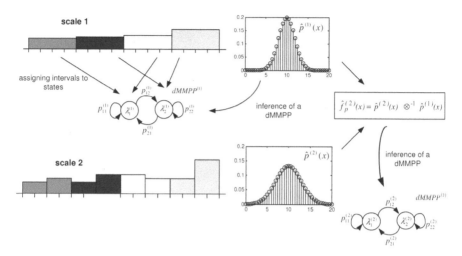

Figure 1.8 Procedure for calculating the empirical PMFs and inferring the partial dMMPPs of the superposition model.

and the dMMPP that best adjusts it is inferred (dMMPP$^{\emptyset}$). Each time interval of the data sequence is then assigned to each dMMPP state and the next step consists on estimating the empirical PMFs associated to each state. For the next immediate scale (scale 2), the empirical PMFs associated to each state will also be estimated and then they are deconvolved from the PMFs corresponding to time scale 1 and to the same states. The dMMPPs that describe the contribution of time scale 2 for the arrival process are calculated based on the PMFs that result from these deconvolution operations.

1.5.3 Construction of the Equivalent dMMPP Model

Due to the construction processes that are inherent to each one of the proposed traffic models, this step will be completely different for both approximations.

Considering the superposition model, matrices that characterize the equivalent dMMPP are obtained using equations that correspond to the Kronecker sum and product:

$$\mathbf{P} = \mathbf{P}^{(1)} \otimes \mathbf{P}^{(2)} \otimes \ldots \otimes \mathbf{P}^{(L)} \tag{1.20}$$

$$\mathbf{\Lambda} = \mathbf{\Lambda}^{(1)} \oplus \mathbf{\Lambda}^{(2)} \oplus \ldots \oplus \mathbf{\Lambda}^{(L)} \tag{1.21}$$

where matrices $\mathbf{\Lambda}^{(l)}$ are $\mathbf{P}^{(l)}$, $l = 1, \ldots, L$, have been calculated in the previous subsection. However, it is also necessary to remove the additional factors

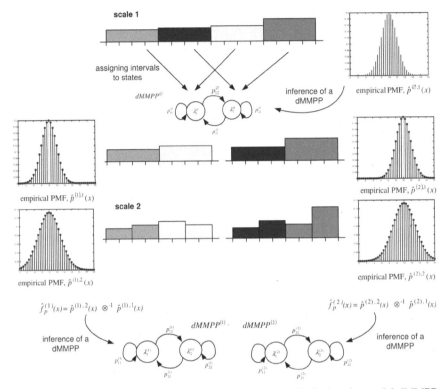

Figure 1.9 Procedure for calculating the empirical PMFs and inferring the partial dMMPPs of the decomposition model.

that were introduced in Section 1.5.2.1. In this way, the final $\mathbf{\Lambda}$ matrix will be given by

$$\mathbf{\Lambda} = \mathbf{\Lambda} - \sum_{l=2}^{L} e^{(l)} \cdot \mathbf{I} \tag{1.22}$$

where $\mathbf{I}$ represents the identity matrix.

For the decomposition model, it is necessary to build a dMMPP that is equivalent to the tree structure of dMMPPs that was obtained on the previous sections. The objective is to incorporate on the model the level of detail that is given by the finest time scale, so the equivalent dMMPP will have a number of states equal to the number of states that exist in the set of dMMPPs belonging to the finest time scale, L, of the tree structure. The states that belong to the finest time scale can be identified by paths on the tree structure having the form $\mathbf{s} = (s_1, s_2, \ldots, s_L)$. Note that each state $\mathbf{s}$ results from the states that are associated to it and belong to path $s_{i+1]} = (s_1, s_2, \ldots, s_{i+1})$ corresponding to the dMMPP$^{s_{i]}}$, $i = 0, 1, \ldots, L - 1$. In this way, the states of the equivalent dMMPP will have Poisson arrival rates that are equal to the sum of the Poisson arrival rates of their predecessor states in the tree structure, that is,

$$\lambda_{\mathbf{s}} = \sum_{j=0}^{L-1} \lambda_{s_{j+1}}^{s_{j]}} \tag{1.23}$$

The transition between each pair of states is determined by the shortest path in the tree structure that passes through the root dMMPP and is able to connect both states. There are two distinct cases: the case where both states belong to different dMMPPs located at the finest time scale and the case where both states belong to the same dMMPP. Anyway, each pair of states can descend from a single common dMMPP (that will be necessarily the root dMMPP) or from more common dMMPPs. The common dMMPP that is located at the time scale with the highest value of l will be designated by $\mathbf{s} \wedge \mathbf{t} = (s_1, s_2, \ldots, s_k)$, where $k = \max\{i : s_j = t_j, j = 1, 2, \ldots, i\}$.

So, starting with the case where $\mathbf{s} \neq \mathbf{t}$, the transition probability from $\mathbf{s}$ to $\mathbf{t}$, $p_{\mathbf{s},\mathbf{t}}$, is given by the product of three factors. The first one takes into account the time scales where $\mathbf{s}$ and $\mathbf{t}$ are associated to the same states and is given by

$$\phi_{\mathbf{s},\mathbf{t}} = \begin{cases} \prod_{j=0}^{|\mathbf{s}\wedge\mathbf{t}|-1} p_{s_{j+1},s_{j+1}}^{s_{j]}}, & |\mathbf{s} \wedge \mathbf{t}| \neq 0 \\ 1, & |\mathbf{s} \wedge \mathbf{t}| = 0 \end{cases} \tag{1.24}$$

The second factor takes into account the transition that occurs at the time scale where vectors $\mathbf{s}$ and $\mathbf{t}$ are associated to different states of a same dMMPP and corresponds to the term $p_{s_{|\mathbf{s}\wedge\mathbf{t}|+1}, t_{|\mathbf{s}\wedge\mathbf{t}|+1}}^{\mathbf{s}\wedge\mathbf{t}}$. The third factor takes into account the steady-state probabilities of the states that are associated to $\mathbf{t}$ on the time scales that are not common to vector $\mathbf{s}$, being given by

$$\psi_{\mathbf{s},\mathbf{t}} = \prod_{j=|\mathbf{s}\wedge\mathbf{t}|+1}^{L-1} \pi_{t_{j+1}}^{t_{j]}} \tag{1.25}$$

where one considers that an empty product is equal to one.

So, for the case where $\mathbf{s} \neq \mathbf{t}$ we have, finally, that the transition probability is given by

$$p_{\mathbf{s},\mathbf{t}} = \phi_{\mathbf{s},\mathbf{t}} \; p_{s_{|\mathbf{s}\wedge\mathbf{t}|+1}, t_{\mathbf{s}\wedge\mathbf{t}+1}}^{\mathbf{s}\wedge\mathbf{t}} \; \psi_{\mathbf{s},\mathbf{t}} \tag{1.26}$$

Table 1.1 Main characteristics of the traffic traces/captures.

Traffic trace	Capure period	Capture size (packets)	Average rate (bytes/s)	Average packet size (bytes)
pOct	Bellcore capture	1 million	322790	568
ISP	10.26pm a 10.49pm, October 18, 2002	1 million	583470	797
P2P file sharing	10.26pm to 11.31pm, October 18, 2002	0.5 million	131140	1029

For the case where $\mathbf{s} = \mathbf{t}$, the transition probability is simply given by

$$P_{\mathbf{s},\mathbf{t}} = \phi_{\mathbf{s},\mathbf{t}} \tag{1.27}$$

1.6 Traffic Traces

In order to evaluate and compare the efficiency of both proposed traffic models and their respective inference procedures, three traffic traces were selected. The first one is the well-known Bellcore *pOct* trace, publicly available [1]. The second one is a capture of aggregated IP traffic corresponding to Internet access; it was measured on the downstream direction and corresponds to approximately 65 simultaneous users. This capture was made on the ADSL network of a Portuguese ISP. The third capture corresponds to traffic of a peer-to-peer file sharing application (specifically, Kazaa), includes 10 simultaneous users and was made on the same access link used to measure the second capture. For all measurements, the traffic analyzer was a 1.2 GHz Athlon AMD PC, having 1.5 Gbytes of RAM and running the WinDump application; for every packet, the arrival instant and the IP header were saved to hard disk. The main characteristics of the traffic traces are described in Table 1.1.

All captures exhibit self-similarity characteristics, since the arrival process exhibits similar fluctuations over a wide range of time scales. Considering the Bellcore traffic trace, for example, we can see from Figure 1.1 that fluctuations over the 0.01 s, 0.1 s and 1 s time scales are in fact similar. For the other traces, the behavior is analogous.

Regarding LRD properties, they can also be easily identified on the different traffic captures: looking at capture *ISP*, for example, the analysis of its

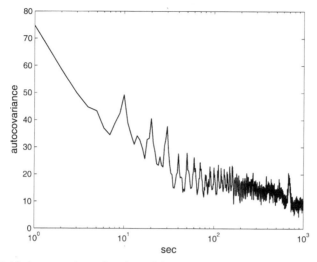

Figure 1.10 Autocovariance function of the packet count process, *ISP* traffic trace.

autocovariance function (Figure 1.10) lead us to suspect that it exhibits LRD properties, due to its slow decay for large time lags. This is confirmed by the scaling analysis made through the second-order logscale diagram, since the energy values are aligned between a medium octave (7) and octave 13, the highest one present in data (Figure 1.11). A similar analysis was made for the other traces, which also reveal similar LRD behaviors.

1.7 Results

The evaluation of the efficiency/suitability and the comparison between the proposed dMMPP models, and their respective parameter inference procedures, is made based on several criteria: (i) by comparing the Hurst parameters of the original and synthesized (according to the parameters inferred for the resulting dMMPP) data traces; (ii) by comparing the PMFs of the average packet counts at different time scales, calculated also from the original and synthesized traces; and (iii) by comparing the queuing behavior, in terms of packet loss ratio (PLR) and average waiting time in queue (AWT), of the original and synthesized traces, through a trace-driven simulation where these traces are used as inputs. Queuing behavior evaluation was based on simulation since the packet processing times are not necessarily multiples of the sampling interval (in fact, these two quantities are not related at all), as it should happen in order to use the theoretical results that are widely available

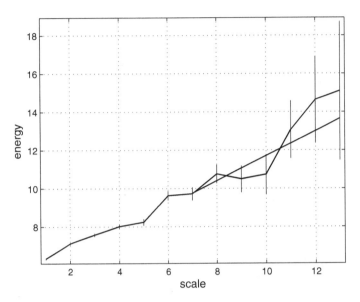

Figure 1.11 Second-order logscale diagram, *ISP* traffic trace.

for dMMPPs. All simulations were conducted using a fixed packet length equal to the mean packet length of the original trace. In order to get statistically significant values, 10 traffic replicas were generated for each inferred model and only the respective average values are shown in the results. The 95% confidence intervals are relativelly small, so we decided not to show them in the graphs due to readability issues.

For all traces, the sampling interval of the counting process was chosen to be 0.1 s and three different time scales were considered: 0.1 s, 0.2 s and 0.4 s. Larger aggregation levels were also considered, with good fitting results. For each trace, the estimation procedures of the superposition and decomposition traffic models took less than 2 minutes to complete, using a MATLAB implementation running on the same PC that was described above, which shows that the procedures are very efficient, from a computational point of view.

In order to verify that the proposed fitting procedures are able to capture the traffic self-similar behavior, Table 1.2 compares the Hurst parameters estimated based on the original and fitted traffic, for each one of the three selected traffic traces. This table also includes (inside brackets) the range of time scales over which energy values follow a straight line. As we can see,

Table 1.2 Comparison between the Hurst parameters estimated according to both proposed traffic models, for all considered traffic traces.

Traffic trace	Original traffic	Traffic fitted by the superposition model	Traffic fitted by the decomposition model
pOct	0.846 (4,11)	0.859 (4,11)	0.851 (4,11)
ISP	0.954 (4,10)	0.956 (4,10)	0.952 (5,10)
P2P file sharing	0.917 (8,12)	0.897 (6,12)	0.901 (6,12)

there is a very good agreement between the Hurst parameter values of the original and fitted traffic, so LRD behavior is indeed well captured by the proposed models.

The second evaluation criteria is based on the comparison between the PMFs of the original and fitted traces, for different time scales and for both considered traffic models. Starting with trace *pOct*, it can be seen from Figures 1.12, 1.13 and 1.14 that there is a good agreement between the probability functions of the original and fitted traces, for all time scales and for both considered models. These results were obtained for equivalent dMMPPs having 344 states for the superposition model case and 81 states for the decomposition model. For the *ISP* trace, Figures 1.15, 1.16 and 1.17 compare the PMFs of the original and fitted traces, again for the different time scales and for both considered traffic models. Finally, for the *P2P file sharing* trace the comparisons are illustrated in Figures 1.18, 1.19 and 1.20. For these cases, there is also a good agreement between the PMFs of the original and fitted traces, for both considered models and for the three considered time scales. The performance of both models is quite similar, although the number of states of the resulting dMMPP is much higher for the superposition model: for the *ISP* traffic trace, the resulting number of states is 245 for the superposition model and 74 for the decomposition model, while for the *P2P file sharing* trace the resulting number of states is 288 for the superposition model and 38 for the decomposition model.

Next, we will verify if the good results that was possible to achieve on the fitting of the Hurst parameter and the PMFs at the different time scales is enough to guarantee similar queuing behaviors for the original and fitted traffic. Regarding queuing behavior, we will compare the Packet Loss Ratio (PLR) and the Average Waiting Time in queue (AWT) obtained, through trace-driven simulation, by considering as inputs the different original and

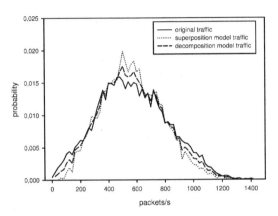

Figure 1.12 Comparison of the PMFs at the finest time scale, *pOct* trace.

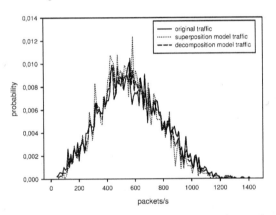

Figure 1.13 Comparison of the PMFs at the intermediate time scale, *pOct* trace.

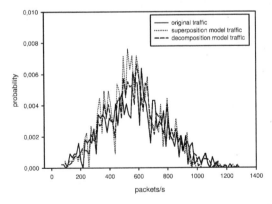

Figure 1.14 Comparison of the PMFs at the largest time scale, *pOct* trace.

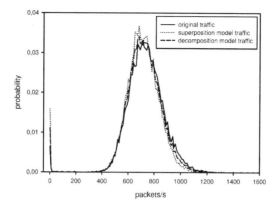

Figure 1.15 Comparison of the PMFs at the finest time scale, *ISP* trace.

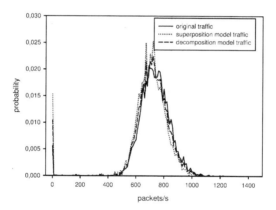

Figure 1.16 Comparison of the PMFs at the intermediate time scale, *ISP* trace.

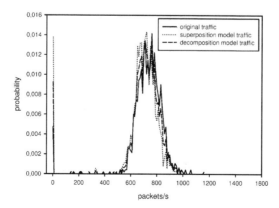

Figure 1.17 Comparison of the PMFs at the largest time scale, *ISP* trace.

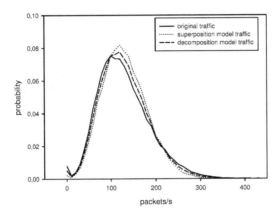

Figure 1.18 Comparison of the PMFs at the finest time scale, *P2P file sharing* trace.

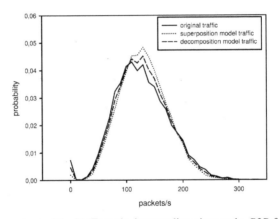

Figure 1.19 Comparison of the PMFs at the intermediate time scale, *P2P file sharing* trace.

fitted (according to the proposed inference procedures) traces. Two different sets of utilization ratios were used in the simulations: for traces *pOct* and *P2P file sharing*, we used the values $\rho = 0.6$, $\rho = 0.7$, $\rho = 1.0$ and $\rho = 1.2$, while for trace *ISP* the selected values were $\rho = 0.8$, $\rho = 0.9$, $\rho = 1.0$ and $\rho = 1.2$. The choice of these two different sets is due to the lower burstiness of the *ISP* traffic that leads to lower packet losses for the same queuing system utilization factors.

From Figures 1.21 and 1.23 it can be seen that, for the *pOct* trace, PLR and AWT are very well approximated by the equivalent dMMPPs that were estimated, for the utilization ratios $\rho = 0.6$ and $\rho = 0.7$. Figures 1.22 and

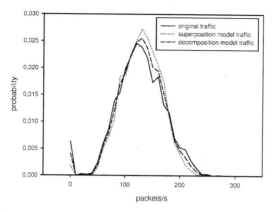

Figure 1.20 Comparison of the PMFs at the largest time scale, *P2P file sharing* trace.

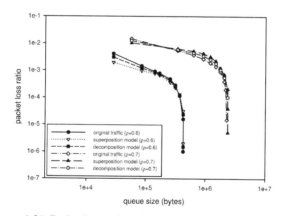

Figure 1.21 Packet loss ratio, *pOct* trace ($\rho = 0.6$ and $\rho = 0.7$).

1.24 show that the fitting accuracy degrades for the higher utilization ratios, $\rho = 1.0$ and $\rho = 1.2$. However, the fitting is still quite reasonable.

For the *P2P file sharing* trace, the results corresponding to $\rho = 0.6$ and $\rho = 0.7$ are depicted in Figures 1.25 and 1.27, while the results corresponding to $\rho = 1.0$ and $\rho = 1.2$ are depicted in Figures 1.26 and 1.28. In this case, the agreement between the curves is good for the smaller utilization ratios, $\rho = 0.6$ and $\rho = 0.7$, beginning to degrade as the utilization ratio increases. This behavior is similar to the one observed for trace *pOct*. Note that for traces *pOct* and *P2P file sharing* the two higher utilization ratios are not realistic since they imply very high packet loss ratios. At this range of values, any small peak that happens on the number of arrivals and is not

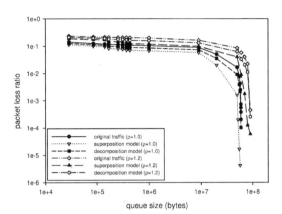

Figure 1.22 Packet loss ratio, *pOct* trace ($\rho = 1.0$ and $\rho = 1.2$).

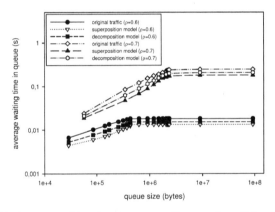

Figure 1.23 Average waiting time in queue, *pOct* trace ($\rho = 0.6$ and $\rho = 0.7$).

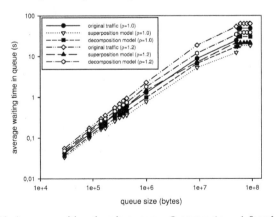

Figure 1.24 Average waiting time in queue, *pOct* trace ($\rho = 1.0$ and $\rho = 1.2$).

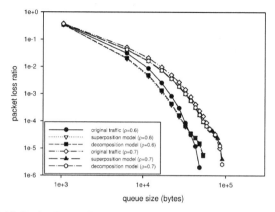

Figure 1.25 Packet loss ratio, *P2P file sharing* trace ($\rho = 0.6$ and $\rho = 0.7$).

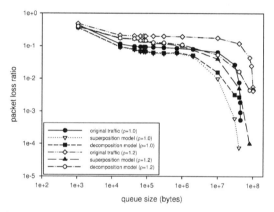

Figure 1.26 Packet loss ratio, *P2P file sharing* trace ($\rho = 1.0$ and $\rho = 1.2$).

conveniently fitted by the inferred models results in significant differences on PLR and AWT values that, however, do not correspond to traffic flows having significantly different characteristics.

For trace *ISP*, the results corresponding to $\rho = 0.8$ and $\rho = 0.9$ are illustrated in Figures 1.29 and 1.31. The results corresponding to $\rho = 1.0$ and $\rho = 1.2$ are represented in Figures 1.30 and 1.32. In all cases, the agreement between the curves corresponding to the original and fitted traces is good for both performance metrics, for all utilization ratio values and for both proposed traffic models.

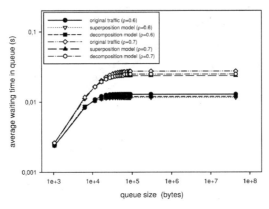

Figure 1.27 Average waiting time in queue, *P2P file sharing* trace ($\rho = 0.6$ and $\rho = 0.7$).

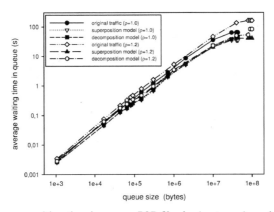

Figure 1.28 Average waiting time in queue, *P2P file sharing* trace ($\rho = 1.0$ and $\rho = 1.2$).

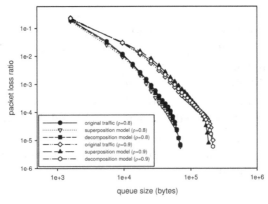

Figure 1.29 Packet loss ratio, *ISP* trace ($\rho = 0.8$ and $\rho = 0.9$).

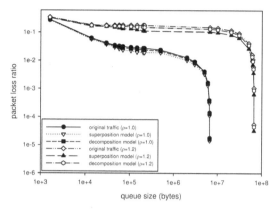

Figure 1.30 Packet loss ratio, *ISP* trace ($\rho = 1.0$ and $\rho = 1.2$).

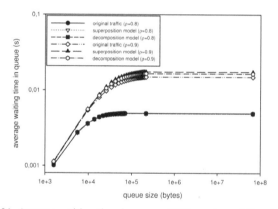

Figure 1.31 Average waiting time in queue, *ISP* trace ($\rho = 0.8$ and $\rho = 0.9$).

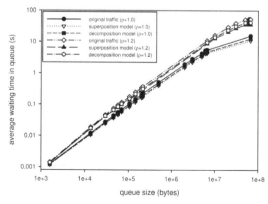

Figure 1.32 Average waiting time in queue, *ISP* trace ($\rho = 1.0$ and $\rho = 1.2$).

Note that, as the utilization factor increases the deviation between the different curves also increases, since the sensibility of the metrics variations to any difference on the queue size (even if it is a small one) is higher.

We can thus say that both proposed traffic models, and their corresponding parameter inference procedures, are able to achieve good results on fitting the Hurst parameters and the PMFs at each time scale, and this efficiency reveals itself enough to lead to good performance results in terms of PLR and AWT. Besides, the computational complexity of the inference procedures is very small. Note that this complexity, as well as the number of states of the resulting dMMPPs, are directly related to the precision level that is used on the approximation of the empirical PMFs by weighted sums of Poisson probability functions.

The performance of both inference procedures is very similar. So, it is not easy to recommend one of the approaches based only on their relative performances. One argument that clearly favors the decomposition approach is that the numbers of states of the resulting dMMPPs are smaller than the corresponding numbers for the superposition approach. This may be due to the fact that in the decomposition approach, as the time scale increases, dMMPPs are fitted to successively smaller sets of intervals whose arrivals characteristics tend to increase in homogeneity and, thus, tend to be associated to a smaller number of states. However, the contribution of each time scale for the characterization of the aggregate traffic characteristics is interpreted in an easier and more natural way through the superposition approach. Note also that, for the same number of states, a smaller number of dMMPPs (and their associated parameters) are needed to compute the final dMMPP using the superposition approach.

1.8 Conclusions

We proposed two discrete-time MMPPs, and their associated parameter fitting procedures, which are able to capture self-similarity over a range of time scales. The first model has a fitting procedure that matches the complete distribution of the arrival process at each time scale of interest, while the second proposed model is constructed using a hierarchical procedure that, starting from a MMPP that matches the distribution of packet counts at the coarsest time scale, successively decomposes each MMPP state into new MMPPs that incorporate a more detailed description of the distribution at finner time scales. The traffic process is then represented by a MMPP equivalent to the constructed hierarchical structure. The accuracy of the fitting procedures was evaluated by comparing the Hurst parameter, the probability mass function at each time scale and the queuing behavior (as assessed by the loss probability and average waiting time), corresponding to the measured and to synthetic traces generated from the inferred models. Three traffic traces exhibiting self-similar behavior were considered: the well-known *pOct* Bellcore trace, a trace of aggregated IP WAN traffic, and a trace corresponding to a popular file sharing application. Our results show that the proposed models and parameter fitting procedures are very effective in matching the main characteristics of the measured traces over the different time scales present in data.

References

[1] W. Leland, M. Taqqu, W. Willinger and D. Wilson, On the self-similar nature of Ethernet traffic (extended version), *IEEE/ACM Transactions on Networking*, vol. 2, no. 1, pp. 1–15, February 1994.

[2] J. Beran, R. Sherman, M. Taqqu and W. Willinger, Long-range dependence in variable-bit rate video traffic, *IEEE Transactions on Communications*, vol. 43, nos. 2–4, pp. 1566–1579, 1995.

[3] V. Paxson and S. Floyd, Wide-area traffic: The failure of Poisson modeling, *IEEE/ACM Transactions on Networking*, vol. 3, no. 3, pp. 226–244, June 1995.

[4] B. Ryu and A. Elwalid, The importance of long-range dependence of VBR video traffic in ATM traffic engineering: Myths and realities, *ACM Computer Communication Review*, vol. 26, pp. 3–14, October 1996.

[5] M. Crovella and A. Bestavros, Self-similarity in World Wide Web traffic: Evidence and possible causes, *IEEE/ACM Transactions on Networking*, vol. 5, no. 6, pp. 835–846, December 1997.

[6] W. Willinger, M. Taqqu, R. Sherman and D. Wilson, Self-similarity through high-variability: Statistical analysis of Ethernet LAN traffic at the source level, *IEEE/ACM Transactions on Networking*, vol. 5, no. 1, pp. 71–86, February 1997.

[7] W. Willinger, V. Paxson and M. Taqqu, *Self-similarity and Heavy Tails: Structural Modeling of Network Traffic*, A Practical Guide to Heavy Tails: Statistical Techniques and Applications. Birkhauser, 1998.

[8] D. P. Heyman and T. V. Lakshman, What are the implications of long range dependence for VBR video traffic engineering?, *IEEE/ACM Transactions on Networking*, vol. 4, no. 3, pp. 301–317, June 1996.

[9] A. Neidhardt and J. Wang, The concept of relevant time scales and its application to queuing analysis of self-similar traffic, in *Proceedings of SIGMETRICS'1998/PERFORMANCE'1998*, pp. 222–232, 1998.

[10] M. Grossglauser and J. C. Bolot, On the relevance of long-range dependence in network traffic, *IEEE/ACM Transactions on Networking*, vol. 7, no. 5, pp. 629–640, October 1999.

[11] A. Nogueira and R. Valadas, Analyzing the relevant time scales in a network of queues, in *SPIE's International Symposium ITCOM 2001*, August 2001.

[12] K. Meier-Hellstern, A fitting algorithm for Markov-modulated Poisson process having two arrival rates, *European Journal of Operational Research*, vol. 29, pp. 370–377, 1987.

[13] P. Skelly, M. Schwartz and S. Dixit, A histogram-based model for video traffic behaviour in an ATM multiplexer, *IEEE/ACM Transactions on Networking*, pp. 446–458, August 1993.

[14] R. Grünenfelder and S. Robert, Which arrival law parameters are decisive for queueing system performance, in *ITC 14*, 1994.

[15] S. Kang and D. Sung, Two-state MMPP modelling of ATM superposed traffic streams based on the characterisation of correlated interarrival times, *IEEE GLOBECOM'95*, pp. 1422–1426, November 1995.

[16] S. Li and C. Hwang, On the convergence of traffic measurement and queuing analysis: A statistical-match and queuing (SMAQ) tool, *IEEE/ACM Transactions on Networking*, vol. 5, no. 1, pp. 95–110, February 1997.

[17] A. Andersen and B. Nielsen, A Markovian approach for modeling packet traffic with long-range dependence, *IEEE Journal on Selected Areas in Communications*, vol. 16, no. 5, pp. 719–732, June 1998.

[18] C. Nunes and A. Pacheco, Parametric estimation in MMPP(2) using time discretization, *Proceedings of the 2nd Internation Symposium on Semi-Markov Models: Theory and Applications*, December 1998.

[19] P. Salvador and R. Valadas, A fitting procedure for Markov modulated Poisson processes with an adaptive number of states, in *Proceedings of the 9th IFIP Working Conference on Performance Modelling and Evaluation of ATM & IP Networks*, June 2001.

[20] T. Yoshihara, S. Kasahara and Y. Takahashi, Practical time-scale fitting of self-similar traffic with Markov-modulated Poisson process, *Telecommunication Systems*, vol. 17, no. 1-2, pp. 185–211, 2001.

[21] P. Salvador, A. Pacheco and R. Valadas, Multiscale fitting procedure using Markov modulated Poisson processes, *Telecommunications Systems*, vol. 23, no. 1-2, pp. 123–148, June 2003.

[22] D. Veitch and P. Abry, A wavelet based joint estimator for the parameters of LRD, *IEEE Transactions on Information Theory*, vol. 45, no. 3, pp. 878–897, April 1999.

2

Statistical Inspection and Analysis Techniques for Traffic Data Arising from the Internet

Natalia M. Markovich[1] and Udo R. Krieger[2]

[1]*Institute of Control Sciences, Russian Academy of Sciences, Profsoyuznaya str. 65, 117997 Moscow, Russia; e-mail: markovic@ipu.rssi.ru*
[2]*Information Systems and Applied Computer Science Department, Otto-Friedrich University, D-96045 Bamberg, Germany; e-mail: udo.krieger@ieee.org*

Abstract

The statistical characterization of measurements arising from Internet traffic requires specific analysis and estimation techniques since the underlying traffic characteristics are often distributed with heavy tails. Here we provide a survey of fast and simple methods to detect the heaviness of tails and dependence in traffic data. We consider the dependence of both univariate and bivariate data. Important practical problems regarding the dependence between the transmission rate and file size or the time of transmission are discussed as well. Finally, the recommended tools are illustrated by applications to real Web, TCP and video data.

Keywords: Traffic characterization, heavy tails, data dependence, extremal index, Pickands' function.

Abbreviations

ACF	autocorrelation function
a.s.	almost sure

D. D. Kouvatsos (ed.), Traffic and Performance Engineering for Heterogeneous Networks, 41–60.

DF	distribution function
EVI	extreme value index
HTML	hypertext markup language
i.i.d.	independent, identically distributed
LRD	long range dependence
PDF	probability density function
r.v.	random variable
TCP	transmission control protocol

2.1 Introduction

Currently, traffic measurements of high-speed packet-switched networks that are taken with high resolution at different time scales constitute an important topic of data mining and teletraffic engineering in the Internet. Associated issues such as the statistical data analysis of the measurements at the session and flow levels, the on-line estimation of the underlying random characteristics of the developed Web and Internet traffic models and network management actions based on on-line traffic analysis are studied intensively, too. In this regard it is necessary to analyze the traffic characteristics of the gathered data in a rigorous mathematical manner and to cope very carefully with the related tasks of model selection, estimation and assessment.

Considering the transported traffic in IP networks, besides the dominant peer-to-peer traffic a large portion of the flows are currently still generated by classical client-server applications of the Web (cf. [7, 15, 16]). From the statistical perspective, many observed Web traffic characteristics are determined by independent random variables and often distributed with heavy-tails (see [4, 14, 24]).

Examples of independent univariate data are given by file sizes and session volumes transmitted to an individual customer. Inter-arrival times of packets in aggregated traffic may be dependent. The pairs (file size, duration of its transmission) generated by an individual client-server relationship are independent, but components of these pairs are evidently dependent. Generally, the dependence is not always obvious and requires testing.

Considering the data analysis, the specific features of heavy-tailed distributions are the following:

(1) the tail of the distribution tends to zero at infinity slower than an exponential tail;

(2) the possible non-existence of all or some higher moments;

(3) sparse data at the tail domain of the distribution.

Examples of heavy-tailed distributions are provided by Pareto, Lognormal, and Cauchy distributions as well as Weibull distributions with shape parameter less than 1. They require specific statistical methods for the rigorous investigation of the data (see [18]).

The reconstruction of such heavy-tailed distributions is a difficult issue. The histogram, for instance, provides a good estimate of the controlling probability density function (PDF). In the case of distributions with heavy tails, however, it shows a misleading estimate in the tail domain or over-smoothes the main part of the PDF. The same is true for kernel estimators (cf. [28]). To improve the estimation of a heavy-tailed PDF at the tail, a preliminary transformation of the data from a Pareto to a triangular distribution has been used in [18]. Kernel estimates with variable bandwidth kernels provide another way to reconstruct such PDFs (cf. [28]).

In practice important traffic characteristics like transfer delays often require the analysis of quantiles of the corresponding distribution. Usually, quantiles can be estimated by means of an empirical distribution function (DF). However, high quantiles like 99%, 99.9%, cannot be calculated in such a way since the empirical DF is equal to one outside the range of the empirical sample. In [21] a new estimate for high quantiles has been proposed and its asymptotic normality was proved in [17].

These examples show that it is important to recognize the heavy tails in the traffic characteristics before their further analysis. To resolve it, formal nonparametric tests (cf. [12, 25]) or rough statistical methods for heavy-tailed features can be applied (cf. [6, 18]).

Considering the analysis of heavy-tailed data, the extremes, e.g., large file sizes, long durations of transmission etc., influence significantly on the rate of transmission (see [10, 20]). In this situation the asymptotic distribution (as the sample size tends to infinity) of the maximum of the sample called extreme value distribution is used as a model of the DFs determining the extremes of traffic characteristics (see [18]). Such models include a parameter called tail index. The tail index $\alpha \in \mathbb{R}$ (or the extreme value index (EVI) $\gamma = 1/\alpha$) shows the shape of the tail.

In this context, one has to test the dependence in the underlying univariate data since statistical methods usually require independence and stationarity of the distribution. Often, it is also interesting to know the dependence between pairs of traffic characteristics.

For this purpose we present in this paper several procedures that may help us to detect heavy tails and to analyze the dependence structure of a gathered sample with Internet data.

The material is organized as follows. In Section 2.2 we sketch simple inspection techniques for heavy-tailed traffic characteristics. We discuss statistical procedures to test the dependence in univariate data in Section 2.3 and to determine the dependence in bivariate data in Section 2.4. Finally, we present some conclusions.

2.2 Testing the Heaviness of Tails

2.2.1 Definitions

The formal definitions of heavy-tailed distributions, their subclasses and theoretical properties can be found in [1, 6, 18]. Here we only define the most important and widest class of regularly varying heavy-tailed distributions and formulate its important property.

Definition 1. *The class $\Re_\alpha$ of distributions with regularly varying tails and the tail index $\alpha = 1/\gamma$, $\gamma > 0$, is defined by*

$$\Re_\alpha = \{F : \mathbb{R} \to [0, 1] \mid 1 - F(x) = x^{-\alpha}\ell(x), 0 < x, 0 < \alpha \in \mathbb{R}\},$$

where $\ell(x)$ is a slowly varying function that satisfies the condition $\lim_{x \to \infty} \ell(tx)/\ell(x) = 1$ for all $t > 0$.

Definition 2. *In the case $\alpha = 0$ $\Re_\alpha$ determines a subclass of distributions called super heavy-tailed. These distributions have no finite moments of any order.*

The theoretical property of distributions with regularly varying tails that is important for practice concerns the finiteness of moments. More exactly, the pth moment of the random variable (r.v.) X_1 exists, i.e., $E|X_1|^p < \infty$ holds, if the tail index α satisfies $0 < p < \alpha$ and the distribution belongs to class $\Re_\alpha$ (cf. [6, p. 330]).

The detection of heavy tails and super heavy tails may be provided by formal tests, see [12] and [25], respectively.

2.2.2 Rough Preliminary Methods

There are some simple statistical procedures that allow us to detect heavy tails. These include the ratio of the maximum to the sum, the plot of the empirical mean excess function, the QQ-plot and the tail index estimation.

Let $X_1, \ldots, X_n$ be i.i.d. r.v.s. The statistic (cf. [6, p. 308]) $R_n(p) = M_n(p)/S_n(p)$, $n \geq 1$, $p > 0$, where $S_n(p) = |X_1|^p + \cdots + |X_n|^p$, $M_n(p) = \max(|X_1|^p, \ldots, |X_n|^p)$, $n \geq 1$ indicates the amount of finite moments. More exactly, if $R_n(p)$ is small for large n, then $E|X|^p < \infty$, otherwise it suggests that the pth moment is infinite, i.e. $E|X|^p = \infty$. This suggestion is based on the theoretical property

$$R_n(p) \xrightarrow{\text{a.s.}} 0 \qquad \Leftrightarrow \qquad E|X|^p < \infty.$$

It is the idea of a QQ-plot ("quantiles against quantiles"-plot) to draw the dependence $\{(X_{(k)}, F^{\leftarrow}((n - k + 1)/(n + 1))) : k = 1, \ldots, n\}$, where $X_{(1)} \geq \ldots \geq X_{(n)}$ are the order statistics of the sample, and $F^{\leftarrow}$ is an inverse function of the DF F. Usually, the QQ-plot is built as a dependence of exponential quantiles against the order statistics of the underlying sample. Generally, one can select any distribution instead of an exponential one. The QQ-plot looks close to linear if the model of the distribution F is selected properly.

The mean excess function $e(u) = E(X - u | X > u)$, $0 \leq u < X_F \leq \infty$, where $X_F = \sup\{x \in \mathbb{R} : F(x) < 1\}$ is the finite right endpoint of the distribution, can be tested in practice by its empirical analogue

$$e_n(u) = \sum_{i=1}^{n} (X_i - u)\mathbf{1}\{X_i > u\} \Big/ \sum_{i=1}^{n} \mathbf{1}\{X_i > u\}.$$

Here $\mathbf{1}\{A\}$ denotes the indicator of an event A. The plot of the mean excess function tends to infinity for heavy-tailed distributions (e.g., for a Pareto distribution it is linear), decreases to zero for light-tailed distributions and remains constant for an exponential distribution.

The tail index or its inversion EVI is the most important characteristic of a heavy-tailed distribution. There are numerous procedures to estimate it. The most popular estimators are given by the Hill's estimator [11]

$$\widehat{\gamma}^H(n, k) = \frac{1}{k} \sum_{i=1}^{k} \log X_{(n-i+1)} - \log X_{(n-k)},$$

which is used to estimate the positive EVI γ of a heavy-tailed r.v. X, and the moment estimator [1]

$$\hat{\gamma}^M(n, k) = \hat{\gamma}^H(n, k) + 1 - 0.5 \left(1 - (\hat{\gamma}^H(n, k))^2 / S_{n,k}\right)^{-1}, \qquad (2.1)$$

where $S_{n,k} = (1/k) \sum_{i=1}^{k} (\log X_{(n-i+1)} - \log X_{(n-k)})^2$. $\hat{\gamma}^M(n, k)$ is also valid for real valued EVIs, although it has a larger asymptotic variance than $\hat{\gamma}^H(n, k)$. For both estimators the parameter k indicates the largest order statistics $X_{(n-k)} \leq \ldots \leq X_{(n)}$ of the underlying sample $X_1, \ldots, X_n$ of size n.

Here we present a rather new estimator proposed in [5] that is only valid for positive EVIs. In [18] it has been called the group estimator. Its advantage is determined by the recursive property that gives rise to use it for on-line calculations. According to this estimator the sample of i.i.d. r.v.s. $X_1, \ldots, X_n$ is divided into l groups $V_1, \ldots, V_l$, each group containing m r.v.s., i.e. $n = l \cdot m$. Let

$$M_{li}^{(1)} = \max\{X_j : X_j \in V_i\}$$

and let $M_{li}^{(2)}$ denote the second largest element in the same group V_i. Then the statistic

$$\gamma_l = 1/z_l - 1, \qquad z_l = (1/l) \sum_{i=1}^{l} M_{li}^{(2)} / M_{li}^{(1)}$$

is suggested as an estimate of γ.

All estimators of the tail index are very sensitive to the choice of the smoothing parameters. The latter are determined by k in the cases of the Hill's and moment estimators and the amount of observations m in each group in case of the group estimator. The use of plots where the estimator is represented against a smoothing parameter is the easiest way to select these parameters. One can plot $\{(m, z_m), m_0 < m < M_0\}$, $m_0 > 2$, $M_0 < n/2$ or draw a Hill plot $\{(k, \hat{\gamma}^H(n, k)), 1 \leq k \leq n - 1\}$ and then choose the estimate of z_m or $\hat{\gamma}^H(n, k)$ from an interval in which these functions demonstrate stability.

2.2.3 Application to Web Data

In this subsection we present an example of the application of EVI estimators to Web data. This data have been gathered in the Ethernet segment of the Department of Computer Science at the University of Würzburg and have been analyzed in [14, 18, 21]. The data comprise superimposed traffic flows

Table 2.1 Estimation of the extreme value index for Web traffic characteristics.

Estimate	s.s.s.	d.s.s.	s.r.	i.r.t.
$\hat{\gamma}^H(n,k)$	0.92	0.7	0.92	0.65
γ_l	0.84	0.7	0.8	0.5
$\hat{\gamma}^M(n,k)$	0.8	0.7	0.92	0.6

of client-server sessions monitored at the client site and contain basic characteristics of sub-sessions, i.e., the size of a sub-session (s.s.s) in bytes and its duration (d.s.s.) in seconds, as well as the characteristics of the transferred Web pages, i.e., the size of the response (s.r.) in bytes and the inter-response time (i.r.t.) in seconds. The sample sizes n of both d.s.s. and s.s.s. are given by 373 whereas $n = 7107$ is used for both i.r.t. and s.r.. A detailed description of these data can be found in [18].

In Table 2.1 and Figure 2.1 one can see the estimation of the EVI for the data samples s.s.s., d.s.s., i.r.t. and s.r. by means of the Hill's, group and moment estimators. The parameter m of the group estimate γ_l and the parameter k of both the Hill's and moment estimates were selected by the plots of these estimates against the corresponding parameters. The straight horizontal lines in the plots correspond to the intervals of stability of the EVI estimates and, therefore, state the estimated values of the EVI. Indeed, one may select several stability intervals. Regarding s.s.s., for example, the value 0.69 corresponds to the first stability interval of the moment estimate. Regarding d.s.s. one could also select 0.6 as appropriate Hill's and moment estimates. This feature demonstrates the obvious disadvantage of the selection by a plot and the necessity to use other data-driven methods. The bootstrap procedure is such an alternative method (see [1, 18]).

Observing the estimates of γ, one may conclude that the estimates of the tail index $\alpha = 1/\gamma$ are all positive. It implies that the distributions of all considered Web characteristics are heavy-tailed. Moreover, the α's are less than 2 for all considered data sets. It follows from extreme value theory [6] that the βth moments, $\beta \geq 2$, of the distributions of s.s.s., s.r., d.s.s. and i.r.t. are not finite. This feature follows subject to the assumption that the distributions of all characteristics are regularly varying. The tails of the s.s.s. and s.r. distributions are heavier than the tails of d.s.s. and i.r.t. since their EVIs γ are larger.

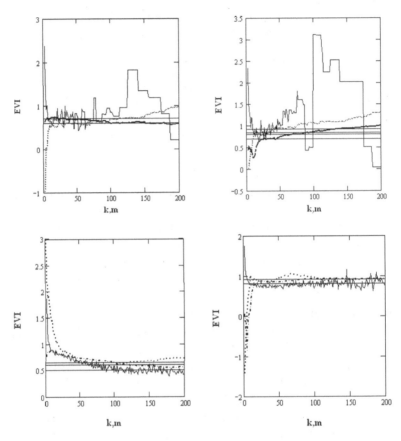

Figure 2.1 *EVI* estimation by the Hill's estimator (dotted line), moment estimator (dashed line) and the group estimator (thin solid line) for the data sets d.s.s., s.s.s., i.r.t., s.r. (from left to right).

2.3 Testing the Dependence of Univariate Data

2.3.1 Classical Measures

The methods stated in Section 2.2 mostly assume that the measurements are independent. Testing independence constitutes a significant part of the analysis. Unfortunately, it is impossible in practice to check formal independence conditions like, e.g., a strong mixing condition, since it requires the knowledge of the marginal and bivariate distributions. Hence, the autocorrelation function (ACF)

$$\rho_X(h) = \rho(X_t, X_{t+h}) = E\left((X_t - EX_t)(X_{t+h} - EX_{t+h})\right) / \mathrm{Var}(X_t)$$

is considered as an important indicator of the dependence structure of a time series.

The standard sample autocorrelation function at lag $h \in \mathbb{Z}$ is determined by

$$\rho_{n,X}(h) = \frac{\sum_{t=1}^{n-h}(X_t - \overline{X}_n)(X_{t+h} - \overline{X}_n)}{\sum_{t=1}^{n}(X_t - \overline{X}_n)^2}, \qquad \overline{X}_n = \frac{1}{n}\sum_{t=1}^{n}X_t. \qquad (2.2)$$

The naive analysis of the ACF includes the comparison of values of the ACF at different lags h and the Bartlett's bounds $\pm 1.96/\sqrt{n}$ (see [2]). Then the observations are assumed to be independent with probability 95% if the ACF falls inside the Bartlett's interval. The Bartlett's bounds are valid only for linear processes with Gaussian noise. The use of this convenient interval is meaningless if the underlying r.v.s. are not Gaussian or a time series belongs to a non-linear process. Moreover, the bounds for heavy-tailed regularly varying distributions that are typical for telecommunications have a complicated form (see [27, §9.5]).

$\rho_{n,X}(h)$ may be inaccurate if the sample size n is small or h is close to n. The relevance of this estimate is determined by its rate of convergence to the real ACF. If the variance is infinite (we have observed this case for Web data in Section 2.2.3), $\rho_{n,X}(h)$ cannot be calculated. In [27, p. 342] it was therefore proposed to use the modified ACF estimate

$$\widetilde{\rho}_{n,X}(h) = \frac{\sum_{t=1}^{n-h} X_t X_{t+h}}{\sum_{t=1}^{n} X_t^2} \qquad (2.3)$$

without the centering by the sample mean. This estimate may not be relevant for non-linear processes. In this context the selection of a proper type of process is a complicated problem.

Hence, we see that the conclusions regarding independence by the observation of the ACF may be unreliable. One can just assume that the observations may be independent if the values of the ACF are small at large lags and fall inside the Bartlett's interval.

Sometimes, the dependence in the time series $\{X_t, t \geq 0\}$ is long-range, i.e. it remains significant over a long time interval. This implies that even if the values of the ACF are small for large lags their cumulative sum may be large, namely

$$\sum_{h=0}^{\infty} |\rho_X(h)| = \infty. \qquad (2.4)$$

Let the ACF be represented for some constant $c_\rho > 0$ by the model

$$\rho_X(h) \sim c_\rho h^{2(H-1)} \quad \text{for large } h,$$

that satisfies the condition (2.4) for $H \in (0.5, 1)$. Then the closer the Hurst parameter H is to one, the deeper is the possible long-range dependence. The Hurst parameter can be calculated by a formula

$$\hat{H}_n = 0.5 \left(1 + \log_2(1 + \rho_{n,X}(1))\right)$$

proposed in [13]. It is obtained under the assumption that the process X_t is second-order self similar with the ACF (see [18, p. 47])

$$\rho_X(h) = \frac{1}{2} \left(|h+1|^{2H} - 2|h|^{2H} + |h-1|^{2H}\right).$$

2.3.2 Estimation of the Extremal Index

Instead of the ACF, we consider here the extremal index as an alternative dependence measure. It arises from the theory of extreme values.

Let $X_1, X_2, \ldots, X_n$ be n (not necessarily independent) r.v.s. arising from some stationary process with marginal DF $F(x)$. Denote by $\widetilde{X}_1, \widetilde{X}_2, \ldots, \widetilde{X}_n$ an associated independent sequence with the same DF. Then it follows for large n and u_n that

$$P\{\max(X_1, \ldots, X_n) \le u_n\} \approx P^\theta\{\max(\widetilde{X}_1, \ldots, \widetilde{X}_n) \le u_n\} = F^{n\theta}(u_n). \tag{2.5}$$

Here, $\theta \in [0, 1]$ is a constant called the extremal index (see [1, 29]). It is clear that $\theta = 1$ holds for i.i.d. sequences. Formula (2.5) implies that the extremal index shows the change in the limiting distribution of the maxima of the sample due to the dependence.

The extremal index has a deeper meaning than only the indication of the dependence. First, it characterizes a cluster structure in the data. The clustering can also be visible on the ACF plots, which implies dependence in the data. For example, the extremal index allows us to divide video data into classes according to the different dependence of frames in the scenes (see [22]).

If, on the other hand, exceedances $\{X_i - u\}, i = 1, \ldots, n$ over a threshold u are considered, θ characterizes the distribution of the inter-exceedance times. A representation of the exponential distribution of the inter-exceedance times with an intensity equal to θ was proved in [8].

Among several estimators of the extremal index we want to mention here the blocks and runs estimators. These are only distinguished by the definition of the cluster. In a way, the extremal index is close to the mean excess function. But the latter shows the sample mean excess over the whole sample and the extremal index over the cluster.

One can define a cluster as a block of the data with at least one exceedance over the threshold. Then the blocks estimator is calculated by the formula:

$$\overline{\theta}^{B}(u) = \frac{k^{-1}\sum_{j=1}^{k}\mathbf{1}(M_{(j-1)r,jr} > u)}{rn^{-1}\sum_{i=1}^{n}\mathbf{1}(X_i > u)}, \qquad (2.6)$$

where $M_{l,j} = \max(X_{i+1}, \ldots, X_j)$, k is the number of blocks, $r = \lfloor n/k \rfloor$ is the number of observations in each block, and $[\cdot]$ denotes the integer part of the number. $1/\overline{\theta}^{B}(u)$ can be simply interpreted as the ratio of the number of observations that exceed the threshold u to the number of clusters. It shows the mean number of exceedances in a cluster.

If we define a cluster as a block of data with some number of exceedances over the threshold and at the same time the r subsequent observations are all below the threshold u, we get the runs estimator:

$$\overline{\theta}^{R}(u) = \frac{(n-r)^{-1}\sum_{i=1}^{n-r}\mathbf{1}(X_i > u, M_{i,i+r} \le u)}{n^{-1}\sum_{i=1}^{n}\mathbf{1}(X_i > u)} \qquad (2.7)$$

In this case no block structure is required. The runs estimate has a better asymptotic bias than the blocks estimate (see [29]).

The selection of the smoothing parameters k in (2.6) or r in (2.7) constitutes a common problem of both estimators. Data-driven methods of this selection pose an open problem. It is the underlying idea to select those parameters which make the clusters independent. The theory is stated in [6, sect. 8.1]. Very roughly speaking, clusters should be sufficiently far away from each other.

The simplest way is to estimate a θ that corresponds to the stable interval of the plot $(1/\overline{\theta}(u), u)$ over a range of thresholds for a fixed parameter k (or r). The reason is that both considered estimates are consistent, i.e. $\theta = \lim_{n\to\infty}\overline{\theta}$.

Sometimes, it is possible to select these parameters by the nature of the problem. In [22], it has been proposed to take scenes of video data as blocks. This selection was motivated by the scene changes with large variations in the bit rate among different scenes. This feature is typical for video traffic due to the visual shifting between scenes. Hence, such blocks have no equal sizes. In this respect the blocks estimator has been modified to a scene blocks estimator

$$\overline{\theta}_S^B(u) = \frac{\sum_{j=1}^k \mathbf{1}(M_{\sum_{m=0}^{j-1} r_m, \sum_{m=1}^{j} r_m} > u)}{\sum_{i=1}^n \mathbf{1}(X_i > u)},$$

where r_j is the number of frames in the jth scene, $\sum_{j=1}^k r_j = n$, $r_0 = 0$ and k is the number of scenes.

2.4 Testing the Dependence of Bivariate Data

2.4.1 Dependence between the Rate, File Size, and Duration of TCP Traffic Flows

In teletraffic engineering we often need to detect the dependence of two r.v.s., such as the basic characteristics of TCP flows. The joint behavior of large values of the file size S and the duration D of a transfer and the throughput or rate R of a session, that is $R = S/D$, is considered by many authors due to its practical importance (cf. [10, 20, 23, 26, 27]).

Regarding this investigation one can summarize the following problems:

- There are ties in the data of the flow size and, hence, the distribution of the size is not continuous.
- Physical limitations in the sizes, durations and rates occur due to differences in the access links or due to the impact of the TCP congestion window and self-congestion.
- No simple classical models for the distributions of size, duration and rate arise, but more complicated structures such as mixtures of Lognormal or Pareto distributions, super heavy-tailed Log-Pareto distributions or regularly varying distributions with varying tail indexes.
- A non-homogeneous dependence structure appears.

In [26] truncated univariate and bivariate Lognormal distributions have been proposed to model the flow size, duration and rate distributions. $\log R = \log(S/D) = \log S - \log D$ is investigated instead of S/D. Further the applied regression models $E(\log S | \log D = y)$ and $E(\log D | \log S = x)$ have been built under the assumption that $\log D$ and $\log S$ are governed by truncated univariate normal distributions. Fortunately, the truncation of the distribution does not reflect much on the regression analysis.

In [27, p. 239] the size, duration and rate are assumed to follow regularly varying distributions with tail indices α_S, α_D and α_R and the independence of

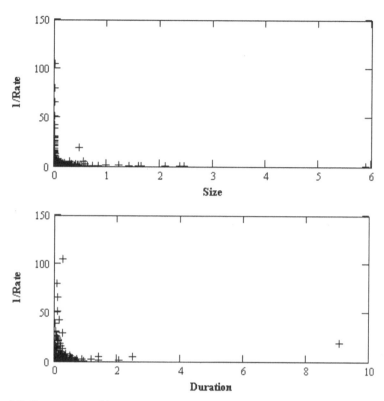

Figure 2.2 Scatter plots of inverse Web session throughput against the size of sub-sessions (top) and duration of sub-sessions (bottom). The "axis hugging" is visible on the top plot suggesting the independence of large sizes and throughputs. For large values the independence of durations and throughputs is not so strict.

R and D is shown by a comparison of the tail indices of the size, α_S, and of $D \cdot R$, $\alpha_D \alpha_R / (\alpha_D + \alpha_R)$, based on Breiman's theorems.

In [10] the independence of R and S and some dependence between R and D are illustrated using data of HTTP responses. The authors have recognized that different strengths of dependence arise between large values and moderate values of a bivariate vector, i.e. the probability of both r.v.s. being large is negligible in case of asymptotic independence. As a consequence, the "axis hugging" by data points at a scatter plot implies extremal independence. The same conclusions regarding the dependence of the pairs (R, S) and (R, D) follow from Figure 2.2. It shows scatter plots of the Web data described in Section 2.2.3.

In [23] the dependence of (D, R) is shown for 80% of the investigated aggregated flows by the examination of the relation $ED \cdot ER/ES = 1$.

In [20] the weak dependence of (S, R) and almost independence of (D, R) of the analyzed TCP-flow data is revealed by the examination of a Pickands' dependence A-function. Subsequently, we shall consider this approach in more detail.

2.4.2 Testing the Dependence by Pickands' Function

The Pickands' function stems from the representation of the limiting distribution of bivariate maxima. It is a convenient form to detect the dependence of extremes of two underlying r.v.s. More exactly, let $(X_1, Y_1), \ldots, (X_n, Y_n)$ be a bivariate i.i.d. sample with a bivariate extreme value distribution G. It implies that for some normalizing constants $0 < a_{j,n} \in \mathbb{R}$ and $b_{j,n} \in \mathbb{R}$, $j = 1, 2$,

$$P\{(M_{1,n} - b_{1,n})/a_{1,n} \le x, (M_{2,n} - b_{2,n})/a_{2,n} \le y\} \to G(x, y), \quad n \to \infty$$

holds, where $M_{1,n} = \max\{X_1, \ldots, X_n\}$, $M_{2,n} = \max\{Y_1, \ldots, Y_n\}$ are the component-wise maxima.

We consider the representation

$$G(x, y) = \exp\left(\log\left(G_1(x)G_2(y)\right) A\left(\frac{\log G_2(y)}{\log\left(G_1(x)G_2(y)\right)}\right)\right),$$

where A is Pickands' function, G_1 and G_2 are univariate extreme value DFs of the maxima $M_{1,n}$ and $M_{2,n}$, i.e. they are limiting DFs of these maxima themselves. In general, the vector $(M_{1,n}, M_{2,n})$ will not be present in the original data.

Let the random pair (X^*, Y^*) have the DF $G(x, y)$. In the bivariate case, the function $A(t)$ satisfies $A(0) = A(1) = 1$, it is convex and lies inside the triangle determined by the points $(0, 1)$, $(1, 1)$ and $(0.5, 0.5)$. The case $A \equiv 1$ corresponds to total independence and the case $A(t) = \max\{(1-t), t\}$ to total dependence of X^* and Y^*.

The correlation between X^* and Y^*

$$\rho = \int_0^1 \frac{dw}{(A(w))^2} - 1$$

is always nonnegative since $A(w) \le 1$ holds (cf. [30]).

In practice, the A-function is evaluated by component-wise maxima over m blocks of data (X_j^*, Y_j^*), $j = 1, \ldots, m$. The best known estimators of the A-function include

$$\widehat{A}_m^{HT}(t) = \left((1/m) \sum_{j=1}^m \min \left(\frac{\xi_j/\overline{\xi}_m}{1-t}, \frac{\eta_j/\overline{\eta}_m}{t} \right) \right)^{-1}$$

proposed in [9], and

$$\log \widehat{A}_m^C(t) = \frac{1}{m} \sum_{j=1}^m \log \max \left(t\xi_j, (1-t)\eta_j \right)$$

$$- t\frac{1}{m} \sum_{j=1}^m \log \xi_j - (1-t)\frac{1}{m} \sum_{j=1}^m \log \eta_j$$

proposed in [3]. Both estimators require a preliminary transformation of the component-wise maxima to new exponentially distributed r.v.s., i.e. $\xi_j = -\log \widehat{G}_1(X_j^*)$ and $\eta_j = -\log \widehat{G}_2(Y_j^*)$, where we use $\overline{\xi} = \frac{1}{m} \sum_{j=1}^m \xi_j$, $\overline{\eta} = \frac{1}{m} \sum_{j=1}^m \eta_j$. To transform the data, one has to estimate first the marginal DFs G_1 and G_2 by the component-wise maxima X_j^* and Y_j^*, e.g., by empirical DFs or parametric models. As such models one can take the Generalized Pareto distribution with the DF

$$\Psi_{\sigma,\gamma}(x) = \begin{cases} 1 - (1 + \gamma x/\sigma)^{-1/\gamma}, & \gamma \neq 0, \\ 1 - \exp(-x/\sigma), & \gamma = 0, \end{cases}$$

where $\sigma > 0$ and $x \geq 0$ if $\gamma \geq 0$ holds and $0 \leq x \leq -\sigma/\gamma$ if $\gamma < 0$ holds, or the Generalized Extreme Value distribution

$$H_\gamma(x) = \begin{cases} \exp\left(-\left(1 + \gamma \left(\frac{x-\mu}{\sigma}\right)\right)^{-1/\gamma} \right), & \gamma \neq 0 \\ \exp(-e^{-(x-\mu/\sigma)}), & \gamma = 0, \end{cases} \tag{2.8}$$

with $1 + \gamma (x - \mu)/\sigma > 0$.

The amount of these maxima may be moderate which may reflect on the accuracy of the DF estimates. The component-wise maxima may not be jointly observable, i.e. there are such pairs of maxima which do not exist in the original sample (the pairs are artificial). Then the question arises whether they should be excluded or not. Some observations may help to provide an answer:

- Artificial pairs do not affect the asymptotical distribution of the bivariate maxima and its Pickands' representation.
- Artificial pairs like those constructed from TCP-flow sizes and durations in a mobile network considered in [19] hug the vertical axis, see Figure 2.3, and hence, their components are independent (cf. [10]). Thus, they do not contribute to the dependence and can be excluded.
- Components of artificial pairs are not artificial itself and contribute to the margins.
- Excluding artificial pairs may lead to the reduction of the number of component-wise maxima and influences the trade-off between the bias and variance of the estimation.
- The accuracy of the estimation is more sensitive to the number of blocks rather than to the artificial pairs.

To check an estimate $\widehat{A}_m$ and to select the number of blocks m for a given sample size n, it is proposed in [20] to use a PP-plot $(\widehat{F}_\chi(\chi_{(i)}), i/m)$, $i = 1, \ldots, m$. Here

$$\widehat{F}_\chi(z) = \frac{z\left(1 + z - \frac{\widehat{A}_m(\frac{1}{1+z})}{\widehat{A}_m(\frac{1}{1+z})}\right)}{(1+z)^2}$$

is the distribution of $\widehat{\chi} = \widehat{\xi}/\widehat{\eta}$, $\widehat{\xi} = -\log \widehat{G}_1(X^*)$ and $\widehat{\eta} = -\log \widehat{G}_2(Y^*)$. If the estimators of the A-function are not convex, they may be improved by taking a convex hull.

To test the required independence, the measure $2(1 - A(1/2))$ proposed in [30] can be used. For the estimator $\widehat{A}_m^C(t)$ it has the following approximate form

$$T_n \approx -\sqrt{n/0.342} \log \widehat{A}_m^C(1/2)$$

(see [3]). The null hypothesis on the independence of two r.v.s. is rejected at level α if T_n exceeds the quantile of order $1 - \alpha$ of the standard normal distribution. The problem of this test is determined by its slow convergence. In other words, its accuracy improves very slowly as the sample size increases. Hence, it can be unreliable for moderate sample size. Applying this test to the TCP-flow data of the mobile network considered in [20], for example, the null hypothesis regarding the independence of R and S, R and D has to be accepted with probability 99%.

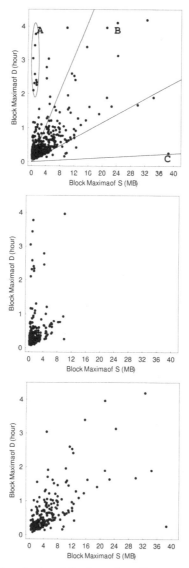

Figure 2.3 Top: a scatter plot of pairs of block maxima $(S_{(n),j}, D_{(n),j})$, $j = 1, \ldots, m$ when the block size is given by $n = 1\,000$ and $m = 610$. The lines, from bottom to top, indicate the access rates 384 kb/s (EDGE), 40.2 kb/s (a typical GPRS rate) and 9.05 kb/s (a minimum GPRS rate), respectively. Middle and bottom: artificial and true data points resulted from the maximization procedure, respectively.

2.5 Conclusions

Considering the statistical characterization of Internet traffic, three items have to be investigated:

(1) the preliminary detection of heavy tails;
(2) the dependence structure of the data;
(3) traffic modeling, model selection, estimation and evaluation.

For heavy-tailed models and, in particular, models with infinite variance, the classical statistical methods are not adequate or applicable and flexible enough. An example is provided by the sample autocorrelation function that has a specific form for heavy-tailed distributions with infinite variance and requires special confidence bounds. Moreover, we should further distinguish between methods that are valid for independent and dependent data.

In this paper we have presented simple methods to detect the heaviness of tails and dependence in data arising from Internet traffic. The exploratory techniques introduced in Section 2.2 like "the ratio of the maximum to the sum" or the group estimator of the tail index started from an i.i.d. assumption on the underlying data. Their interpretation may become hazardous when they are applied to the non-iid case.

In Section 2.3 we have discussed some methods to test the dependence in univariate data and focussed on the estimation of the extremal index. In Section 2.4 techniques to determine the dependence in bivariate data have been stated. They have been illustrated by the application of Pickands' function to TCP-flow data. The estimation of Pickands' function assumes that the underlying (size, duration) pairs of flows are independent, too.

In conclusion, we are convinced that the presented statistical techniques and examples illustrating their application to real Internet data provide a useful guideline for the rigorous teletraffic analysis of various features arising from the measurements of Internet traffic at the time scales of the session and flow levels.

Acknowledgement

The research was partly supported by the FP6-NoE-project EuroFGI under contract 028022.

References

[1] J. Beirlant, Y. Goegebeur, J. Teugels and J. Segers, *Statistics of Extremes: Theory and Applications*, Wiley, Chichester, West Sussex, 2004.

[2] P. J. Brockwell and R. A. Davis, *Time Series: Theory and Methods*, 2nd edition, Springer, New York, 1991.

[3] P. Capéraà, A.-L. Fougères and C. Genest, A nonparametric estimation procedure for bivariate extreme value copulas, *Biometrika*, vol. 84, pp. 567–577, 1997.

[4] M. E. Crovella and A. Bestavros, *Self-Similarity in World Wide Web Traffic: Evidence and Possible Causes*, ACM Sigmetrics, pp. 226–244, 1996.

[5] Y. Davydov, V. Paulauskas and A. Račkauskas, More on P-stable convex sets in Banach spaces, *J. Theoret. Probab.*, vol. 13, no. 1, pp. 39–64, 2000.

[6] P. Embrechts, C. Klüppelberg and T. Mikosch, *Modeling Extremal Events*, Springer, Berlin, 1997.

[7] Eurescom Project P1112, New dimensions – Network dimensioning based on modeling of Internet traffic. Traffic characteristics and statistical estimation, Technical Report D8, Heidelberg, June 2003.

[8] C. A. T. Ferro and J. Segers, Inference for clusters of extreme values, *J. Royal Stat. Soc., Ser. B*, vol. 65, pp. 545–556, 2003.

[9] P. Hall and N. Tajvidi, Distribution and dependence-function estimation for bivariate extreme-value distributions, *Bernoulli*, vol. 6, pp. 835–844, 2000.

[10] F. Hernandez-Campos, J. S. Marron, S. I. Resnick, C. Park and K. Jeffay, Extremal dependence: Internet traffic applications, *Stochastic Models*, vol. 21, no. 1, pp. 1–35, 2005.

[11] B. M. Hill, A simple general approach to inference about the tail of a distribution, *Ann. Statist.*, vol. 3, pp. 1163–1174, 1975.

[12] J. Jurečková and J. Picek, A class of tests on the tail index, *Extremes*, vol. 4, pp. 165–183, 2001.

[13] H. Kettani and J. A. Gubner, A novel approach to the estimation of the hurst parameter in self-similar traffic, in *Proceedings of IEEE Conference on Local Computer Networks*, Tampa, Florida, 2002.

[14] U. R. Krieger, N. M. Markovitch and N. Vicari, Analysis of World Wide Web traffic by nonparametric estimation techniques, in *Performance and QoS of Next Generatione Networking*, K. Guto et al. (Eds.), Springer, London, pp. 67–83, 2001.

[15] U. R. Krieger (Ed.), *New Mathematical Methods, Algorithms and Tools for Measurement, IP Traffic Characterization and Classification*, IST-FP6 NoE EuroFGI, Contract No. 028022, Deliverable D.WP.JRA.5.1.1, December 2007.

[16] U. R. Krieger (Ed.), *Achievements on Measurements, IP Traffic Characterization, Classification and Statistical Methods*, IST-FP6 NoE EuroFGI, Contract No. 028022, Deliverable D-WP-JRA-5.1-2, March 2008.

[17] N. M. Markovich, High quantile estimation for heavy-tailed distributions, *Performance Evaluation*, vol. 62, nos. 1–4, pp. 178–192, 2005.

[18] N. M. Markovich, *Nonparametric Estimation of Univariate Heavy-Tailed Data. Research and Practice*, J. Wiley & Sons, Chichester, 2007.

[19] N. M. Markovich and J. Kilpi, Bivariate statistical analysis of TCP-flow sizes and durations, in *Proceedings Stochastic Performance Models for Resource Allocation*

in *Communication Systems*, Amsterdam, 8–10 November 2006, pp. 47–50, 2006, http://www.cwi.nl/events/2006/StoPeRa/.

[20] N. Markovich and J. Kilpi, Bivariate statistical analysis of TCP-flow sizes and durations, *Annals of Operations Research*, to be published, 2008.

[21] N. M. Markovitch and U. R. Krieger, The estimation of heavy-tailed probability density functions, their mixtures and quantiles, *Computer Networks*, vol. 40, no. 3, pp. 459–474, 2002.

[22] N. M. Markovich, A. Undheim and P. Emstad, Classification of slice-based VBR video traffic and estimation of link loss by exceedance, *Computer Network*, 2009.

[23] R. van de Meent and M. Mandjes, Evaluation of 'user-oriented' and 'black-box' traffic models for link provisioning, in *Proceedings of 1st Conference on Next Generation Internet Design and Engineering*, Rome, Italy, 2005, IEEE, Piscataway, NY, pp. 380–387, 2005.

[24] M. Nabe, M. Murata and H. Miyahara, Analysis and modeling of World Wide Web traffic for capacity dimensioning of Internet access lines, *Performance Evaluation*, vol. 34, pp. 249–271, 1998.

[25] C. Neves and I. Fraga Alves, The ratio of maximum to the sum for testing super heavy tails, in *Advances in Mathematical and Statistical Modeling*, B. C. Arnold, N. Balakrishnan, J. M. Sarabia and R. Minguez (Eds.), Birkhäuser, Boston, pp. 181–194, 2008.

[26] A. Pacheco, C. Pascoal and M. R. de Oliveira, Analysis of internet traffic flows using the truncated bivariate normal distribution, Technical Report, 2007, see [15].

[27] S. I. Resnick, *Heavy-Tail Phenomena. Probabilistic and Statistical Modeling*, Springer, New York, 2006.

[28] B. W. Silverman, *Density Estimation for Statistics and Data Analysis*, Chapman & Hall, New York, 1986.

[29] R. L. Smith and I. Weissman, Estimating the extremal index, *J. Royal Stat. Soc., Ser. B*, vol. 56, no. 3, pp. 515–528, 1994.

[30] J. A. Tawn, Bivariate extreme value theory: Models and estimation, *Biometrika*, vol. 75, no. 3, pp. 397–415, 1988.

3

BitTorrent Session Characteristics and Models
Extended Version

David Erman, Dragos Ilie and Adrian Popescu

*Department of Telecommunication Systems, School of Engineering,
Blekinge Institute of Technology, 371 79 Karlskrona, Sweden;
e-mail: {david.erman, dragos.ilie, adrian.popescu}@bth.se*

Abstract

This paper reports on a modeling and evaluation study of session characteristics of BitTorrent traffic. BitTorrent is a second generation Peer-to-Peer (P2P) application recently developed as an alternative to the classical client-server model to reduce the load burden on content servers and networks. Results are reported on measuring, modeling and analysis of application layer traces collected at the Blekinge Institute of Technology (BTH) and at a local Internet Service Provider (ISP). For doing this, a measurement infrastructure has been developed at BTH to collect P2P traffic. A dedicated modeling methodology has been put forth as well. New results are reported on session characteristics of BitTorrent, and it is observed that session interarrival times can be accurately modeled by the hyper-exponential distribution while session durations and sizes can be reasonably well modeled by the lognormal distribution.

Keywords: BitTorrent, traffic measurements, traffic modeling, traffic self-similarity.

*D. D. Kouvatsos (ed.), Traffic and Performance Engineering for Heterogeneous
Networks,* 61–84.

3.1 Introduction

Over the last years, P2P file sharing systems have evolved to be some of the major traffic contributors in the Internet [19]. Although an exact definition of "P2P systems" is still debatable, such a system typically represents a distributed computing paradigm where a spontaneous, continuously changing group of collaborating computers act as equals in supporting applications such as resource redundancy, content distribution, and other collaborative actions.

There are currently several architectural designs for P2P systems, which follow different strategies used for resource discovery and content distribution [8, 18, 21]. For instance, resource discovery can be done either with the help of a centralized directory (e.g., Napster [22]) or with the help of a decentralized directory (e.g., KaZaa/FastTrack [28]) or with the help of query flooding (e.g., Gnutella [5]). Content distribution is usually performed between peers directly, without further server interaction. In early P2P systems, such as Napster, files were transferred in their entirety. More recent systems such as BitTorrent and later versions of Gnutella employ *swarming*, i.e., the peers download non-intersecting parts of the content from different peers. Specific advantages and drawbacks are associated with every architectural design, as reported in [18, 21].

There are several important consequences related to the appearance of P2P systems. One of the more significant is the high traffic volumes caused by these systems, which are due to both signaling traffic and data traffic. Furthermore, another serious consequence is the high variability introduced by P2P systems in the Internet traffic patterns, with fluctuations that strongly variate in both time and space. For instance, recent measurement studies showed that P2P traffic may come up to 80% of the total traffic in a high speed IP backbone link carrying TCP traffic towards several ADSL areas [3]. In the same study, the authors also observe that both Long-Range Dependence (LRD) properties and the degree of traffic self-similarity seem to reduce with the predominance of P2P traffic. Other open problems in P2P systems are related to the appearance of "mice" (short transfers) and "elephants" (long transfers) phenomena in Internet traffic, scalability, expressiveness, efficiency and robustness of search mechanisms as well as security issues.

Measurement studies and analysis of P2P traffic have been rather limited so far. This is because of the complexity of this task, which involves answering difficult questions related to data retrieval and content location, storage, data analysis and modeling of traffic and topological characteristics as well as privacy and copyright issues.

BitTorrent, a P2P replication and distribution system, has become extremely popular over the last years. According to Cachelogic, the BitTorrent traffic volume has increased from 26% to 52% of the total P2P traffic volume during the first half of 2004 [2]. BitTorrent relies on swarming techniques in combination with the "tit-for-tat" mechanism for creating incentive in content distribution. No search functionality is built into the protocol, and the signaling is geared towards an efficient dissemination of data [13].

There are only a few measurement studies of BitTorrent [15, 26, 27]. This is primarily because the protocol is quite new, only a few years old. In these studies, traffic has been collected from trackers as well as with the help of modified clients. The drawback with using modified clients is related to the accuracy of the timestamps at the application level, which is directly dependent on the type and version of the computer hardware, Operating System (OS) and application software.

The main goals of the paper are towards an understanding of the characteristics of BitTorrent sessions, to be further used in a P2P simulation environment. To this end, we have designed a dedicated measurement system for P2P environments [14]. Detailed results are reported on measuring, modeling and analysis of BitTorrent traffic collected from BTH, Karlskrona as well as at a local ISP with a 5 Mbps link. Our results show that BitTorrent session interarrival times can be accurately modeled by the hyper-exponential distribution while session durations and sizes can be reasonably well modeled by the lognormal distribution. It is also important to mention that additional work has been done on further modeling and analysis of the collected traces, which model the BitTorrent traffic at the message level [11].

The rest of the paper is organized as follows. In Section 3.2 we provide a short overview of the BitTorrent protocol. In Section 3.3 we describe the P2P measurement infrastructure developed at BTH. Section 3.4 reports on the BitTorrent traffic measurements done at BTH and at a local ISP. Section 3.5 presents the traffic metrics used in the evaluation of BitTorrent. In Section 3.6 we describe the modeling methodology used in our experiments. Section 3.7 reports the BitTorrent session characteristics with summary statistics. Finally, Section 3.8 concludes the paper.

3.2 The BitTorrent Protocol

BitTorrent is a P2P protocol for content distribution and replication designed to quickly, efficiently and fairly replicate data [6]. In contrast to other P2P protocols, the BitTorrent protocol does not provide any resource query or lookup functionality, but rather focuses on fair and effective replication and distribution of data. The signaling is geared towards an efficient dissemination of data only. The protocol is fair in the sense that peers exchange content in a tit-for-tat fashion. Non-uploading peers are only sporadically allowed to download. The protocol operates over TCP and uses swarming, i.e., peers download parts, so-called *pieces*, of the content from several peers simultaneously. The consequence of this is efficient network utilization. The size of the pieces is fixed on a per-resource basis and cannot be changed.

A peer interested in downloading some content by using BitTorrent must first obtain a set of metadata, the so-called *torrent* file, to be able to join a set of peers engaging in the distribution of the specific content. In the following we use the term *swarm*, or *distribution swarm*, to define a set of metadata together with the associated network entities.

A BitTorrent distribution swarm can be partitioned into three network entities and two protocols. The first network entity is a centralized software entity, the so-called *tracker*, which keeps lists of connected peers as well as information about their evolution. The tracker replies to peer requests for other peer addresses and ports as well as records simple statistics about the evolution of the swarm. The second entity is the set of active peers, which can be further divided into *seeds* and downloading peers, or *leechers*. A seed is defined to be a peer that already possesses all the content of the swarm, and has stopped downloading data from other peers. A seed may however continue to serve other peers. Also, an initial seed is necessary for peers to be able to start replicating the content. Finally, the third network entity is a server, usually a web server, which provides the metadata required for joining a specific swarm. The distribution of the metadata is not necessarily done via HTTP, but it can be done in any manner. Any way of distributing the torrent file is valid.

The metadata needed to join a BitTorrent swarm consists of the network address information of the tracker (in BitTorrent terminology called the *announce URL*) and resource information such as file size and piece size. An important part of the resource information is a set of Secure Hash Algorithm One (SHA-1) hash values, each corresponding to a specific piece of the resource. These hash values are used to verify the correct reception of a piece. The resource information is also used to calculate a separate SHA-1 hash value, the *info* field, used as an identification of the current swarm. The hash value appears in both the tracker and peer protocols. The metadata does not contain any information regarding the peers participating in a swarm.

The BitTorrent protocols are the tracker protocol and the peer wire protocol. The tracker protocol uses HTTP. Peers make HTTP GET requests and the tracker sends responses in the returning HTTP response data. The purpose of the peer request to the tracker is to locate other peers in the distribution swarm and to allow the tracker to record simple swarm statistics. The peer sends a request containing information about itself and some basic statistics to the tracker, which responds with a randomly selected subset of all peers engaged in the swarm.

The peer wire protocol operates over TCP, and uses in-band signaling for peer communication. Signaling and data transfer are done in the form of a continuous bi-directional stream of fixed-size protocol messages. A P2P session is equivalent to a TCP session, and there are no protocol entities for tearing down a BitTorrent session beyond the TCP teardown itself. Connections between peers are thus single TCP sessions, carrying both data and signaling traffic. Once a TCP connection between two peers is established, the initiating peer sends a handshake message containing the *peer_id* and *info_field* hash (Figure 3.1a). If the receiving peer replies with the corresponding information, the BitTorrent session is considered to be opened and the peers start exchanging messages across the TCP streams. In other cases, the TCP connection is closed. Immediately following the handshake procedure, each peer sends information about the pieces of the resource it possesses. This is done only once, and only by using the first message after the handshake. The information is sent in a *bitfield* message, consisting of a stream of bits, with each bit index corresponding to a piece index.

A peer maintains two states for each peer relationship: *interested* and *choked*. If a peer is choked, then it will not receive any data unless unchoking occurs. Usually, unchoking is equivalent to uploading. The *interested* state indicates whether other peers have parts of the sought content. Interest should be expressed explicitly, as should lack of interest. This means that a peer wishing to download notifies the sending peer (where the sought data is) by sending an *interested* message, and as soon as the peer no longer needs any other data, a *not interested* message is issued. Similarly, for a peer to be allowed to download, it must have received an *unchoke* message from the sending peer. Once a peer receives a *choke* message, it will no longer be permitted to download. This allows the sending peer to keep track of the peers that start downloading when unchoked. A new connection starts out choked and not interested. A peer with all data, i.e., a seed, is never interested.

The choke/unchoke and interested/not interested mechanism provides fairness in the BitTorrent protocol. As it is the transmitting peer that decides

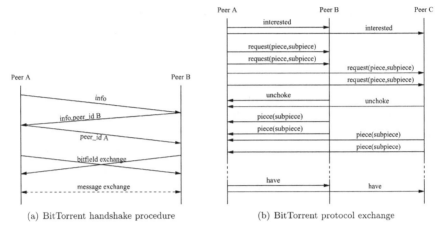

(a) BitTorrent handshake procedure   (b) BitTorrent protocol exchange

Figure 3.1 BitTorrent message exchanges.

whether to allow a download or not, peers not sharing content will be re-
ciprocated in the same manner. To allow peers that have no content to join
the swarm and start sharing, a mechanism called *optimistic unchoking* is employed. From time to time, a peer with content will allow even a non-sharing
peer to download.

Data transfer is done in parts of a *piece* (called *sub-piece*) at a time,
by issuing a *request* message. The sub-pieces are typically of size 16384 or
32768 bytes. To allow TCP to increase the throughput, several requests are
usually sent back-to-back. Each request should result in the corresponding
sub-piece to be transmitted. If the sub-piece is not received within a certain
time (typically one minute), the non-transmitting peer is snubbed, i.e., it is
punished by not being allowed to download, even if unchoked. Data transfer
is done by sending a piece message, which contains the requested sub-piece
(Figure 3.1b). Once the entire piece, i.e., all sub-pieces, has been received,
and the SHA-1 hash of the piece has been verified to be correct, a *have*
message is sent to all connected peers.

3.3 Traffic Measurements

A mixed methodology for traffic measurements of P2P systems has been developed at BTH. Our procedure is based on a combination of instrumentation
at the application layer with transport flow identification and extraction of

Figure 3.2 Measurement procedures.

packets captured at the link-layer (Figure 3.2). This solution allows accurate measurements on both generations of P2P protocols.

The P2P measurement infrastructure developed at BTH consists of peer nodes and protocol decoding software [12]. Tcpdump [16] and tcptrace [23] are used for traffic recording and protocol decoding. Although the infrastructure is currently geared towards P2P protocols, it can be easily extended to measure other protocols running over TCP. Furthermore, we plan to develop similar modules to measure UDP-based applications.

The BTH measurement nodes run the Gentoo Linux 1.4 operating system, with kernel version 2.6.5. Each node is equipped with an Intel Celeron 2.4 GHz processor, 1 GB RAM, 120 GB hard drive, and 10/100 FastEthernet network interface. As shown in Figure 3.3, the network interface is connected to a 100 Mbps switch in the BTH networking lab, which is further connected through a router to the GigaSUNET backbone.

Our experience with the current setup has been that the traffic recording step alone accounts for about 70% of the total time taken by measurements. Protocol decoding is not possible when the hosts are recording traffic. The main reason is the protocol decoding phase, which is I/O intensive and requires large amounts of CPU power and RAM memory. To overcome this problem a distributed measurement infrastructure (shown in Figure 3.4) will be developed in the future.

When used in the distributed infrastructure the P2P nodes are equipped with an additional network interface, which we refer to as the *management* interface. P2P traffic is recorded from the primary interface and stored in a directory on the disk. The directory is exported using the Network File System (NFS) over the management interface. Data processing workstations can read recorded data over NFS as soon as it is available. Optionally, the data processing workstations can be located in private LAN or VPN in order to increase security, save IP address space and decrease the number of collisions on the Ethernet segment. In this case, the router provides Internet access to the workstations, if needed.

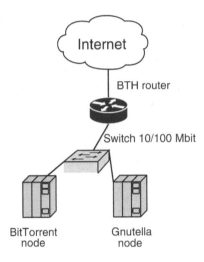

Figure 3.3 Measurement setup.

3.4 BitTorrent Measurements

Two sets of BitTorrent measurements have been performed. The first set used the instrumented version of the reference BitTorrent client as the main measurement tool, with only partial packet capture to determine timestamp accuracy. The second set involved full packet capture and stream reassembly in addition to application logging.

The traffic for the first set of measurements was collected at two different locations over a three-week time period starting on May 3rd, 2004. One location was the BTH networking lab and the other was at a local ISP with a 5 Mbps link. The measurements represent 12 different runs (with lengths of 2 to 12 days) of the instrumented client, 3 of which were run as the only active application. This was done so as to establish a point of reference without applications competing for available bandwidth. To measure more realistic scenarios, the rest of the runs were done with some temporal overlap [13]. A total of 20 GB of uncompressed XML logs were collected in the first set of measurements. After postprocessing, the amount of logs was over 25 GB. The logs contain approximately 100 million protocol messages from almost 300000 individual sessions. The BitTorrent log files contain a list of client software states, e.g., tracker announcements, new connections, choke, unchoke, interested, uninterested, along with the timestamps when the state change took place.

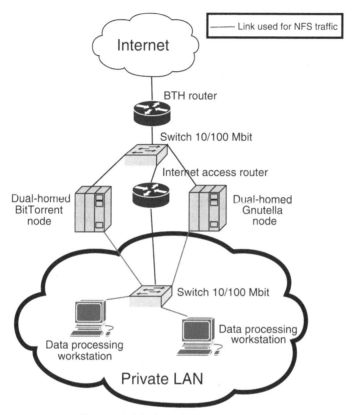

Figure 3.4 Distributed measurement setup.

The second set of traces were collected as `tcpdump` traces at the BTH networking lab during one week, starting June 4th, 2004. A single instance of the reference client was run as the only application on the measurement node. The set contains 150 GB of data, out of which 143 GB are `tcpdump` traces. The rest of the data are application logs and postprocessed logs. Approximately 22 million messages were transmitted in 53000 sessions during the second measurement set.

An important issue regarding traffic measurements in P2P networks is the copyright issue. The most popular content in these networks is often copyrighted material. To circumvent this problem, we joined BitTorrent swarms distributing several popular Linux operating system distributions.

3.5 Traffic Metrics

The BitTorrent client application logs are in essence timestamped protocol events. This means that metrics like interarrival and interdeparture times are readily available by simple calculations. Furthermore, it is possible to compute detailed statistics on several levels of aggregation, which offers the advantage of being able to look into potential burstiness on timescales determined by the timestamp accuracy.

Specific software has been written to extract several important statistics and metrics, to characterize the peer behavior only, and not the entire swarm [11]. The goal is to use accurate characterization and modeling of the behavior of a peer in modeling entire swarms. However, to measure the true size of the swarm, active probing of the tracker is necessary. This is subject for future work.

A number of metrics have been used for the characterization of the BitTorrent signaling traffic [11]. The most important ones are as follows:

Download time. This is the time it takes for the modified client to do a complete download. This metric also provides information about the peer changes from being both a downloading and uploading peer to being a seed, thus offering the possibility to collect statistics about the seed and leecher states.

Session duration and size. A BitTorrent session is equivalent to a TCP session, given that the BitTorrent handshake is completed. As BitTorrent protocol messages are fixed-length messages, there is a one-to-one mapping between the messages sent and received during a session and the session size. A BitTorrent session time is given by the TCP session time, whereas the session size is given by the amount of data transmitted during the TCP session.

Number and type of messages. We count the number of messages of each type in both upstream and downstream directions. Together with the session duration and size, this gives us valuable insights into the behavior of a peer.

Host persistence. We also count the number of unique host IP addresses and peer client IDs. If a given host IP address has a one-to-one mapping to a peer ID and we have a long session time, the peer is considered to be persistent. Persistent peers indicate a healthy swarm in the sense that new peers are more likely to find a larger number of seeds in a swarm with many persistent peers than in swarms with less persistent peers.

Peer swarm size. The peer swarm size refers to the number of peers observed by the measuring client at any given time. This is not the size of the entire swarm, i.e., the total number of collaborating peers, but the number of peers to which the measuring peer is connected. Information about the total swarm size is only available at the tracker, and therefore it is not considered in the reported measurements.

Piece response times. The piece response time is defined to be the time elapsed between the moment of the initial *request* for any subpiece belonging to a given piece to the moment of the transmission of the associated *have* message. This parameter gives the possibility to estimate the downstream bandwidth usage.

Piece popularity. The popularity of a piece is given by the number of requests for any subpiece of a given piece. This gives an indication of the effectiveness of the piece selection algorithms of the requesting peers.

3.6 Modeling Methodology

Detection and estimation of heavy-tailed properties in the distribution of application layer objects is an important part in performance modeling of applications. It may for instance reveal the presence of infinite mean or variance. Accurate estimation of these properties is also important in order to capture the degree of LRD inherent in the objects. Such estimates are also useful in building simulation models that can reproduce traffic conditions as observed in real networks.

Often the random variable possessing heavy tail appears hidden behind another distribution. While the two distributions may have very different tail behavior in a mathematical sense, it may be quite difficult to segregate the two in a practical fitting problem. The crux of the problem lies therefore in determining the cutoff point between the two distributions [7, 17, 29].

The modeling process for mixture models is partitioned into three separate activities: distribution selection, parameter estimation and fitness assessment.

3.6.1 Distribution Selection

The first step is to do a visual inspection of various plots such as histogram (or experimental Probability Density Function (PDF)), Empirical Distribution Function (EDF), complementary cumulative distribution function (CCDF), Hill plots and α-estimation plots [7]. We inspect the lower quantiles of the data using the PDF and CCDF for the upper tail. The CCDF is useful for discerning potentially heavy tail behavior in the distribution such as for file sizes and session durations [20]. The histogram is more suitable for observing metrics in situations where higher frequency behavior is to be modeled, such as for interarrival times. Hill plots give an indication of the amount of heavy tail behavior, and also potential cutoff points in the mixture model case. The α-estimation provides indications of the degree of self-similarity in the data.

The visual inspection helps in eliminating many candidate distributions, and indicates whether a single distribution will suffice or if a mixture model is required. For this work, we have primarily considered single distributions and mixtures of two distributions, as the number of measurements makes the heuristics involved in calculating more cutoff points prohibitively complex.

3.6.2 Parameter Estimation

Based on the candidate distributions selected for modeling, we employ Maximum Likelihood Estimation (MLE) to obtain parameter estimates. Given the large number of sessions available for the measurements, we assume that the obtained parameter estimations are accurate enough to consider the associated distribution fully specified, provided that the confidence intervals for the estimated parameters are within acceptable boundaries.

In the case of single distributions, the parameter estimation is a straightforward procedure, and estimates are obtained from the complete set of data. In the mixture model case, we use successive right censoring as employed in [17] together with an error percentage assessment (described in the following section) to find out the cutoff points for the mixture model.

3.6.3 Fitness Assessment

To determine whether a distribution is representative of the observed data, we employ visual procedures, formal hypothesis tests, and an error percentage assessment. We use visual procedures like histogram and CCDF overplots and Quantile-Quantile (QQ) plots. Overplots give insight in the fitness of the lower and upper tails respectively of a single distribution. We use the QQ plot as a visual aid to assess the representativeness of the chosen model to several measurements simultaneously.

Formally defined goodness-of-fit hypothesis tests such as the Kolmogorov–Smirnov (KS), λ^2 and Anderson–Darling (AD) tests are used to test the null hypothesis H_0 : "*The samples $X_1 \ldots X_n$ are drawn from a distribution $F(x; \Theta)$*" [9]. A major drawback with these types of tests is that they tend to reject the null hypothesis in the case of large sample sets [4]. A possible reason could be the parametrization errors, even if these errors are ever so slight. This is especially true for EDF tests that need modified test statistics, which depend on the size of the data set, e.g., the KS and Cramér–von Mises tests. These tests react to very small discrepancies in the model, thus failing the formal hypothesis test and rejecting the hypothesized model. However, for many purposes (such as simulation), it is sufficient to have a model that is "good enough". Therefore, we only use the AD statistic as an additional goodness-of-fit measure for the metrics where the number of samples is relatively low (less than about 1000 samples). For larger sample sets we use a different method, as described below.

To assess the quality of the fitted distributions in a more quantitative manner, we employ a method similar to the EDF test but that does not suffer as much with increasing size of sample space. For the case of single distribution, the fitness assessment is the final step of the modeling, as we accept the MLE estimated parameters and do not perform any further parameter optimisation. On the other hand, in the case of mixture models, we use this step as part of the process of locating a suitable cutoff point between the distributions making up the mixture.

The method is based on the EDF test for a fully specified distribution, as described in [9]:

1. Obtain the ordered statistics $X_{(1)} < X_{(2)} < \cdots < X_{(n)}$ from the measured data $X_1, X_2, \cdots, X_n$.
2. Transform the ordered statistics by using the probability integral transform (PIT) method with the selected distribution $F(\cdot)$ and estimated parameters $\hat{\Theta}$. If the samples $X_1 \cdots X_n$ are IID samples from some distribution F, then $\hat{U}_i = F(X_i; \hat{\Theta})$, where $i = 1, 2 \ldots n$, are uniformly IID on $[0, 1]$. $\hat{U}_{(i)} = F(X_{(i)}; \hat{\Theta})$ are then ordered samples from the uniform distribution.
3. Obtain the error percentage by using the following expression:

$$E_\% = 100 \times \frac{\sum_{i=1}^{n} |U_{(i)} - \hat{U}_{(i)}|}{n E_{\max}} \tag{3.1}$$

Table 3.1 Fitness quality boundaries.

$E_\%$ ≈	0	1	2	3	4
Degree	perfect	very good	good	fair	poor

where $E_{\max}$ is defined as $\int_0^1 \sup \{U_{(x)}, 1 - U(x)\}\, dx = 0.75$ or, in plain terms, the maximum discrepancy from a true $U[0, 1]$ distribution that may occur, and $U_{(i)}$ are ordered samples from a true uniform distribution.

4. Accept or discard the fitted distribution as "good enough" according to some predefined criteria. In our case, we choose $E_\% \approx 5$ as an upper limit for accepting the fitted match. It is important to mention that this is not a statistical significance level, but rather an acceptable margin of error.

Additionally, fuzzy classification or rough set theory may be employed in quantifying the goodness-of-fit in a more formal way. We use the informal degrees of fitting quality presented in Table 3.1. More formally defined measures, e.g., proper membership functions, are subject of future research.

3.7 Session Characteristics

In this section we report the modeling results for the distributions of session interarrival times, upstream session sizes and durations. Table 3.2 provides a summary of the number of sessions in each of the 13 measurements, except for number 9, which was lost due to hardware failure. It is observed that measurement 6 is different with regards to both mean session size and mean session duration. Further, the maximum session size for this measurement is more than twice that of any other measurement. The mean session size is observed to be about twice that of the corresponding measurement of the same content (measurement 10). As measurements 6 and 10 have large session sizes, it is likely that the session size in this case is related to the total content size (4.3 GB).

The minimum session durations are all set to 0, indicating that all of them are shorter than the accuracy provided by the application logs. These very short sessions are also indicated in the minimum session sizes, and they correspond to a session containing only a handshake or an interrupted handshake. More detailed information is available in [11].

Table 3.2 Session and peer summary.

Measurement number	Sessions	Session duration (s)				Session size (MB)			
		Mean	Max	Min	Std	Mean	Max	Min[a]	Std
1	29712	343	98991	0	2741	27.49	647.26	73	70.65
2	46022	233	117605	0	2316	27.15	646.03	73	64.05
3	28687	465	171074	0	3614	28.54	539.20	73	61.70
4	13493	750	143707	0	3942	49.88	671.99	73	100.65
5	12354	910	180298	0	4504	57.08	668.53	73	116.10
6	10685	1207	223235	0	7016	74.25	3117.79	73	247.74
7	4444	218	46478	0	1642	49.96	431.13	78	76.48
8	17287	231	87026	0	1972	33.11	695.94	73	109.31
10	9701	652	267497	0	5907	37.78	1499.85	73	109.08
11	43939	448	141509	0	3791	17.22	475.86	73	52.73
12	68288	197	292241	0	2580	8.31	987.89	73	30.63
13	52833	465	483996	0	4036	32.2	1652.83	73	99.4

[a]This column measured in bytes.

3.7.1 Session Interarrival Times

The reported distributions refer to interarrival times for remotely initiated sessions during the seeding phase of our measurement peer. We do not consider the leech phase, partly because it is short compared to the seed phase and the number of non-locally initiated sessions is fairly low, and partly because the peer is more active during this phase than during the seed phase. The combination of active peer status and low number of samples that is present during the leech phase (e.g., only 10–20 sessions) makes the analysis more difficult.

We have modeled the session interarrival times by using a two-stage hyperexponential distribution, denoted by H_2. The associated probability density function is

$$H_2(x) = u(x) \left\{ p\lambda_1 e^{-\lambda_1 x} + (1 - p)\lambda_2 e^{-\lambda_2 x} \right\} \qquad (3.2)$$

where λ_1 and λ_2 are the arrival rates for the two exponential terms, p is the probability of an arrival being drawn from the first exponential term, and $u(x)$ is the unit step function. In Figure 3.5 we present examples of visual assessment tools. Figures 3.5a and 3.5b show PDF and CCDF overlay plots for measurement 3. Both indicate a very good fitting for up to 99% probability

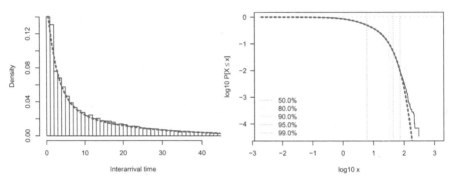

(a) Empirical PDF for measurement 3 with estimate overlaid

(b) CCDF for measurement 3 with estimate overlaid

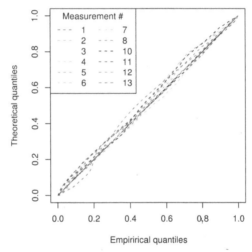

(c) QQ-plot of all measurements subject to $H_2(\hat{\lambda}_1, \hat{\lambda}_2, \hat{p})$

Figure 3.5 Fitness assessment plots.

mass, with most of the errors in the tail of the distribution. Figure 3.5c shows a QQ plot with all measurements.

Parameter estimates for each of the measurements have been obtained by using a maximum likelihood estimator implemented in the R language [1]. The estimation procedure is part of the MASS package for R [30]. It uses the built-in optimization function of R and is based on a gradient algorithm. Table 3.3 reports the parameter estimates and the associated standard devi-

Table 3.3 Fitted hyperexponential parameters.

Measurement number	$\hat{\lambda}_1$	$\hat{\sigma}_{\lambda_1}$	$\hat{\lambda}_2$	$\hat{\sigma}_{\lambda_2}$	$\hat{p}$	$\hat{\sigma}_p$	$E_\%$	Comment
1	0.0593	0.0046	0.1696	0.0085	0.2215	0.0467	2.07367	Pass, fair
2	0.1158	0.0009	0.7556	0.0279	0.7936	0.0066	0.41535	Pass, very good
3	0.0566	0.0006	0.3653	0.0099	0.6575	0.0077	0.49009	Pass, very good
4	0.5372	0.0178	0.0168	0.0002	0.2533	0.0052	2.79455	Pass, fair
5	0.5538	0.0212	0.0162	0.0002	0.2156	0.0052	2.79722	Pass, fair
6	0.4798	0.0174	0.0127	0.0002	0.2879	0.0060	3.93588	Pass, poor
7	0.4188	0.0143	0.0052	0.0001	0.3014	0.0076	2.05430	Pass, good
8	0.5142	0.0113	0.0168	0.0002	0.4252	0.0050	2.79291	Pass, fair
10	0.5581	0.0205	0.0128	0.0002	0.3276	0.0064	3.76412	Pass, poor
11	0.0140	0.0009	0.0802	0.0005	0.0219	0.0024	2.20763	Pass, good
12	0.0935	0.0004	5.8224	0.1380	0.8252	0.0021	3.84606	Pass, poor
13	0.0563	0.0004	0.4175	0.0065	0.5897	0.0048	1.87389	Pass, good

ations obtained in the fitting procedure. Also presented is the $E_\%$ value and the resulting fitness decision and degree.

Summarizing the results for session interarrival times during the seeding phase we observe that all measurements pass according to the selected error criteria. Furthermore it is observed that measurements 2 and 3 have low $E_\%$ values, and they show significance levels of ≈ 0.005 when using the Anderson–Darling test. This is an indication for good quality of fitting for the selected distributions.

The appearance of a hyper-exponential model for session interarrival times is interesting, though not very surprising as Paxson and Floyd showed that network user session interarrival times are exponentially distributed [25]. A BitTorrent session arrival process is in effect filtered through the tracker, since a new peer first needs to contact the tracker to obtain a subset of the total number of peers in the swarm. This provides one hint as to why the hyper-exponential model fits. An additional reason could be that the arrival rates are slightly varying with time. As the model applies to several measurements with different content type, size and popularity, we expect this model to apply to BitTorrent in general, regardless of content characteristics. Current work is being done to verify this assumption.

Table 3.4 Correlation coefficients for session duration and sizes.

Measurement	1	2	3	4	5	6	7	8	10	11	12	13
ρ_{xy}	0.32	0.36	0.29	0.30	0.30	0.34	0.47	0.40	0.67	0.43	0.38	0.25

Table 3.5 Percentages of session sizes exceeding 0 bytes and 1 piece size.

	Measurement	1	2	3	11	12	13
> 0 bytes	Sessions	1558	1619	1795	3092	3793	3438
	% of sessions	5	4	6	7	6	7
$\geq$ 1 pieces	Sessions	1392	1356	1564	1769	2612	3017
	% of sessions	5	3	5	4	4	6

3.7.2 Session Duration and Size

In this section we report the modeling results for the size and duration of remotely initiated peer sessions. We observe that they show fairly high correlation, as shown in Table 3.4.

For reasons similar to those considered at session interarrival times, we consider for modeling the following:

- Measurements with more than 20000 sessions.
- Sessions that have been initiated after the start of the seeding phase.
- Sessions that actually request and receive at least one piece.

The reason for this is threefold:

- As observed in Table 3.5, most sessions do not transfer any data after the initial TCP handshake, with the consequence of a fairly low number of samples (3–6% of the total number of sessions) left for parameter estimation. By including the measurements with fewer sessions, the remaining number of sessions would be inadequate for proper parameter estimations.
- The α-estimations for measurements (Table 3.6) indicate that there could be some heavy tail behavior present in the distributions, as observed in the CCDF plots. The shape in Figure 3.6b is representative for the CCDFs of session duration for all measurements. We observe

Table 3.6 Session α-estimates.

Measurement		1	2	3	11	12	13
duration	$\hat{\alpha}$	1.335	1.264	1.523	1.379	1.272	1.435
	$\hat{\sigma}_\alpha^2$	0.149	0.163	0.116	0.134	0.060	0.176
size	$\hat{\alpha}$	1.176	1.147	1.233	0.961	0.902	1.289
	$\hat{\sigma}_\alpha^2$	0.353	0.339	0.320	0.222	0.147	0.207

clear multi-modal behavior, which means that the heuristic approach of locating the cutoff points, as described in [17], should be used.

The α-estimations in Table 3.6 were obtained using the software described and implemented by Crovella in [7].

- Both session sizes and durations appear to be drawn from a single, similar distribution when inspecting only sessions that have transmitted at least one piece (Figure 3.6).

The models for session size and durations are reported in Tables 3.7 and 3.8, respectively. Only the sessions that actually receive data have been modeled. Lognormal distributions with parameters μ and σ_{LN} have been used for modeling. The lognormal density is given in Eq. (3.3), where μ is the mean and σ_{LN} is the standard deviation.

$$
f(x) = \begin{cases} \dfrac{1}{x\sqrt{2\pi\sigma_{LN}^2}} e^{-(\ln x - \mu)^2/2\sigma_{LN}^2} & x > 0, \sigma > 0, \\ 0 & \text{otherwise.} \end{cases} \tag{3.3}
$$

The second to fifth columns show the estimated parameters, together with the associated estimated standard deviations, for which the best value of $E_\%$ was obtained. The value of $E_\%$ is given in column 8. The sixth column indicates the tail probability mass for which the fitting passed the 5% fitness limit of $E_\%$, while the seventh column shows the tail probability mass for which the *best* value was obtained.

Since the number of samples is substantially smaller than for the hyperexponential models shown in Section 3.7.1, we also calculate the Anderson-Darling statistic for the fitted distribution. Column 9 shows the significance levels obtained in the Anderson–Darling test, under the assumption that the parameter estimates are good enough to assume a fully specified distribution.

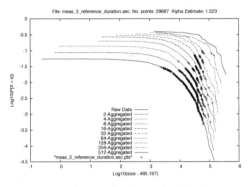

(a) α-estimate plot for session duration

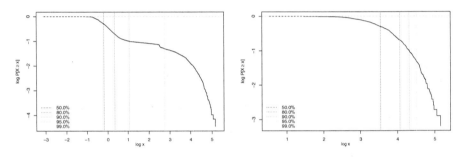

(b) Session duration CCDF for all sessions   (c) Session duration CCDF for sessions with $\geq$ 1 piece

Figure 3.6 α-estimates and CCDF for measurement 3.

The last column shows the fitting decision, together with the result of the AD test passing at the critical level.

Although we expected that a Pareto distribution or a mixture of the Pareto and log-normal distributions would provide a better fitting model, we found that this was not the case. We believe this is due to the limitation in the amount of data available in a BitTorrent swarm. There is no point in a peer downloading more data once the entire content is obtained, and no incentive for a peer to remain in the swarm once it has become a seed.

For sessions not included in the model, i.e., sessions that do not receive any content, we have observed that session durations and sizes can be fairly accurately modeled by the inverted gamma distribution.

Table 3.7 Upstream size parameters.

Measurement number	$\hat{\mu}$	$\hat{\sigma}$	$\hat{\sigma}_{LN}$	$\hat{\sigma}$	Pass	Tail mass	$E_\%$	AD sign.	Comment
1	18.7	0.04	0.62	0.02	0.45	0.21	2.1	> 0.25	Pass, good AD: Pass
2	17.8	0.04	0.99	0.03	1	0.4	2.9	> 0.025	Pass, fair AD: Fail
3	18.4	0.04	0.60	0.02	1	0.24	3.3	> 0.05	Pass, fair AD: Pass
11	14.1	0.06	2.44	0.04	1	0.99	2.4	≈ 0.001	Pass, good AD: Fail
12	13.6	0.05	2.36	0.04	0.86	0.74	3.4	< 0.001	Pass, fair AD: Fail
13	19.0	0.03	0.69	0.02	1	0.17	3.0	> 0.025	Pass, fair AD: Fail

The models obtained for session sizes and durations largely reflect those reported in [24] in that they are in general well-described by a log-normal distribution. In [24], the distributions for the sizes of bulk transfers such as FTP, SMTP and NNTP are shown to closely resemble a log-normal distribution. Since BitTorrent is a bulk transfer application, the similarity of results is not surprising either. Since the models are slightly poorer than the corresponding interarrival time models, we believe that the session sizes and durations are more dependent on the content and swarm characteristics than the individual user behaviours. The relation of the models to these characteristics is also part of ongoing work.

As BitTorrent is purely a data transfer P2P system, we do not expect the results to hold for P2P systems in general, such as e.g. KaZaA or eDonkey [10, 28]. However, we do expect them to be relevant for similar applications or sub-sets of P2P systems, primarily those making use of swarming downloads. Several, if not most, P2P systems currently do this. Studies of these systems with respect to the data transfer-related sessions would therefore be a valuable addition to our work.

Table 3.8 Duration parameters.

Measurement number	$\hat{\mu}$	$\hat{\sigma}$	$\hat{\sigma}_{LN}$	$\hat{\sigma}$	Pass	Tail mass	$E_\%$	AD sign.	Comment
1	8.55	0.03	1.08	0.02	1	0.74	2.2	≈ 0.01	Pass, good AD: Fail
2	8.16	0.04	1.33	0.03	1	0.99	1.5	> 0.15	Pass, good AD: Pass
3	8.17	0.04	1.38	0.02	1	0.98	1.6	> 0.05	Pass, good AD: Pass
11	8.09	0.04	1.56	0.03	1	1	2.4	> 0.001	Pass, good AD: Fail
12	7.2	0.03	1.57	0.02	1	1	3.9	$\ll 0.001$	Pass, poor AD: Fail
13	7.94	0.03	1.52	0.02	1	1	2.3	< 0.001	Pass, good AD: Fail

3.8 Conclusions

A characterization study of BitTorrent application traffic collected at two locations has been reported. Detailed results have been reported on measuring, modeling and analysis of the traffic collected. The modeling activity focuses on typical characteristics of remotely initiated peer sessions during the seeding phase of the measurement peer. New results have been reported on modeling BitTorrent session interarrival times, sizes and durations. Session interarrival times have been observed to be accurately modeled by the hyper-exponential distribution while session durations and sizes have been observed to be reasonably well modeled by the log-normal distribution.

Future work includes modeling of other relevant parameters like peer swarm size and piece popularity, to be further used in a P2P simulation environment as well as verifying our results by new measurement sets.

Acknowledgements

The authors wish to acknowledge the support of The Swedish Governmental Agency for Innovation Systems (VINNOVA) and the EuroNGI and EuroFGI Networks of Excellence.

References

[1] The R Project, http://www.r-project.org, August 2005.

[2] C. A. Parker, The true picture of peer-to-peer file sharing, http://www.cachelogic.com/research/slide9.php, May 2005.

[3] N. Ben Azzouna and F. Guillemin, Experimental analysis of the impact of peer-to-peer applications on traffic in commercial IP networks, *European Transactions on Telecommunications*, Special Issue on P2P Networking and P2P Services, vol. 15, 2004.

[4] J. Beran, *Statistics for Long-Memory Processes*, Chapman & Hall, 1994.

[5] Clip2, *The Annotated Gnutella Protocol Specification v0.4*. The Gnutella Developer Forum (GDF), 1.8th edition, July 2003. http://groups.yahoo.com/group/the_gdf/files/Development/.

[6] B. Cohen, BitTorrent protocol specification, 2005.

[7] M. E. Crovella and M. S. Taqqu, Estimating the heavy tail index from scaling properties. *Methodology and Computing in Applied Probability*, vol. 1, no. 1, pp. 55–79, 1999.

[8] D. Tsoumakos and N. Roussapoulos, A comparison of peer-to-peer search methods, in *Proceedings International Workshop on the Web and Databases (WebDB)*, San Diego, California, USA, 2003.

[9] R. B. D'Agostino and M. A. Stephens (Eds.), *Goodness-of-Fit Techniques*, Dekker, 1986.

[10] eDonkey, http://www.edonkey.com, February 2005.

[11] D. Erman, Bittorrent traffic measurements and models, Licentiate Thesis, Blekinge Institute of Technology, October 2005.

[12] D. Erman, D. Ilie and A. Popescu, Peer-to-peer traffic measurements, Technical Report, Blekinge Institute of Technology, Karlskrona, Sweden, 2005.

[13] D. Erman, D. Ilie, A. Popescu and A. A. Nilsson, Measurement and analysis of BitTorrent traffic, in *Nodic Teletraffic Seminar (NTS) 17*, August 2004.

[14] D. Ilie, D. Erman and A. Popescu, Traffic measurements of P2P systems, in *Proceedings Swedish National on Computer Networking Workshop (SNCNW04)*, November 2004.

[15] M. Izal, G. Urvoy-Keller, E. W. Biersack, P. A. Felber, A. Al Hamra and L. Garcés-Erice, Dissecting BitTorrent: Five months in a torrent's lifetime, in *Proceedings Passive and Active Measurements (PAM2004)*, 2004.

[16] V. Jacobsen, C. Leres and S. McCanne, Tcpdump, http://www.tcpdump.org, August 2005.

[17] A. K. Jena, A. Popescu and A. A. Nilsson, Modeling and evaluation of internet applications, in *Proceedings International Teletraffic Conference (ITC18)*, Berlin, Germany, August 2003..

[18] J. F. Kurose and K. W. Ross, *Computer Networking, A Top-Down Approach Featuring the Internet*, Addison-Wesley, 2003.

[19] T. Karagiannis, A. Broido, N. Brownlee, K. C. Claffy and M. Faloustos, File sharing in the Internet: A characterization of P2P traffic in the backbone, Technical Report, University of California, Riverside, USA, 2003.

[20] B. Krishnamurty and J. Rexford, *Web Protocols and Practice*, Addison Wesley, 2001.

[21] L. L. Peterson and B. S. Davie, *Computer Networks: A Systems Approach*, Morgan Kaufmann, 2003.

[22] Napster, http://www.napster.com, August 2005.

[23] S. Ostermann, Tcptrace, http://www.tcptrace.org, August 2005.

[24] V. Paxson, Empirically derived analytic models of wide-area tcp connections, *IEEE/ACM Transactions on Networking*, vol. 2, pp. 316–336, 1994.

[25] V. Paxson and S. Floyd, Wide area traffic: The failure of Poisson modeling, *IEEE/ACM Transactions on Networking*, vol. 3, no. 3, pp. 226–244, 1995.

[26] J. A. Pouwelse, P. Garbacki, D. H. J. Epema and H. J. Sips, The BitTorrent P2P file-sharing system: Measurements and analysis, in *Proceedings 4th International Workshop on Peer-to-Peer Systems (IPTPS'05)*, February 2005.

[27] D. Qiu and R. J. Srikant, Modeling and performance analysis of bittorrent-like peer-to-peer networks, Technical Report, University of Illinois at Urbana-Champaign, USA, 2004.

[28] Sharman Networks, KaZaA, http://www.kazaa.com, February 2005.

[29] D. M. Titterington, A. F. M. Smith and U. E. Makov, *Statistical Analysis of Finite Mixture Distributions*, John Wiley & Sons, 1985.

[30] W. N. Venables and B. D, Ripley, *Modern Applied Statistics with S*, 4th edition, Springer, http://www.stats.ox.ac.uk/pub/MASS4/, 2002.

4

Adaptive Load Balancing with OSPF

Riikka Susitaival and Samuli Aalto

*Networking Laboratory, Helsinki University of Technology, 02150 Espoo, Finland;
e-mail: {riikka.susitaival, samuli.aalto}@hut.fi*

Abstract

The objective of load balancing is to move traffic from congested links to other parts of the network. If the traffic demands are known, load balancing can be formulated as an optimization problem. The resulting traffic allocation can be realized in the networks that use explicit routes, such as MPLS-networks. It has recently been found that similar load balancing is possible to be implemented even in the IP networks based on OSPF-routing by adjusting OSPF-weights of the links and traffic splitting ratios in the routers. However, if the traffic demands are unknown or they may change rapidly, another approach is needed. In this paper we study adaptive load balancing in OSPF-networks based on measured link loads. We propose an adaptive and distributed algorithm that gradually balances the load by making small changes in the traffic splitting ratios in the routers. We develope different traffic scenarios for testing the algorithm numerically. The results show that the performance of OSPF-networks can significantly be improved by our simple algorithm as compared to the equal splitting.

Keywords: OSPF, Traffic Engineering, adaptive routing, load balancing.

4.1 Introduction

Traditionally traffic is routed along the minimum-hop paths in IP networks. By this approach the usage of link resources is minimized. However, some

D. D. Kouvatsos (ed.), Traffic and Performance Engineering for Heterogeneous Networks, 85–107.

links may become congested while others remain underloaded. Recent studies have shown that although most of the links of networks are clearly under-utilized, the load of a link can be over 90% during traffic spikes [1]. In addition, many emerging applications, such as peer-to-peer applications, are bandwidth intensive in their nature. Thus efficient use of network resources necessitates implementation of some of traffic engineering mechanisms.

The idea of traffic-aware routing, and specially load balancing, is to avoid congested links when traffic is routed from a router to another. There are two distinct methodologies to implement traffic-aware routing in IP networks. The first one uses current routing protocols like Open Shortest Path First (OSPF) [2] but the link weights and the traffic splitting ratios in the routers are defined differently from the traditional approach. The second one takes advantage of some explicit routing protocol like Multi Protocol Label Switching (MPLS) [3] and defines the used paths beforehand.

One of the first proposals to tune OSPF-weights to achieve an optimal load distribution is presented by Fortz and Thorup in [4]. They assume that the routers are bound to split the traffic to a fixed destination equally to the admissible[1] next hops. Under this assumption, however, it is not possible to select the OSPF-weights so that the load distribution is optimal. Even finding the best possible weight setting is shown to be a NP-hard problem. Instead, Fortz and Thorup propose a local heuristic search method for setting the OSPF-weights. In [5] the same authors point out that OSPF-weight changes should be avoided as much as possible, because they confuse the active routing and the performance of TCP goes down.

The load balancing problem in IP networks based on OSPF-routing (OSPF-networks) is also investigated by Wang et al. in [6]. They show that optimal routing in terms of any objective function can be converted to a shortest-path routing with positive link weights if the traffic of each ingress-egress pair can be split arbitrarily to the shortest paths. However, in current IP routing with OSPF, only equal splitting is possible and, furthermore, the splitting in each router is done based on the destination address only.

Sridharan et al. [7] solve the problems appeared in [6]. The source-destination based splitting is easy to convert to destination based splitting by dividing the sums of incoming and outgoing traffic at the node. The problem of the unequal splitting ratios is solved by taking advantage of the existence of multiple prefixes to a certain destination. For a particular prefix only part of

---

[1] A next hop is defined here to be admissible if it belongs to some shortest path to the destination.

next hops are available. As the size of the routing table increases this approach approximates well the arbitrary splitting ratios of the optimal routing.

The problem of the approaches presented above is that the traffic demands are assumed to be known. Defining the traffic demand matrix describing the point-to-point demands in the network is not straightforward task. If the traffic demands are not known, or the traffic conditions may change unexpectedly, another approach is needed. One possibility is to adaptively react to changes in the traffic detected by measurements, such as end-to-end monitoring or monitoring of each link individually. One of mechanisms to to improve the performance of the network adaptively uses a bandwidth estimation method to compute new link weights to be distributed [8] and another attempts to minimize the packet loss rate by optimizing the OSPF weights by online simulation [9]. However, it is not desirable to modify the link weights too frequently, since this can lead to undesired effects in the network performance and in addition, the optimization of link weights requires lots of computation.

Some adaptive mechanisms without link weight optimization developed recently are presented in [10, 11]. In OSPF Optimized Multipath (OSPF-OMP) [10] load is balanced by moving traffic from the paths that include critically loaded segments to other paths according to "move increments", which are defined dynamically. As OSPF-OMP, Adaptive Multi-Path (AMP) establishes multiple paths to each destination by allowing also the use of non-shortest paths and balances load by changing the splitting ratios at each router. However, in AMP a node has only local information of its adjacent links. In paper [11] it remains unclear how much the use of only local information affects the performance of AMP. Another problem is that the studies above do not provide a proper comparison of the algorithms to minimum-hop routing and optimal routing, and the analysis of the stability issues is missing. Thus development of adaptive load balancing in OSPF-networks is required.

In [12] we studied adaptive load balancing in MPLS-networks. In the present paper we study how similar ideas can be applied in OSPF-networks. Our assumption is that the link loads are measured periodically and the information on the measured loads is distributed to all routers. We suggest an adaptive and distributed algorithm to improve the performance of the network without knowledge of the traffic demands. The idea is that, based on the measured link loads, the routers make independently small changes in the load distribution by adjusting their own traffic splitting ratios. The algorithm is simple and does not need any special protocol to distribute measurements.

We develop a numerical evaluation method and a measurement based traffic demand model to test functioning of the algorithm properly.

The rest of the paper is organized as follows. In Section 4.2 we first review a static load balancing problem for off-line optimization of OSPF-weights and then formulate another optimization problem for adjusting the splitting ratios when the paths are fixed. The adaptive and distributed algorithm to optimize the splitting ratios is presented in Section 4.3, and the performance of the proposed algorithm in different test networks and under various traffic conditions is evaluated numerically in Section 4.4. Section 4.5 concludes the paper.

4.2   Load Balancing Based on Known Traffic Demands

In this section we consider a static load balancing problem, in which the traffic demands are assumed to be known. We start with an OSPF-network model. Then we consider the optimization problem in general, after which we review how the OSPF-weights can be determined so that the optimal performance is achieved using shortest path routing. Finally we consider the case where the paths are fixed and only the traffic splitting ratios in the routers may be optimized.

4.2.1   Network Model

Consider an IP network based on OSPF-routing (OSPF-network). Let $\mathcal{N}$ denote the set of nodes (routers) n and $\mathcal{L}$ the set of links l of the network. Alternatively we use notation (i, j) for a link from node i to node j. The capacity of link l is denoted by b_l. The set of ingress-egress (IE) pairs $k = (s_k, t_k)$ is denoted by $\mathcal{K}$ with s_k referring to the ingress node and t_k referring to the egress node of IE-pair k. Let $\mathcal{P}_k$ denote the set of all possible paths p from node s_k to node t_k. We use notation $l \in p$ if link l belongs to path p. The traffic demand of IE-pair k is denoted by d_k.

In the link state based routing protocols like OSPF, each link l is associated with a fixed weight w_l and the traffic is carried along shortest paths. Let $\mathcal{P}_k^{SP}$ denote the set of shortest paths from node s_k to node t_k with respect to link weights w_l,

$$\mathcal{P}_k^{SP} = \left\{ p \in \mathcal{P}_k \ \middle| \ \sum_{l \in p} w_l = \min_{p' \in \mathcal{P}_k} \sum_{l' \in p'} w_{l'} \right\}.$$

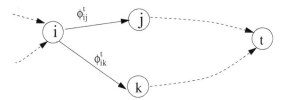

Figure 4.1 The network model.

The standard choice $w_l = 1$ for all l results in minimum-hop paths and, thus, minimizes the total required link bandwidth.

In each node i, the incoming traffic with the same destination t is aggregated and then split to links (i, j) that belong to some shortest path of the ingress-egress pair (i, t). Such adjacent nodes j are called admissible next hops. Let ϕ_{ij}^t denote the corresponding splitting ratios. Thus, ϕ_{ij}^t refers to the fraction of overall traffic passing node i and destined to node t that is forwarded on link (i, j). It is required that

$$\sum_{j:(i,j)\in p \text{ for some } p\in\mathscr{P}_{(i,t)}^{SP}} \phi_{ij}^t = 1.$$

For an illustration, see Figure 4.1. As, e.g., in [4], it is usually assumed that these splitting ratios ϕ_{ij}^t are equal,

$$\phi_{ij}^t = \frac{1}{|\{j' : (i, j') \in p \text{ for some } p \in \mathscr{P}_{(i,t)}^{SP}\}|}.$$

This choice is referred to as Equal Cost Multiple Path (ECMP). However, as mentioned in Section 4.1, there is a method that allows unequal splitting ratios [7].

It may happen that the load for the bottleneck links becomes too high. Then, one naturally asks whether it is possible to choose the link weights w_l and the splitting ratios ϕ_{ij}^t in such a clever way that the maximum link utilization is minimized. This question is answered in the following subsections.

4.2.2 Static Load Balancing Problem

Load balancing[2] can be solved in terms of various objective functions. Minimizing the maximum link utilization and minimizing the mean delay of the

---

[2] Also known as optimal routing.

network are two examples of them. We note that the first one may result in long paths consuming much resources, while the latter one both balances the load and gives preference to short paths. However, since minimizing the maximum link utilization can be formulated as linear programming problem, we concentrate on that in this paper.

The optimal solution to the minimization problem of the maximum link utilization is not unique in general. Among the optimal solutions, the one that minimizes the overall usage of the resources is the most reasonable. Thus it is convenient to formulate an Linear Programming -problem (LP-problem) that minimizes the maximum link utilization with a greater weight but also takes into account the overall usage of the resources with a smaller weight as, e.g., in [6]:

$$
\begin{aligned}
&\text{Minimize } \alpha + r \sum_{k \in \mathcal{K}} \sum_{l \in \mathcal{L}} x_l^k \text{ subject to the constraints} \\
&\alpha \geq 0, \; x_l^k \geq 0, &&\text{for each } k \in \mathcal{K} \text{ and } l \in \mathcal{L}, \\
&\sum_{k \in \mathcal{K}} x_l^k \leq \alpha b_l, &&\text{for each } l \in \mathcal{L}, \\
&Ax^k = R^k, &&\text{for each } k \in \mathcal{K},
\end{aligned}
\tag{4.1}
$$

where α and x_l^k are the free variables describing the minimum of the maximum link utilization and the traffic load of IE-pair k on link l, respectively, and r is some small constant. Furthermore, $A \in \mathbb{R}^{N \times L}$, where $N = |\mathcal{N}|$ and $L = |\mathcal{L}|$, denotes the matrix for which $A_{nl} = -1$ if link l directs to node n, $A_{nl} = 1$ if link l leaves from node n, and $A_{nl} = 0$ otherwise; $x^k \in \mathbb{R}^{L \times 1}$, $k \in \mathcal{K}$, refers to the link load vector with elements x_l^k; and $R^k \in \mathbb{R}^{N \times 1}$, $k \in \mathcal{K}$, denotes the vector for which $R_{s_k}^k = d_k$, $R_{t_k}^k = -d_k$, and $R_n^k = 0$ otherwise.

From the optimal traffic loads x_l^k it is possible to determine the set $\mathcal{P}_k^{\text{LB}}$ of paths p that are used to carry the traffic demand d_k from node s_k to node t_k,

$$
\mathcal{P}_k^{\text{LB}} = \{ p \in \mathcal{P}_k \mid x_l^k > 0 \text{ for all } l \in p \}.
$$

4.2.3 Load Balancing in OSPF-Networks

Wang et al. [6] proved that there is a set of positive link weights w_l so that the optimal paths in the load balancing problem (4.1) are shortest paths with respect to these link weights. In other words, $\mathcal{P}_k^{\text{LB}} \subseteq \mathcal{P}_k^{\text{SP}}$ for all k. The procedure to define these link weights is given below.

Let $\tilde{y}_l = \sum_k \tilde{x}_l^k$ denote the traffic load allocated to link l in the optimal solution $\tilde{x}_l^k$ of the load balancing problem (4.1). Formulate then another LP-problem (primal) and its dual. In the primal LP-problem the induced traffic loads $\tilde{y}_l$ serve as new capacity constraints:

$$\text{Minimize} \sum_{k \in \mathcal{K}} \sum_{l \in \mathcal{L}} x_l^k \text{ subject to the constraints}$$

$$\begin{aligned} x_l^k &\geq 0, & \text{for each } k \in \mathcal{K} \text{ and } l \in \mathcal{L}, \\ \sum_{k \in \mathcal{K}} x_l^k &\leq \tilde{y}_l, & \text{for each } l \in \mathcal{L}, \\ Ax^k &= R^k, & \text{for each } k \in \mathcal{K}, \end{aligned} \qquad (4.2)$$

The dual of the problem above is:

$$\text{Maximize} \sum_{k \in \mathcal{K}} d_k U_{t_k}^k - \sum_{l \in \mathcal{L}} \tilde{y}_l W_l \text{ subject to the constraints}$$

$$\begin{aligned} W_l &\geq 0 & \text{for each } l \in \mathcal{L}, \\ U_{s_k}^k &= 0, & \text{for each } k \in \mathcal{K} \\ U_j^k - U_i^k &\leq W_{(i,j)} + 1, & \text{for each } k \in \mathcal{K} \text{ and } (i, j) \in \mathcal{L}. \end{aligned} \qquad (4.3)$$

The required link weights are then given by $w_l = W_l + 1$, where the variables W_l are determined as the solution to the dual problem.

In addition, optimal destination based traffic splitting ratios ϕ_{ij}^t are determined from the link loads x_l^k of the solution of the primal problem. These splitting ratios are calculated as follows [7]:

$$\phi_{ij}^t = \frac{\displaystyle\sum_{k:t_k=t} x_{(i,j)}^k}{\displaystyle\sum_{j':(i,j')\in\mathcal{L}} \sum_{k:t_k=t} x_{(i,j')}^k} \qquad (4.4)$$

To summarize, the answer to the question posed at the end of Section 4.2.1 is positive: it is possible to choose the link weights w_l and the traffic splitting ratios ϕ_{ij}^t in such a way that the maximum link utilization is minimized.

4.2.4 Optimization of the Splitting Ratios

The traffic control reacting to changes in traffic demands by changing the link weights is often too time consuming or impractical. In such a case with fixed link weights, we can still affect the traffic distribution by optimizing the traffic splitting ratios used in the routers.

We present a procedure to determine the splitting ratios that minimize the maximum link utilization with the given link weights. As before, let $\mathcal{P}_k^{\text{SP}}$ denote the set of shortest paths for IE-pair k with respect to these link weights. Let ϕ_p denote the fraction of traffic demand d_k that uses path $p \in \mathcal{P}_k^{\text{SP}}$. We start by solving these splitting ratios for each IE-pair k from the following LP-problem:

$$\text{Minimize } \alpha + r \sum_{k \in \mathcal{K}} \sum_{l \in \mathcal{L}} \sum_{p \in \mathcal{P}_k^{\text{SP}}:l \in p} d_k \phi_p \text{ subject to the constraints}$$

$$\alpha \geq 0, \ \phi_p \geq 0, \qquad\qquad \text{for each } p \in \bigcup_{k \in \mathcal{K}} \mathcal{P}_k^{\text{SP}},$$

$$\sum_{k \in \mathcal{K}} \sum_{p \in \mathcal{P}_k^{\text{SP}}:l \in p} d_k \phi_p \leq \alpha b_l, \qquad \text{for each } l \in \mathcal{L}, \qquad\qquad (4.5)$$

$$\sum_{p \in \mathcal{P}_k} \phi_p = 1, \qquad\qquad \text{for each } k \in \mathcal{K}.$$

Let ϕ_p be the optimal traffic share on path p. This induces the following link loads:

$$x_l^k = \sum_{p \in \mathcal{P}_k^{\text{SP}}:l \in p} d_k \phi_p.$$

The destination based splitting ratios for each node i can then be calculated as in (4.4).

4.3 Adaptive Load Balancing

The static load balancing problem presented in the previous section is possible to be formulated and solved only if the traffic demands d_k are known. It may well be the case that such information is either imprecise, outdated, or totally missing. In such a case, another approach is needed. In this section we first formulate the corresponding dynamic load balancing problem for OSPF-networks and then describe an adaptive and distributed algorithm to solve the dynamic problem.

4.3.1 Dynamic Load Balancing Problem

Our assumptions are as follows. The traffic demands of the network are un-known. The link loads are periodically measured at times t_n at each node adjacent to the link. Let $\hat{y}_l(n)$ denote the measured link load of link l from the measurement period (t_{n-1}, t_n). The information on the measured loads is distributed to all nodes in the network. This is done by flooding mechanism of OSPF routing protocol and will be described in more detail later. The time needed to distribute the information is negligible in comparison to the length of the measurement period.

In general, the objective of our dynamic load balancing problem is as follows. Based on the measured link loads, the link weights w_l and the traffic splitting ratios ϕ_{ij}^t should be adjusted so that they converge, as soon as possible, to the (unknown) optimal values of the corresponding static load balancing problem presented in Section 4.2.3.

However, as mentioned in section 4.1, it is not desirable to modify the link weights too frequently. Therefore, we consider the dynamic problem in two time-scales. In the shorter time-scale, only the traffic splitting ratios are adjusted but the link weights are kept fixed. The objective in this case is to adjust the traffic splitting ratios so that they converge to the (unknown) op-timal values of the corresponding restricted optimization problem presented in Section 4.2.4. In the longer time-scale, also the link weights should be adjusted so that the optimal load distribution is finally achieved. One way to do it is to estimate the traffic demands from the measurement data and then determine the link weights as a solution to the dual problem presented in Section 4.2.3.

In this paper we focus on dynamic load balancing at the shorter time scale. A destination-based, adaptive and distributed algorithm to solve this problem is described in the following subsection.

4.3.2 Adaptive and Distributed Algorithm for Load Balancing

We assume that the link weights w_l are fixed. For each IE-pair k, let $\mathcal{P}_k^{\mathrm{SP}}$ denote the set of shortest paths from node s_k to node t_k with respect to these link weights w_l.

Let $\phi_{ij}^t(n)$ denote the traffic splitting ratios that are based on the measured link loads $\hat{y}(n) = (\hat{y}_l(n); l \in \mathcal{L})$. We note that, since the measured link loads are distributed to all nodes, the decisions concerning the traffic splitting ratios can be done in a distributed way. Thus, in our adaptive and distributed algorithm, each node i independently determines the traffic splitting ratios ϕ_{ij}^t for all destination nodes t and admissible next hops j.

The decisions in the algorithm are based on a cost function $D_p(y)$ defined for each path $p \in \mathcal{P}_{(i,t)}^{SP}$ by

$$D_p(y) = \max_{l \in p} \frac{y_l}{b_l},$$

where $y = (y_l; l \in \mathcal{L})$. This is a natural choice as the objective is to minimize the maximum link utilization. The idea in the algorithm is simply to alleviate the congestion on the most costly path by reducing the corresponding traffic splitting ratio. This should, of course, be compensated by increasing the splitting ratio related to some other path. A problem in adaptive adjustment of the splitting ratios in a short time-scale is the possible disorder of the packets. However, this can be solved by changing only a part of the splitting ratios at a time, for example.

Since the algorithm is adaptive, we have a closed-loop control problem: the splitting ratios that depend on measured loads have a major effect on the upcoming load measurements. It is well-known that feedback control systems are prone to instability if the gain in the loop is too large. Thus, to avoid harmful oscillations, we let the splitting ratios change only with minor steps. The step size is determined by the granularity parameter g. A finer granularity is achieved by increasing the value of g. The measurement period should be short enough to obtain reasonably fast convergence.

Algorithm. At time t_n, after receiving the information $\hat{y}(n)$ concerning all the measured loads, node i adjusts the traffic splitting ratios for all destination nodes t as follows:

1. Calculate the cost $D_p(\hat{y}(n))$ for each path $p \in \mathcal{P}_{(i,t)}^{SP}$.
2. Find the path $q \in \mathcal{P}_{(i,t)}^{SP}$ with maximum cost, i.e. $D_q(\hat{y}(n)) = \max_{p \in \mathcal{P}_{(i,t)}^{SP}} D_p(\hat{y}(n))$, and decrease the splitting ratio of the first link (i, j) of that path as follows:

$$\phi_{ij}^t(n) = \phi_{ij}^t(n-1) - \frac{1}{g}\phi_{ij}^t(n-1).$$

3. Choose another path $r \in \mathcal{P}_{(i,t)}^{SP}$ randomly and increase the splitting ratio of its first link (i, k) as follows:

$$\phi_{ik}^t(n) = \phi_{ik}^t(n-1) + \frac{1}{g}\phi_{ij}^t(n-1).$$

4. For all other admissible next hops j', keep the old splitting ratio,

$$\phi_{ij'}^{t}(n) = \phi_{ij'}^{t}(n-1).$$

On computational complexity. An important question of the proposed algorithm is its computational complexity, since the feasibility and scalability of the algorithm depends greatly on that feature. In the algorithm each node adjusts its routing parameters in a distributed manner. Thus it is sufficient to estimate the algorithmic complexity at each node separately.

Let $|\mathcal{N}|$ be the size of the network. Given node i proceeds the steps of the algorithm for all other $|\mathcal{N}| - 1$ destination nodes. Let P be the number of parallel shortest paths and L be the length of a path. In the first step of the algorithm, the path cost, which includes L sum operations, is calculated for P paths. Thus the number of the operations in the first step is $L \times P$ per destination.

In the second step the path with maximum cost is selected. The calculation can actually be done in the first step already, if during path cost calculation the momentary maximum of the path costs is stored in memory. Decreasing of splitting ratio in the second step is a standard time operation as well as choosing a random path and decreasing a splitting ratio of that in the third step. Finally, the fourth step does not need calculation at all.

In total, each iteration round of the algorithm at each node requires $|\mathcal{N}| - 1 \times P \times L$ calculations. OSPF specification [2] does not limit the number of parallel paths, but in implementations of IP routers P is typically bounded from 4 to 8 for practical reasons. Path length L depends on the size of the network. The work case is the chain of nodes, where L equals to $|\mathcal{N}| - 1$. However, in random networks L is closer to $\log |\mathcal{N}|$. As the size of the network increases, the number of calculations $|\mathcal{N}| - 1 \times P \times L$ approaches $|\mathcal{N}| \log |\mathcal{N}|$.

4.3.3 Flooding Measurement Information

In OSPF routing the routers exchange information about the link states and the metrics [2]. The flooding protocol is activated when the link state changes occur. The idea of the protocol in its simplicity is the following: Router A detects a change in the status of some of its adjacent links and updates the link state database accordingly. After that, router A sends link state advertisements (LSA) to its neighbors, which forward them again. The routing by the OSPF shortest path calculation does not provide sufficient capabilities to serve different types of traffic properly and to avoid congestion situations in the network. Thus in RFC 3630 [13] new LSAs, also called traffic engineering LSAs, have deployed. These LSAs have a standard header but the payload enables transmission of additional information.

We use these traffic engineering LSAs to spread information of the measured link loads in our adaptive load balancing algorithm. Thus implementation any additional flooding protocol is avoided and signalling overhead of load balancing is negligible. Time needed to flood information is principally the same as the end-to-end network delay, which is short compared to the time scale of load balancing.

4.4 Numerical Performance Evaluation

In this section we evaluate numerically the performance of the proposed adaptive load balancing algorithm. First, in Section 4.4.1, a simple but efficient numerical evaluation method is described. The method is similar to that developed in [12]. Then we propose traffic demand and link failure models used by the evaluation method. Finally, in Section 4.4.4, the results of this evaluation method applied to two different test networks are presented.

4.4.1 Evaluation Method

The evaluation method is iterative and runs as follows. The test network (including the nodes n, links l, IE-pairs k, and paths p), and the link weights w_l. Let d_k denote the deterministic traffic demand of IE-pair k and $d_k(n)$ time-dependent traffic demand of the same IE-pair.[3] Traffic of each IE-pair is initially allocated to the shortest paths with respect to the fixed link weights w_l. If multiple shortest paths exist, traffic is initially split equally in each node (ECMP).

At each iteration n, the measured link loads $\hat{y}_l(n)$ induced by the splitting ratios $\phi_{ij}^t(n-1)$ are calculated as follows. First we calculate, for each IE-pair k, the induced traffic splitting ratios $\phi_p(n-1)$ for each path $p \in \mathcal{P}_k^{\mathrm{SP}}$ by

$$\phi_p(n-1) = \prod_{(i,j) \in p} \phi_{ij}^{t_k}(n-1).$$

---

[3] Note that the traffic demands are used only for the *evaluation* purposes. The algorithm itself does *not* use any information on these demands.

Then the measured link loads $\hat{y}_l(n)$ are determined by

$$\hat{y}_l(n) = \sum_{k \in \mathcal{K}} d_k(n) \sum_{p \in \mathcal{P}_k^{\text{SP}}:l \in p} \phi_p(n-1),$$

After this, the new traffic splitting ratios $\phi_{ij}^t(n)$ are determined from the measured loads $\hat{y}_l(n)$ as presented in Section 4.3.2.

4.4.2 Traffic Demand Models

We use two different approaches to generate traffic demands $d_k(n)$ to the network. The first one is simple deterministic model, and second is Gaussian IID model. We assume that these models capture essential traffic characteristics of the current IP backbones and using more complex models such as autoregressive ARIMA process is not required at the time-scale of traffic engineering.

Deterministic traffic. In the deterministic traffic model the traffic demands $d_k(n)$ are constant over time, that is, $d_k(n) = d_k$. In this model random traffic fluctuations in the time scale of measurements are ignored. The model is rather simplified but gives a possibility to study the convergence of the algorithm well. The length of measurement period is not fixed to any specific time period.

Gaussian traffic with diurnal traffic pattern. Many measurement studies, such as [14, 15], have shown that traffic of IP backbone network has both clear diurnal traffic pattern and noisy fluctuations around the mean. Cao et al. [14] propose a *moving IID Gaussian model* for IE-flows, in which flows of IE-pair k are considered as realizations of corresponding stochastic process $(X_k(n); n = 1, \ldots, N)$. The flow model consists of a deterministic demand $E[X_k(n)]$ and random fluctuating term $D[X_k(n)]Z_k(n)$:

$$X_k(n) = E[X_k(n)] + D[X_k(n)]Z_k(n), \tag{4.6}$$

where $Z_k(n)$ is referred to as *standardized residual*. The residuals are assumed to be independent Gaussian random variables with mean 0 and unit variance.

We apply the above Gaussian model to our study and compose a measurement-based model for the time-dependent traffic demands. Let $m(n)$ be a moving sample-average and $s(n)$ be a moving sample-standard-deviation of measured traffic trace $x(n)$. Time-dependent traffic demand $d_k(n)$ is derived from the measurements by assuming that variation coefficient $c = s(n)/m(n)$ remains constant for the original traffic trace and a derived traffic demand. Thus, traffic demand $d_k(n)$ of IE-pair k can be formulated as

$$d_k(n) = \left[\frac{m(n)}{\max_i m(i)} + \frac{s(n)}{m(n)} z_k(n) \right] d_k, \tag{4.7}$$

where $z_k(n)$ is standardized residual corresponding to $Z_k(n)$ of Cao's model. The measured traffic trace $x(n)$ from one day is obtained from our recent link measurement study [16]. The moving sample-average and standard-deviation using one hour averaging period of the trace are presented in Figure 4.2. We can see that the traffic trace is really unstationary and has diurnal variation.

The relevance of the moving IID Gaussian model is studied more in [16] and its follow-up study. To be short, the model might be valid for the traffic demands in IP backbone networks, if traffic is aggregated from sufficiently many sources.

4.4.3 Link Failures

In addition to variation in traffic demands, the proposed load balancing algorithm should react also to unexpected changes in the traffic capacity. Such event could be a failure of a link, for example. According to the measurement study of Iannaccone et al. [17] about 50% of failures takes less than one minute. Recovery from these short-term failures should be handled by other approaches as load balancing. However, same study shows that 20% of link failures last over 10 minutes. In these long-term link failures by load balancing the performance of the network can be improved. From the measurements of [17] we conclude that median for time between link failures is 4 days. In our evaluation we use very simple link failure model where each link failures during one day with probability $p = 0.1$. The time moment for the failure is random and duration is uniformly distributed between 10 and 100 minutes.

4.4.4 Numerical Results

Two different test networks (see Figure 4.3) are used with the following characteristics:

1. 10 nodes, 52 links, and 72 IE-pairs;
2. 20 nodes, 102 links, and 380 IE-pairs.

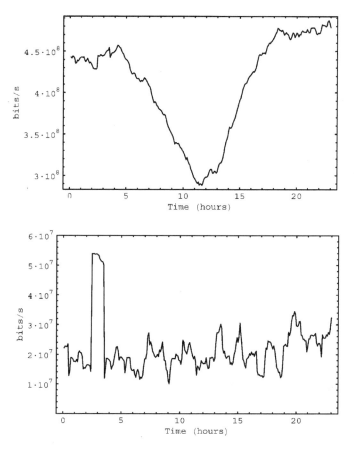

Figure 4.2 Top: Standardized moving sample-average of a link measurement. Bottom: Standardized moving sample-standard-deviation of a link measurement.

From the test-networks we can define the nodes, the links between the nodes, IE-pairs and their deterministic traffic demands d_k. These fixed demands are then used to develop time-dependent traffic demands as explained above. The test networks are random networks generated by the mechanism described in [10].

The results of the adaptive algorithm are compared with the following approaches to forward traffic:

1. "Min-hop": traditional approach where traffic routed to a path that minimizes hop count.

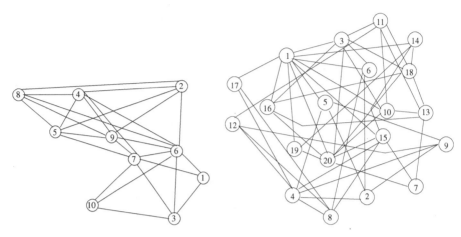

Figure 4.3 Left: 10-node network. Right: 20-node network.

2. "ECMP": the standard policy where the traffic is split equally to the shortest paths with the unit link weights.
3. "Sub-optimal": the optimal value of the restricted optimization problem (4.5) with link weights fixed to 1. This gives a lower bound to the load balancing algorithms that change only the splitting ratios but not the link weights.
4. "Optimal": the optimal value of the static load balancing problem (4.1). The lower bound for all heuristic load balancing algorithms, where both the splitting ratios and link weights can be adjusted.

A huge amount of computation effort is required to obtain optimal and suboptimal reference values for every time moment when traffic demands or link capacities change. For this reason we have optimal values only for part of the test scenarios.

We studied also a scenario, where load balancing was done at two time scales. Earlier we changed only the splitting ratios but now we assume that the link weights can be also optimized at a longer time scale (hours to days). At the longer time scale the optimal link weights and the splitting ratios corresponding to the original traffic demands d_k are determined from the dual problem (4.3) and the primal problem (4.2) of Section 4.2.3, correspondingly. After that the link weights remain unchanged. The splitting ratios are there-after balanced at the short time scale using our proposed algorithm. We refer this load balancing approach of two time scales by "OLW" (Optimal Link Weights).

No traffic fluctuations. In the first scenario out traffic model is deterministic. The shortest paths needed for the adaptive algorithm are calculated
using link weights $w_l = 1$ for all links l. Note that in this scenario the length
of measurement period is not fixed. We study the number of iterations required to convergence of the algorithm instead. The actual convergence time
of the algorithm can be calculated by multiplying the number of iterations by
the length of the measurement period. We have noted that the algorithm works
well when granularity parameter is $g > 20$, which means that approximately
5% of the traffic load is moved at a time.

Figure 4.4 shows the resulting maximum link utilization for the 10-node
and 20-node networks as a function of the number of iterations for granularity
parameters $g = 20$ and $g = 50$. The adaptive algorithm is compared to
minimum hop routing, ECMP, sub-optimal (unit link weights) and optimal
results (also link weights optimized). We can see that in 10-node network
the performance of the adaptive algorithm approaches the sub-optimal value
and decreases the maximum link utilization remarkably as compared to the
minimum hop routing and equal splitting. A small step size in the algorithm
ensures that oscillations are insignificant. In 20-node network the optimal and
sub-optimal results equals and thus optimal performance can be obtained by
the adaptive algorithm.

The explanation for the differences in optimal and sub-optimal results of
10-node and 20-node networks is that, in the 10-node network, the number of
shortest paths related to the unit link weights is 116 whereas it is 153 in the
case of the optimal link weights. Thus the number of ϕ-parameters is larger
in the latter case and also the results are better. In the 20-node network, the
corresponding numbers of the shortest paths are 782 and 785 and the optimal
and sub-optimal results are almost similar.

As a conclusion, in both networks the adaptive algorithm converges to
the lowest possible result. The convergence times are only two times longer
in the 20-node network (approx. 200 iterations) than in the 10-node network
(approx. 100 iterations) in spite of the huge growth in the complexity of the
network.

Next we study the effect of link failures. In 10-node networks 5 links
failed at iterations 9, 66, 148, 228, and 241, correspondingly. The links were
working again at iteration rounds 71, 135, 178, 247, and 264. The results
are presented in Figure 4.5 (top). We compare the adaptive algorithm (with
$g = 20$) to minimum hop routing and ECMP. It can be seen that maximum
link utilization changes dramatically in these events. In the iteration steps
from 148 to 178 there is not enough capacity to carry traffic by minimum

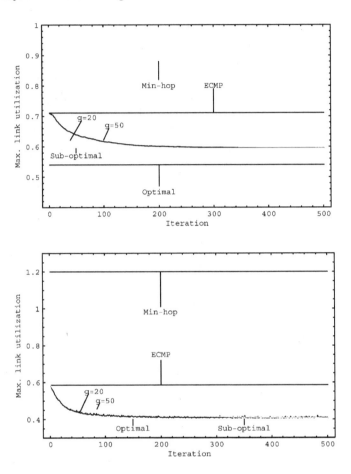

Figure 4.4 The maximum link utilization as a function of the number of iterations, when traffic demand is fixed. Top: 10-node network. Bottom: 20-node network.

hop routing, since the maximum utilization of the links is over 1. By load balancing the maximum utilization remains under 80%. Results of load balancing in the 20-node network are given in Figure 4.5 (bottom). The number of the link failures is 10. Also in this case link failures affects the maximum link utilization greatly. Load balancing reduces the maximum link utilization approximately 55% when compared to minimum-hop routing and 30% when compared to ECMP.

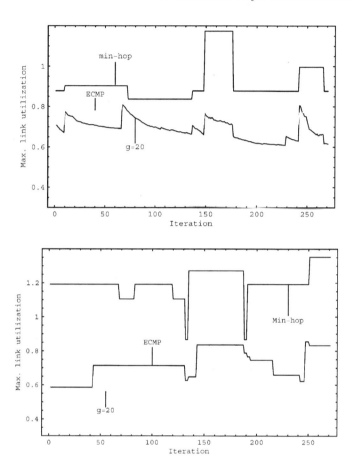

Figure 4.5 The maximum link utilization as a function of the number of iterations, when traffic demand is fixed and there are link failures. Top: 10-node network. Bottom: 20-node network.

Gaussian traffic and diurnal traffic pattern. In this test scenario we use Gaussian IID traffic demand model. The model utilizes the measurements done in a backbone network and thus we are able to check whether the actual convergence time of the algorithm is short enough. In this case we have to also fix the measurement period. Selection of the length of the period and frequence to flood the measurement information is two-fold: the algorithm itself converges as faster as the measurements period is shorter. On the other hand, a short measurement period induces signaling overhead and

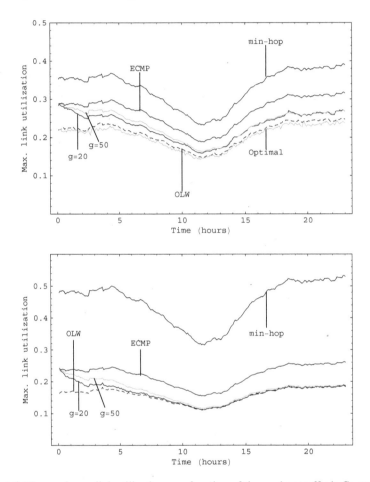

Figure 4.6 The maximum link utilization as a function of time, when traffic is Gaussian and has diurnal pattern. Top: 10-node network. Bottom: 20-node network.

oscillations to the performance of the algorithm. We compromise the period length on 5 minutes, which is also the length of SNMP measurements. The shortest paths needed for the adaptive algorithm are again calculated using link weights $w_l = 1$ for all links l.

Figure 4.6 shows the resulting maximum link utilization during one day for the 10-node and 20-node networks as a function of time for granularity parameters $g = 20$ and $g = 50$. In this case the adaptive algorithm is compared to minimum hop routing, ECMP, optimal (only in 10-node network) and Optimal Link Weight-approach (OLW). Again the proposed algorithm

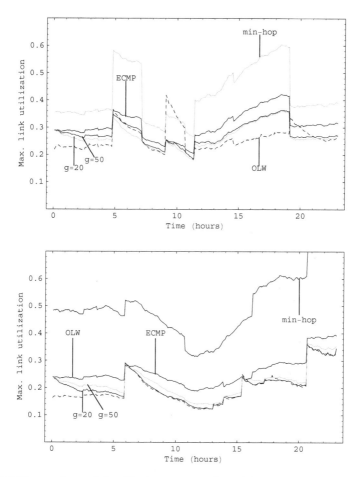

Figure 4.7 The maximum link utilization as a function of time, when traffic is Gaussian and has diurnal pattern and there are link failures. Top: 10-node network. Bottom: 20-node network.

improves the performance of the network as compared to standard policies. The performance of the OLW (dashed line) is almost optimal. The initial convergence time of the adaptive algorithm is 2–3 hours. When the traffic demands change, the algorithm adapts to changes.

Next we combine link failures to Gaussian traffic model. The maximum link utilizations for the 10-node and 20-node networks as a function of time are presented in Figure 4.7. We compare adaptive algorithm ($g = 20$ and $g = 50$) with minimum hop routing, ECMP and OLW. In this scenario the network conditions change both slowly due to diurnal traffic pattern and dramatically due to link failures. Interesting is that although optimizing of the link weights by OLW strategy (dashed line) gives the best performance first, in the time period from 9th to 11th hour the performance of OLW is worse than that of the adaptive algorithm and even EMCP. The reason for poor performance of OLW is that link weights are far from optimal after the link failure. Thus using unit link weights can be a better approach, if the link capacities change much.

Summary of results.   As a conclusion, the proposed adaptive algorithm improved the performance of the network in all test scenarios. When the network capacities changed dramatically, also the adaptive algorithm reacted. The optimization of the link weights at the longer time scale (OLW) improved the performance of the network little bit more but gave also poor performance in some cases. If a 5-minutes measurement period is used, the convergence of the adaptive algorithm is short enough.

4.5 Conclusion

In this paper we studied how load can be balanced adaptively in OSPF-networks using a distributed approach. We have proposed a simple adaptive algorithm for that and tested it in several traffic scenarios. The scenarios included both unstationary traffic demands and link failures. By adaptive algorithm it is possible to decrease the load of the network significantly and thus improve the performance of the network.

We also considered the procedure of optimizing the OSPF-weights by primal-dual methods and how this can be combined to adaptive heuristics. The results show that the optimization of traffic splitting ratios improves the performance of the network when compared to equal splitting. However, when network conditions change much, optimized link weights can produce worse performance than using unit weights.

In the future the approach which combines the shorter and longer time scale optimization has to be developed further. In addition, we have to study if the disorder of packets is really a problem and how this problem can be solved.

Acknowledgements

This work was financially supported by the Academy of Finland (grant No. 74524). We like to also thank anonymous reviewers for valuable comments of the paper.

References

[1] E. J. Anderson and T. E. Anderson, On the stability of adaptive routing in the presence of congestion control, in *Proceedings of IEEE Infocom*, 2003.

[2] J. Moy, OSPF Version 2, IETF RFC2328, April 1998.

[3] E. Rosen, A. Viswanathan and R. Callon, Multiprotocol label switching architecture, IETF RFC3031, January 2001.

[4] B. Fortz and M. Thorup, Internet traffic engineering by optimizing OSPF weights, in *Proceedings of IEEE Infocom*, 2000.

[5] B. Fortz and M. Thorup, Optimizing OSPF/IS-IS weights in a changing world, *IEEE Journal on Selected Areas in Communications*, vol. 20, no. 4, pp. 756–767, 2002.

[6] Y. Wang, Z. Wang and L. Zhang, Internet traffic engineering without full mesh overlaying, in *Proceedings of IEEE Infocom*, 2001.

[7] A. Sridharan, R. Guérin and C. Diot, Achieving near-optimal traffic engineering solutions for current OSPF-IS-IS networks, in *Proceedings of IEEE Infocom*, 2003.

[8] T. B. Pereira and L. L. Ling, Network performance analysis of and adaptive OSPF routing strategy-effective bandwidth estimation, in *Proceedings of ITS*, 2002.

[9] H. T. Kaur, T. Ye, S. Kalyanaraman and K. S. Vastola, Minimizing packet loss by optimizing OSPF weights using on-line simulation, in *Proceedings of the 11th International IEEE/ACM Symposium on Modeling, Analysis and Simulation of Computer and Telecommunication Systems*, MASCOTS-2003, 2003.

[10] C. Villamizar, OSPF optimized multipath (OSPF-OMP), IETF Internet-draft, draft-ietf-ospf-omp-02, February 1999.

[11] I. Gojmerac, T. Ziegler, F. Ricciato and P. Reichl, Adaptive multipath routing for dynamic traffic engineering, in *Proceedings of IEEE Globecom*, 2003.

[12] R. Susitaival, S. Aalto and J. Virtamo, Adaptive load balancing using MPLS, in *Proceedings of MMB&PGTS*, 2004.

[13] D. Katz, K. Kompella and D. Yeung, Traffic engineering (TE) extensions to OSPF version 2, IETF RFC 3630, September 2003.

[14] J. Cao, D. Davis, S. V. Wiel and B. Yu, Time-varying network tomography, *Journal of the American Statistical Association*, vol. 95, pp. 1063–1075, 2000.

[15] M. Roughan, A. Greenberg, C. Kalamanek, M. Rumcewitcz, J. Yates and Y. Zhang, Experience in measuring internet backbone traffic variability: Model, metrics, measurements and meaning, in *Proceedings of ITC*, 2003.

[16] I. Juva, R. Susitaival, M. Peuhkuri and S. Aalto, Traffic characterization for traffic engineering purposes: Analysis of Funet data, in *Proceedings of NGI*, 2005.

[17] G. Iannaccone, C. Chuah, R. Mortier, S. Bhattacharyya and C. Diot, Analysis of link failures in an IP backbone, in *Proceedings of ACM SIGCOMM IMW*, 2002.

5

Modelling of Unconstrained and Constrained H.26x Traffic over IP Networks

S. Kouremenos[1], S. Domoxoudis[2], V. Loumos[2] and A. Drigas[1]

[1]*National Center for Scientific Research "DEMOKRITOS", Institute of Informatics and Telecommunications, P.O. Box 15310 Gr. Ag. Paraskevi, Attiki, Greece*
[2]*National Technical University of Athens "NTUA", School of Electrical and Computer Engineering, Multimedia Technology Laboratory, P.O. Box 15780 Gr. Zographou, Attiki, Greece*

Abstract

In this manuscript, methods for modelling and parameter assessment of unconstrained and constrained videoconference traffic are proposed. In the case of unconstrained traffic the encoder operates in an independent of the network mode (open-loop) while in constrained traffic the encoder has knowledge of the networking constrains and operates using rate-control algorithms (in the loop). The analysis of extensive data that were gathered during experiments with popular videoconference terminals, as well as of traffic traces available in literature, suggested that while the unconstrained traffic traces exhibited high short-term correlations, the constrained counterpart patterns appeared to be mostly uncorrelated, in a percentage not affecting queueing. On the basis of these results, this study discusses methods for accurate modelling and analytical treatment of both types of traffic. Extensive model-based queuing results, in single-source and multiplexed environments, using continuous methods, compared to trace-driven results, confirm the validity of our modelling proposals.

Keywords: Videoconference traffic, unconstrained, constrained, network performance, VBR encoders, modelling, simulation, queueing, H.261, H.263, H.263+, DAR, C-DAR, multiplexing.

D. D. Kouvatsos (ed.), Traffic and Performance Engineering for Heterogeneous Networks, 109–133.

5.1 Introduction

H.26x videoconference traffic is expected to account for large portions of the multimedia traffic in future heterogeneous networks (wire, wireless and satellite). The videoconference traffic models for these networks must cover a wide range of traffic types and characteristics because the type of the terminals will range from a single home or mobile user (low video bit rate), where constrained video traffic is mainly produced, to a terminal connected to a backbone network (high video bit rate), where the traffic is presented to be both constrained and unconstrained. Furthermore, successful videoconference traffic modelling can lead to a more economical network usage (improved traffic policing schemes), leading to lower communication costs and a more affordable and of higher quality service to the end-users.

Partly due to the above reasons, the modelling and performance evaluation of videoconference traffic have been extensively studied in literature and a wide range of modelling methods exist. The results of relevant early studies [2,4–10] concerning the statistical analysis of variable bit rate videoconference streams being multiplexed in ATM networks, indicate that the histogram of the videoconference frame-size sequence exhibits an asymmetric bell shape and that the autocorrelation function decays approximately exponentially to zero. An important body of knowledge, in videoconference traffic modelling, is the approach in [7] where the DAR(1) [1] model was proposed. More explicitly, in this study, the authors noted that AR models of at least order two are required for a satisfactory modelling of the examined H.261 encoded traffic patterns. However, in the same study, the authors observed that a simple DAR(1) model, based on a discrete-time, discrete state Markov Chain performs better – with respect to queueing – than a simple AR(2) model. The results of this study are further verified by similar studies of videoconference traffic modelling [9] and VBR video performance and simulation [8, 12]. In [15], Dr Heyman proposed and evaluated the GBAR process, as an accurate and well performed single-source videoconference traffic model.

The DAR(1) and GBAR(1) models provide a basis for videoconference traffic modelling through the matching of basic statistical features of the sample traffic. On this basis and towards the modelling of videoconference traffic encoded by the Intra-H261 encoder of the ViC tool, Ryu [18] proposed a DAR(p) model using the Weibull instead of the Gamma density for the fit of the sample histogram. In [19], the authors introduced a Continuous Markov chain model, called C-DAR, which is based on the DAR model and is suitable for theoretical analysis. In the same study, the authors concluded that Long Range Dependence (LRD) has minimal impact on videoconference traffic modelling (conclusion also declared in [14]). Looking at the C-DAR model as a Markov modulated rate process, as in [3], the same study's authors applied the fluid-flow method to compare the C-DAR versus a trace-driven simulation. The C-DAR(1) model, via the fluid-flow method, has the advantage of being analytically treated and as a result can be directly applicable to VBR video traffic engineering studies (used as a modelling validation method in the current study).

Relevant newer studies of videoconference traffic modelling reinforce the general conclusions obtained by the above earlier studies by evaluating and extending the existing models and also proposing new methods for successful and accurate modelling. An extensive public available library of frame size traces of unconstrained and constrained MPEG-4, H.263 and H.263+ off-line encoded video was presented in [21] along with a detailed statistical analysis of the generated traces. In the same study, the use of movies, as visual content, led to frames generation with a Gamma-like frame-size sequence histogram (more complex when a target rate was imposed) and an autocorrelation function that quickly decayed to zero (a traffic model was not proposed though in the certain study).

Of particular relevance to our work is the approach in [22], where an extensive study on multipoint videoconference traffic (H.261-encoded) modelling techniques was presented. In this study, the authors discussed methods for correctly matching the parameters of the modelling components to the measured H.261-encoded data derived from realistic multipoint conferences (in "continuous presence" mode).

The above studies certainly constitute a valuable body of knowledge. However, most of the above studies examine videoconference traffic traces compressed by encoders (mainly H.261) that were operating in an unconstrained mode and as a result produced traffic with similar characteristics (frame-size histogram of Gamma form and strong short-term correlations). Today, a large number of videoconference platforms exist, the majority of them operating over IP-based networking infrastructures and using practical implementations of the H.261 [25], H.263 [26, 27] and H.263+ [26, 27] encoders.[1] The above encoders operate on sophisticated commercial software

---

[1] Although a newer encoder, namely, H.264, exists, it is not in a compatible version yet and only two commercial video systems where found to support it, which could not establish a common H.264 communication.

packages that are able of working in both unconstrained and constrained modes of operation. In unconstrained VBR mode, the video system operates independently of the network (i.e. using a constant quantization scale throughout transmission). In the constrained mode, the encoder has knowledge of the networking constraints (either imposed off-line by the user or on-line by an adaptive bandwidth adjustment mechanism of the encoder) and modulate its output in order to achieve the maximum video quality for the given content (by changing the quantization scale, skipping frames or combining multiple frames into one). Furthermore, most of the previous studies have dealt with the H.261-encoding of movies (like Starwars) that exhibit abrupt scene changes. However, the traffic patterns generated by differential coding algorithms depend strongly on the variation of the visual information. For active sequences (movies), the use of a single model based on a few physically meaningful parameters and applicable to a large number of sequences does not appear to be possible. However, for videoconference, this is more probable as the visual information is a typical head and shoulders content that does not contain abrupt scene changes and is consequently more amenable to modelling. Moreover, an understanding of the statistical nature of the constrained VBR sources is useful for designing call admission procedures. Modelling constrained VBR sources, to the best knowledge of these authors, is an open area for study. Our approach towards this direction was to gather video data generated by constrained VBR encoders that used a particular rate control algorithm to meet a predefined channel constraint and then model the resulting trace using techniques similar to those used for unconstrained VBR. The difficulty with this approach is that the resulting model could not be used to understand the behaviour of a constrained VBR source operating with a different rate control algorithm or a different channel constraint. However, given that in constrained VBR the encoder is in the loop, it is more likely that network constraints are not violated and that the source operates closer to its maximum allowable traffic. This may make constrained VBR traffic more amenable to modelling than unconstrained VBR traffic. The basic idea is that we can assume worst case sources (i.e. high motion contents), operating close to the maximum capacity and then characterize these sources.

Taking into account the above, it is important to examine whether the models established in literature are appropriate for handling this contemporary setting in general. It is a matter of question whether all coding strategies result in significantly different statistics for a fixed or different sequence. Along the above lines, this study undertook measurements of the videoconference traffic encoded, during realistic low and high motion head and shoulders experiments, by a variety of encoders of popular commercial software modules operating in both unconstrained and constrained modes. Moreover, the modelling proposal was validated with various traces available in literature [21] (to be referred as "TKN traces" from now on).

The rest of the manuscript is structured as follows: Section 5.2 describes the experiment characteristics and presents the first-order statistical quantities of the measured data. Section 5.3 discusses appropriate methods for parameter assessment of the encoded traffic. In Section 5.3.3, our modelling results are validated through the comparison of model-based and trace-driven simulations. Finally, Section 5.4 culminates with conclusions and pointers to further research.

5.2 The Experimental and Measurement Work

The study reported in this manuscript employed measurements of the IP traffic generated by different videoconference encoders operating in both unconstrained and constrained modes. More explicitly, we measured the traffic generated by the H.26x encoders[2] included in the following videoconference software tools: ViC (version v2.8ucl1.1.6) [32], VCON Vpoint HD [33], France Telecom eConf 3.5 [34] and Sorenson EnVision [35]. These are: H.261, H.263 and H.263+. All traces examined in the current study are representative of the H.26x family video systems. Especially, the ViC video system uses encoders implemented by the open H.323 community [36]. These encoders are based on stable and open standards and as a consequence their examination is more probable to give reusable modelling results.

For all the examined encoders, compression is achieved by removing the spatial (intraframe) and the temporal (interframe) redundancy. In intraframe coding, a transform coding technique is applied at the image blocks, while in interframe coding, a temporal prediction is performed using motion compensation or another technique. Then, the difference or residual quantity is transform coded. Here, we must note that the ViC H.261 encoder [13, 16] performs only intraframe coding oppositely to the H.261 encoders of Vpoint, eConf and EnVision, where blocks are inter or intra coded. The above encoding variations influence the video bit rate performance of the encoders and as a consequence the statistical characteristics of the generated traffic traces.

---

[2] The NV, NVDCT, BVC and CellB encoders [11] were examined in [24] and it was found that they resulted in similar traffic patterns with the H.261 encoder. Thus the modelling proposal for H.261, in the current study, is applicable for these encoders.

At this point, we may discuss about the basic functionality of the examined video systems which is a fundamental factor in the derived statistical features of the encoded traffic and a basic reason of the experiments' philosophy we followed. The rate control parameter (bandwidth and frame rate) sets a traffic policy, i.e. an upper bound on the encoded traffic according to the user's preference (obviously depending on his/her physical link). An encoder's conformation to the rate control of the system is commonly performed by reducing the video quality (and consequently the frame size quantity) through the dynamic modulation of the quantization level. In the case of ViC, a simpler method is applied. The video quality remains invariant and a frame rate reduction is performed when the exhibited video bit rate tends to overcome the bandwidth bound. In fact, in ViC, the video quality of a specific encoder is a parameter determined a priori by the user. In the case of Vpoint, eConf and EnVision, the frame rate remains invariant and a video quality reduction is performed when the exhibited video bit rate tends to overcome the bandwidth bound. This threshold can be set through the network setting of each client. Moreover, Vpoint utilizes adaptive bandwidth adjustment (ABA). ABA works primarily by monitoring packet loss. If the endpoint detects that packet loss exceeds a pre-defined threshold, it will automatically drop to a lower conference data rate while instructing the other conference participant's endpoint to do the same.

Two experimental cases were examined in the current study as presented in Table 5.1 (TKN traces are also included). Case 1 included experiments where the terminal clients were operating in unconstrained mode while Case 2 covered constrained-mode trials. In both cases, two "talking-heads" raw-format video contents were imported in the video systems through a Virtual Camera tool [37] and then peer-to-peer sessions of at least half an hour were employed in order to ensure a satisfactory trace length for statistical analysis. These contents were offline produced by a typical webcam in uncompressed RGB-24 format: one with mild movement and no abrupt scene changes, "listener", (to be referred as VC-L) and one with higher motion activities and occasional zoom/span, "talker" (VC-H). The video size was QCIF (176x144) in both cases and all scenarios (VC-H and VC-L). In Case 1, no constraint was imposed either from a gatekeeper or from the software itself. The target video bit rates that were imposed in Case 2 are shown in Table 5.1. In each case, the UDP packets were captured by a network sniffer

Table 5.1 Statistical quantities of the sample frame-size sequences.

Trace	Client	Encoder	Duration	Content	Target Rate	No Frames	Frame Rate	Rate	Mean	Variance
				Case 1 (UNCONSTRAINED)						
1	ViC	Intra H261	3600	VC-H	700	54006	15	111	921	249670
2				VC-L		53937	15	63	527	174130
3		H263		VC-H		54011	15	72	603	56981
4				VC-L		53453	15	54	457	24588
5		H263+		VC-H		53679	15	39	327	167780
6				VC-L		53633	15	27	224	205200
7	Vpoint	H261	1800	VC-H	700	27275	15	365	3009	731270
8				VC-L		27276	15	193	1592	347080
9	eConf	H263+	3600	VC-H	444	96805	27	298	1387	389300
10				VC-L		91710	25	130	640	147790
11	TKN	H263	2700	OFFICE		33825	13	91	904	107160
12			3600	LECTURE		45459	13	62	618	136850
				Case 2 (CONSTRAINED)						
1	Vpoint	H261	3600	VC-H	155	49596	14	144	1306	48718
2				VC-L		50636	14	143	1275	29285
3	eConf	H263+		VC-H	82	51331	14	84	733	8267
4				VC-L		48848	14	67	619	22758
5	EnVision	H263	1800	VC-H	256	22501	13	239	2394	27275
6	TKN		2700	OFFICE	64	13800	5	64	1565	37316
7			3600	LECTURE		16788	5	64	1715	105530
8			2700	OFFICE	256	13936	5	256	6200	456790
9			3600	LECTURE		17707	5	256	6505	1540500

and the collected data were further post-processed at the frame level[3] by tracing a common packet timestamp. The produced frame-size sequences were used for further statistical analysis.

Specific parameters shown in Table 5.1, for the VC-H and VC-L traces, depend on the particular coding scheme, the nature of the moving scene, and the confidence of the measured statistics. Moreover, traffic traces available in literature where used for further validation. Specifically, the traces used were: "office cam" and "lecture room cam" (from the TKN library). These traces were offline H.263 encoded in a constrained and unconstrained (no target bit rate was set during offline encoding) mode.

Some primary conclusions, as supported by the experiments' results (see Table 5.1), arise concerning the statistical trends of the encoders' traffic patterns. Specifically, H.263+ produces lower video bit rate than H.263 and H.261 do. This was expected, since the earlier encoder versions have improved compression algorithms than the prior ones (always with respect to the rate produced). Finally, for all the encoders, the use of the VC-H content led to higher rate results (as reasonable). Similar results were observed for the mean frame size and variance quantities. In all cases, the variance quantities of the VC-H content were higher than that of VC-L with the exception of the ViC H.263+ encoder (Case 1 – Traces 5, 6) where the opposite phenomenon appeared.

The encoders used for the production of the TKN traces tend to adjust their quality in a "greedy" manner so as to use up as much of the allowed bandwidth as possible. At this point, we must note that Trace 4 of Case 2 is semi-constrained (i.e. the client did not always need the available network bandwidth). However, this particular case can be covered by the "worst-case" Case 2 – Trace 3, where the target rate is reached (full-constrained traffic).

Taking into account the above context, the following questions naturally arise:

- What is the impact of the encoders' differences on the generated videoconference traffic trends?
- Can a common model capture both types of traffic, unconstrained and constrained?
- Are the traffic trends invariant of the constraint rate selected?

---

[3] It is important to note, here, that analysis at the MacroBlock (MB), as in [20], level has been examined and found to provide only a typical smoothing in the sample data. We believe that the analysis at the frame level is simpler and offers a realistic view of the traffic.

- How does the motion of the content influence the generated traffic – for each encoder – and the parameters of the proposed traffic model?
- Can a common traffic model be applied for all the above cases?

The above questions pose the research subject which is thoroughly examined in the context to follow. Their answers will be given along with the respective analysis.

5.3 Traffic Analysis and Modelling Assessment

The measured traffic analysis for all experimental sets confirms the general body of knowledge that literature has formed concerning videoconference traffic. Traffic analysis was employed for all experimental cases. More explicitly, in all cases, the frame-size sequence can be represented as a stationary stochastic process, with a frequency histogram of an approximately bell-shaped (more narrow in the case of H.263 and H263+ encoding) Probability Distribution Function (PDF) form, see Figures 5.1a–c, 5.2a–c more complex in the TKN traces as their content (office and lecture cam) probably contained more scene changes than our contents VC-L and VC-H. Examining more thoroughly the sample histograms, we noted that the smoothed frame-size frequency histograms of the H.261 encoder have an almost similar bell-shape (see Figures 5.1a, b and Figures 5.2a, b) while a more narrow shape appears in the H.263 and H.263+ histograms (Figures 5.1c and 5.2c). The VC-H frequency frame-size histograms appeared to be more symmetrically shaped than the correspondent VC-L histograms. This is reasonable as the rate of the H.26x encoders depends on the activity of the scene, increasing during active motion (VC-H) and decreasing during inactive periods (VC-L).

Furthermore, the AutoCorrelation Function (ACF) of the unconstrained traffic (for all traces of Case 1) appeared to be strongly correlated in the first 100 lags (short-term) and slowly decaying to values near zero (see some indicative Figures 5.3a–c) of the traces of Case 1. On the contrary, the ACFs of the constrained traffic (Case 2) decayed very quickly to zero denoting the lack of short-term correlation (see Figures 5.3d–f). This conclusion is very critical in queueing as the short-term correlation parameter has been found to affect strongly buffer occupancy and overflow probabilities for videoconference traffic. In fact, to verify this assumption, we measured the buffer occupancy of the constrained traces in queueing experiments of different traffic intensities. Buffering was found to be very small at a percentage not affecting queueing. On this basis, it is evident, that for the purpose of modelling of the two types of traffic not a common model can be applied. More explicitly, a correlated model is needed for the case of unconstrained traffic while a simpler non-correlated model is enough for constrained traffic.

The DAR model, proposed in [7], has an exponentially matching autocorrelation and so matches the autocorrelation of the data over approximately hundred frame lags. This match is more than enough for videoconference traffic engineering. Consequently, this model is a proper solution for the treatment of unconstrained traffic. When using the DAR model, it is sufficient to know the mean, variance and autocorrelation decay rate of the source, for admission control and traffic forecasts. A negative feature of the DAR model is that it exhibits "flat spots" which make its sample paths "look" different from those of the data when comparisons are made for a single source (for multiplexed data sources they are indistinguishable). Though these flat spots may not affect traffic engineering, there is another model which is more specialized for modelling accurately the short-term fluctuations of single teleconference sources, namely, GBAR. However, the GBAR model cannot be applied in our study as it is exclusively based on the Gamma density (except from the cases where the Gamma density is proposed).

For the constrained traffic traces, a simple random number generator based on the fit of the sample frame-size histogram can be directly applied. The DAR model with the autocorrelation decay rate value equal to zero can also be a solution. This feature turns constrained videoconference traffic more amenable to traffic modelling than its counterpart unconstrained as only two parameters are needed, the mean and the variance of the sample.

The rest of the paper discusses methods for correctly matching the parameters of the modelling components to the data and for combining these components into the DAR model (to be analytical treated via the C-DAR and the fluid-flow method for unconstrained traffic).

5.3.1 Fitting of the Frame-Size Frequency Histograms of the Traces

A variety of distributions was tested for fitting the sample frame-size frequency histograms. These are the following: Gamma, Inverse Gamma (or Pearson V), Loglogistic, Extreme Value, Inverse Gauss, Weibull, Exponential, Lognormal. The most dominant ones found to be the first three. Even though the Inverse Gauss density performed similarly to the Gamma distribution, it is not included in the analysis to follow, as the Gamma distribution is more popular and simpler. Finally, the Extreme Value distribution performed, in total, worse than the other ones.

For the purpose of fitting the selected distributions' density to the sample frame-size sequence histogram, although various full histogram-based methods (e.g. [22]) have been tried in literature, as well as maximum likelihood estimations (MLE), we followed the approach of the simple moments matching method. This method has the advantage of requiring only the sample mean frame size and variance quantities and not full histogram information. Thus, taking into account that the sequence is stationary – and as a result the mean and the variance values are almost the same for all the sample windows – it is evident that only a part of the sequence is needed to calculate the corresponding density parameters. Furthermore, this method has the feature of capturing accurately the sample mean video bit rate, a property that is not ensured in the case of MLE or histogram-based models. However, in the cases of not satisfactory fit by none of the examined distributions (as in the case of the TKN traces) a histogram-based method can be applied as an unconvential fitting method.

If m is the mean, v the variance of the sample sequence s and m_l the mean and v_l the variance of the logarithm of the sample s, then the distribution functions and the corresponding parameters derived from the moments matching method are given by the following equations, for each distribution correspondingly, Eq. (5.1): Gamma, Eq. (5.2): Inverse Gamma, Eq. (5.3): Loglogistic.

$$f(x) = \frac{1}{\beta \Gamma(\alpha)} \left(\frac{x}{\beta}\right)^{\alpha-1} e^{-x/\beta} \tag{5.1}$$

where $\alpha = m^2/v$, $\beta = v/m$ and $\Gamma(\alpha) = \int_0^\infty u^{\alpha-1} e^{-u} du$;

$$f(x) = \frac{1}{\beta \Gamma(\alpha)} \cdot \frac{e^{-\beta/x}}{\left(\frac{x}{\beta}\right)^{(a+1)}} \tag{5.2}$$

where $a = m^2/v + 2$ and $\beta = m(m^2/v + 1)$;

$$f(z) = \frac{e^{[z(1-\sigma)-m_l]}}{\sigma(1 + e^z)^2} \tag{5.3}$$

where $z = (\ln(x) - m_l)/\sigma$, $m_l = \mathbb{E}[\ln(s)]$ and $\sigma = \sqrt{3\mathbb{E}[\ln(s)]}/\pi$.

Given the dominance of the above distributions, modelling analysis and evaluation will be presented for the above three densities. The numerical results (densities' parameters) from the application of the above parameters-matching methods appear in Table 5.2. At this point, we must note that the Loglogistic density (5.3), although it provides better fits in the H.263+ cases (as will be commented upon later) exhibits mean and variance values that slightly deviate from the sample counterpart values. However, this is negligible with respect to queuing as concluded by the fluid-flow simulations presented in Section 5.4.

The modelling evaluation of the above methods has been performed from the point of queueing. As a consequence, we thoroughly examined fits of cumulative distributions. This was done as follows: we plotted the sample quantiles from the sample cumulative frequency histogram and the model quantiles from the cumulative density of the corresponding distribution. The Q-Q plot of this method refers to cumulative distributions (probabilities of not exceeding a threshold).

Figures 5.1a–c and 5.2a–c present Q-Q plots for all traces of both Cases 1 and 2, respectively. The results suggest that for fitting videoconference data, the coding algorithm used should be taken into consideration. There seems to be a relationship between the coding algorithms and the characteristics of the generated traffic. For instance, for H.261, in most cases, the dominant distribution is Gamma (5.1), as can be verified from the Q-Q plots depicted in Figures 5.1a and b, and for H.263 and H.263+, the Loglogistic density (5.3) has a more "stable" performance than the other two (Q-Q plots shown in Figures 5.1c and 5.2c. The Inverse Gamma density (5.2) seems to be suitable for H.263 traffic (see Figure 5.1c) although it was outperformed by the Loglogistic density in some cases. However, as will be commented upon later, it did not provide a solution in all cases of constrained traffic.

We must note that in Case 2, where a constrain was imposed, the moments matching method for calculating the distribution's parameters did not always provide a good fit, and performed as shown in Figures 5.2a–c (Inverse Gamma and Loglogistic are depicted; the Gamma density provided similar fit). To provide an acceptable fit, a histogram-based method proposed for H.261 encoded traffic in [22], known as C-LVMAX, was used. This method relates the peak of the histogram's convolution to the location at which the Gamma density achieves its maximum and to the value of this maximum. The values of the shape and scale parameters of the Gamma density are derived from: $a = (2\pi x_{max}^2 f_{max}^2 + 1)/2$ and $b = 1/(2\pi x_{max} f_{max}^2)$ where f_{max} is the unique maximum of the histogram's convolution density at x_{max}. Numerical values for this fit appear also in Table 5.2 (for Case 2 only). Figures 5.2a–c show how the three distributions fit the empirical data using the method of moments (Inverse Gamma, Loglogistic) and the C-LVMAX method (Gamma C-LVMAX). The Inverse Gamma density could not be calculated for all the constrained traces (Case 2 – Traces 5, 6, 8, 9), due to processing limitations

Table 5.2 Parameter values of the modeling components.

Trace	Gamma		Inverse Gamma		Loglogistic		C-LVMAX		Exponential		
	α	β	α	β	m_l	σ	α	β	w	λ_1	λ_2
	Case 1 (UNCONSTRAINED)										
1	3.3971	271.1005	5.3971	4049.5	6.648	0.3549			0.5057	0.9925	0.7757
2	1.5944	330.4804	3.5944	1367	5.9603	0.4449			0.5575	0.9905	0.8709
3	6.3878	94.4476	8.3878	4457.1	6.3382	0.1909			0.4368	0.9915	0.7528
4	8.5059	53.7656	10.5059	4347.3	6.0789	0.1609			0.3083	0.9906	0.8269
5	0.6385	512.6355	2.6385	536.2591	5.5827	0.3051			0.0683	0.992	0.5528
6	0.2454	914.5151	2.2454	279.4386	5.1082	0.318			0.0454	0.994	0.5493
7	12.3817	243.024	14.3817	40266	7.9663	0.1661			0.5322	0.991	0.8192
8	7.2998	218.0509	9.2998	13211	7.3199	0.1695			0.6018	0.9935	0.8781
9	4.9385	280.7675	6.9385	8234.1	7.1117	0.3038			0.5373	0.9932	0.7762
10	2.7722	230.8974	4.7722	2414.5	6.3161	0.3023			0.6022	0.9935	0.8059
11	7.6226	118.5656	9.6226	7792.9	6.7613	0.1533			0.7935	0.9785	0.8769
12	2.787	221.5946	4.787	2338.8	6.3424	0.193			0.7793	0.999	0.9316
	Case 2 (CONSTRAINED)										
1	35.017	37.2996	37.017	47043	7.1602	0.0954	5.8689	236.9217			
2	55.4924	22.9723	57.4924	72016	7.1415	0.0752	9.1345	145.5206			
3	65.0183	11.2758	67.0183	48400	6.5901	0.0673	28.1127	25.8396			
4	16.8458	36.7556	18.8458	11050	6.3966	0.1456	2.7951	261.8669			
5	210.0779	22.7889	212.0779	1010500	8.4713	0.0388	294.3836	16.3024			
6	65.6089	23.8487	67.6089	104220	7.3455	0.0845	245.985	6.422			
7	27.8637	61.5424	29.8637	49496	7.4205	0.1407	187.5797	9.7124			
8	84.1382	73.6822	86.1382	527810	8.7237	0.0796	319.6409	19.6355			
9	27.4707	236.8108	29.4707	185210	8.7524	0.1455	249.7543	27.7889			

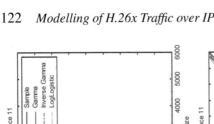
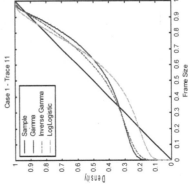
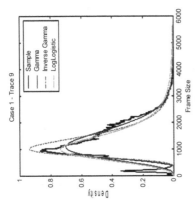
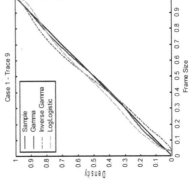
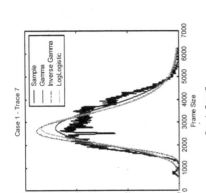
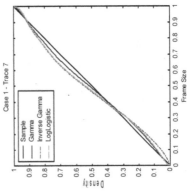

(a): Vpoint, H.261, VC-H

(b): eConf, H.263+, VC-H

(c): TKN, H.263, Office

Figure 5.1   Frame-size histograms versus moment fit and the respective Q-Q plots for unconstrained traces.

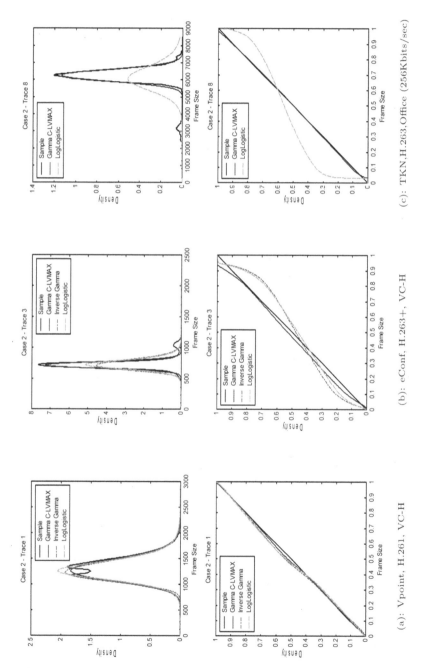

(a): Vpoint, H.261, VC-H

(b): eConf, H.263+, VC-H

(c): TKN.H.263,Office (256Kbits/sec)

Figure 5.2 Frame-size histograms versus moment and C-LVMAX fit and respective Q-Q plots for constrained traces.

(for large a, b parameters the factor $(1/\beta)^{(a+1)}$ in Eq. (5.2) is very small, near zero, and consequently its inverse quantity could not be calculated[4]).

Summing up the above analysis, it is evident that the Gamma density is better for H.261 unconstrained traffic, the Loglogistic for unconstrained H.263, H.263+ traffic and the C-LVMAX method for all cases of constrained traffic. However, if a generic and simple model needed to be applied for all cases then the most dominant would be the Loglogistic density.

5.3.2 Calculation of the Autocorrelation Decay Rate of the Frame-Size Sequences

At this point, we may discuss about the calculation of the autocorrelation decay rate of the frame-size sequence of the unconstrained traces (as denoted in the previous sections, constrained traffic appeared to be uncorrelated and as a result the decay rate of its autocorrelation function can be set to zero). From Figures 5.3a–d, it is observed that the ACF graphs of unconstrained traffic exhibit a reduced decay rate beyond the initial lags. It is evident that unconstrained video sources have very high short term correlation, feature which cannot be ignored for traffic engineering purposes. This is a behaviour also noted in earlier studies [6].

To fit the sample ACF, we applied the model proposed in [22] that is based on a compound exponential fit. This model fits the autocorrelation function with a function equal to a weighted sum of two geometric terms:

$$\rho_k = w\lambda_1^k + (1-w)\lambda_2^k \qquad (5.4)$$

where λ_1, λ_2 are the decay rates with the property: $|\lambda_2| < |\lambda_1| < 1$. This method was tested with a least squares fit to the autocorrelation samples for the first 100, since the autocorrelation decays exponentially up to a lag of 100 frames (short-term behavior) or so and then decays less slowly (long-term behavior). This match is more than enough for traffic engineering, as also noted in [28]. What is notable is that using this model, the autocorrelation parameter ρ is chosen *not* at lag -1, as in the DAR model. For each encoder (in Case 1), the parameter numerical values of the above fit appear in Table 5.2.

In the following section, the discussed modelling components are combined into complete traffic models with the C-DAR method. Furthermore, the different modelling parameters are validated comparing sample-based against

---

[4] However, the values of the parameters of the Inverse Gamma density for Case 2 – Traces 5, 6, 8, 9 are given in Table 5.2.

model-based fluid-flow simulations in a single-server queueing system. This analysis is performed only for unconstrained traffic as for constrained traffic, the Q-Q plots are enough for modelling validation purposes.

5.3.3 Queueing Analysis via the C-DAR Model and the Fluid-Flow Method – Modelling Validation

The C-DAR model that was proposed and used analytically in [19] can be directly applied for full modelling and analytical treatment of H.26x unconstrained (correlated) traffic over IP networks. This model is defined as a continuous-time discrete-state Markov chain with a transition rate matrix Q of the form:

$$Q = f_c(P - I) \tag{5.5}$$

where

$$f_c = \frac{\ln \rho}{\rho - 1} f, \quad P = \rho I + (1 - \rho)A$$

from the DAR(1) model [7] with ρ the autocorrelation decay rate derived from Eq. (5.4), f is the frame rate of the videoconference traffic, I is the identity matrix and A is a rank-one stochastic matrix with all rows equal to the probabilities resulting from the fit of the selected distribution. The C-DAR model demands the representation of the frame-size sequence with a constant number of states, whose probabilities values will fill the rows of the stochastic matrix A. These states can be easily chosen by dividing the interval between the maximum and the minimum frame size of the sequence into M frame-size states. So, if x_{min} is the minimum and x_{max} the maximum frame-size value then a reasonable state step n is $n = (x_{max} - x_{min})/M$, with n rounded to the nearest integer. The rate of each state can be easily calculated by the relative mean rate of a histogram window, as follows: if $\mathbb{P}_i$ is the probability mass of frame size S_i (derived from the corresponding density) then the rate value of the state value is equal to $f \sum_{i=1}^{n} \mathbb{P}_i S_i / \sum_{i=1}^{n} \mathbb{P}_i$. The value of the autocorrelation decay rate ρ should be chosen equal to the parameter λ_1 of the model used to fit the ACF in Eq. (5.4) (see Table 5.2) and the elements for the rows of table A should be determined through the fit produced by the PDF models (with parameters chosen from Table 5.2).

Following the approach in [19], the C-DAR model – as a continuous-time Markov chain model – is suitable for theoretical analysis using the fluid flow method (see also [3, 29, 30]). The above scheme is a very fast and simple queueing analysis method for VBR video traffic. Dr. K. Kontovasilis provided us with a Matlab implementation of the above scheme, namely, "genflow".

The "genflow" program takes as input the characteristics of N statistically identical fluid-flow Markov-Modulated sources (with global matrices $Q_g = Q \otimes Q \otimes \cdots \otimes Q$ from Eq. (5.5) and state space compressed to $\binom{N+M}{M}$ states due to the statistical identical feature of the superposed streams) and solve the congestion problem of those N sources being statistically multiplexed over a multiplexer with infinite buffering capabilities. This program has been used in other studies too (see [31]). This method is analyzed as follows: consider a single server queueing system fed by videoconference traffic $r(t) \geq 0$ as a Markov modulated rate process according to the C-DAR model with a finite number of M states and transition rate matrix Q_g as above. More explicitly, in each state $i = 1 \ldots m$, we correspond a video rate r_i. If Π is the corresponding steady state probability vector, then the mean input rate $\bar{r}$ is calculated as follows: $\bar{r} = \sum_{i=1}^{M} \Pi_i r_i$. The mean rate of the calculated rate vector captures always the mean rate of the sample (with a slight deviation in the case of the Loglogistic density). Let $R = \text{diag}\{r_1 \ldots r_M\}$ and C be the constant server capacity. When $r(t) > C$, the input traffic cannot be served entirely and its excess part is stored into a buffer in order to be served later. Let $\{X(t), t \geq 0\}$ be the stochastic process that represents the buffer occupancy. It is noted that the traffic intensity of the system is equal to $\bar{r}/C$. Define the steady state distribution $F_i(x)$ as the joint probability that the buffer occupancy is less than or equal to x when in the i state of the source model. Let: $F(x) = [F_1(x), F_2(x), \ldots, F_M(x)]^T$.

Then from [29, 30], we have the differential equation:

$$\frac{dF(x)}{dx} D = F(x) Q_g \qquad (5.6)$$

where $D = R - CI$. Given the infinite buffer assumption, we determine a buffer threshold B and define the buffer overflow probability as follows:

$$P_{\text{overflow}} = 1 - F(B)\mathbf{1} \qquad (5.7)$$

where $\mathbf{1} = (1, \ldots, 1)^T$. From Eq. (5.6) and the boundary conditions for the infinite buffer size approach in [3, 29, 30], the following relation holds:

$$F(x) = \sum_{i=1}^{M} \alpha_i e^{z_i x} \phi_i \qquad (5.8)$$

where the coefficients a_i must be calculated from the boundary conditions and z and ϕ are, correspondingly, the eigenvalue and the left eigenvector of the

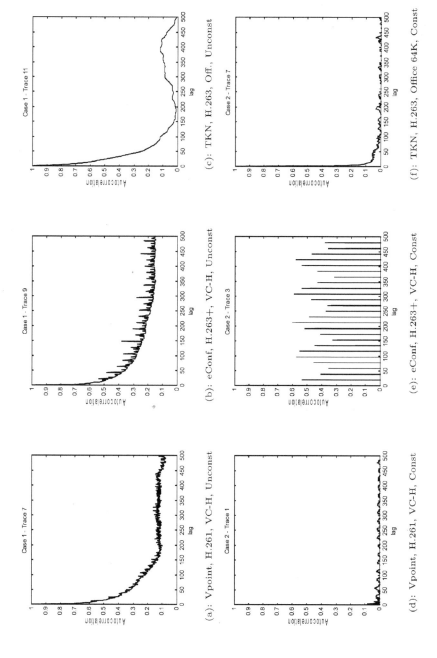

Figure 5.3 Autocorrelation graphs for unconstrained and constrained traces.

matrix $Q_g D^{-1}$. Given the infinite buffer assumption, the solution of Eq. (5.8) is given as follows:

$$F(x) = \Pi + \sum_{i \in S_o} a_i e^{z_i x} \phi_i \qquad (5.9)$$

where $S_o \triangleq \{j | r_j > C\}$, $z_i < 0$ and $z_1 = 0$.

Using the above method (with the assumption of a finite buffer), Xu et al. [19] proved experimentally (comparing the analytical model versus trace-driven simulation) that the C-DAR model provides accurate queueing results (mean cell loss rate, mean queue length) and therefore is suitable for theoretical analysis of videoconference traffic. To validate the modelling proposals of the previous sections, we present experimental queueing results comparing the complementary distribution of the buffer overflow given by the C-DAR Markov chain as derived from the calculation of Eqs. (5.7) and (5.9) for any value of buffer threshold B versus the one given by a discrete-event simulation [17] using the actual traces (trace-driven simulation e.g. [21]). For a variety of Case 1 traces,[5] the complementary buffer size densities from the results of the fluid-flow method for all the examined distribution models (for different values of multiplexed sources N) and the corresponding sample (derived from the discrete simulation of the trace being multiplexed[6] and frame interarrival times equal to $1/f$) are plotted together (see Figures 5.4a–f). The probabilities values are always assigned at the logarithmic scale. The traffic intensity was chosen equal to 0.85,[7] the autocorrelation decay rate properly chosen from Table 5.2 and the number of states of the Markov chain M equal to 5 (increasing the number of states higher than five led to identical results). The comparison between the simulation and the analytical results gives a clear indication of the queueing performance of the proposed models for unconstrained videoconference traffic. As can be seen there is quite good agreement between all curves, apart from the fact that the curve deviates from those derived from by analysis from small buffer sizes. This is physical as with the fluid-flow method the discreteness of the buffer occupancy is neg-

---

[5] Queueing analysis for VC-H and VC-L traces of the same encoder and video system was found to be similar. Consequently, for brevity reasons, only the "worst-case" VC-H traces are examined.

[6] In trace-driven multiplexing, the first frame occurrence of each source was randomized over the interval of a frame and then the source kept its individual frame synchronization.

[7] If the models retain close to the sample at high traffic intensities, their applicability is ensured for lower values of the traffic intensities (as declared in [2]). This property, though, was tested and expected results were found.

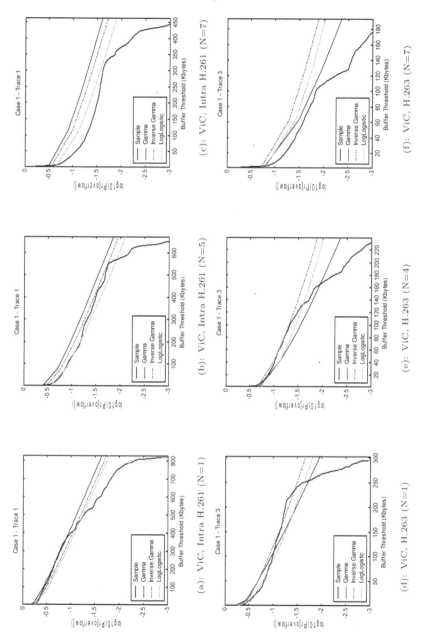

Figure 5.4 Complementary buffer overflow density plots of model versus sample.

lected. Moreover, in all plots where more than one sources were multiplexed the multiplexing gain property led to more conservative results for the models (asymptotically tight though). More explicitly, it is concluded that, in most cases of single source service (e.g. see couples of Figures 5.1a, 5.4a–5.1c, 5.4d) the models that exhibited the best fits in Q-Q plots provided closely accurate queueing results. This fact constitutes the selection of the ACF decay rate at the first 100 lags valid for unconstrained videoconference traffic engineering purposes. However, there are some obvious deviations from the above conclusion where the sample results are roughly fitted by the models. This phenomenon, which is less intense in the case of multiplexing of the same sources due to the multiplexing gain property, claims the existence of notable long-term trends in the ACFs of the respective traces (Case 1 – Traces 9, 11, 12). It is evident that in these particular cases, more than 100 lags (this was also remarked in [22] where the authors proposed a fit in the first 500 lags of the ACF) are needed to capture the strong correlation structure of the traffic. This can be also verified by the bad queueing performance of the Loglogistic density, despite its better fitting behavior in the corresponding Q-Q plots. Though for the majority of the examined cases, where a 100-lag fit was found to be accurate, a network administrator could choose a fit at 500 lags in order to ensure the conservativeness of single-source queueing results. For multiplexing this is already ensured.

Regarding the constrained (uncorrelated) traffic, it is repeated that there is no need to perform simulations, since the buffering is done inside the encoder. The traffic can be captured by the Gamma density calculated via the C-LVMAX method, as can be validated from the Q-Q plots shown in Figures 5.2a–c.

5.4 Conclusions

The current study is a contribution of modelling and simulation results for a variety of existing videoconference encoders for talking heads communication. An extensive analysis of the measured data, a careful but simple modelling of the frame-size sequences and the extensive evaluation of the modelling components, led us to the general conclusion that the traffic can be distinguished into two categories: unconstrained and constrained. In the unconstrained traffic, strong correlations between successive video frames can be found (and a large buffer has to be used for better performance). On the other hand, where bandwidth constraints are imposed during the encoding process, the generated traffic is uncorrelated.

We used the measured data to develop statistical traffic models for unconstrained and constrained traffic. These models were further validated with different videoconference contents (low motion and high motion, TKN library). Different statistical models for fitting the empirical distribution (method of moments and C-LVMAX method) were examined.

For fitting the videoconference data, the coding algorithm used should be taken into consideration. There seems to be a relationship between the coding algorithms and the characteristics of the generated traffic. For instance, for H261, in most cases, the dominant distribution is Gamma, and for H263 and H263+, Loglogistic has a more "stable" performance. Moreover, the Inverse Gamma density could not be calculated for all constrained traces, due to processing limitations. This fact constitutes the Inverse Gamma density as impractical as a generic model for H.263 traffic.

Regarding the unconstrained traces, a careful but simple generalization of the DAR model can simulate conservatively and steadily the measured videoconference data. The model was further verified using the Continuous version of the DAR model, namely, C-DAR model (analytical solution). For the constrained traces, the traffic can be captured by the C-LVMAX method via a random number generator, producing frames at a time interval equal to the sample. On the other hand, if a moments matching method needed to be applied, then the Loglogistic density is a direct solution. Another interesting assumption is that the traffic trends remain invariant when a different network constrained is selected, as evident from the TKN traces. So, the proposed model for the constrained traffic can be applied without taking into account the specific network constraint.

It is evident that if a generic and simple model needed to be applied for all cases of videoconference traffic then the most dominant would be the DAR model based on the fit of the Loglogistic density with a decay rate properly assigned to the fit of the sample ACF at the first 100 lags (although a 500 lags fit would lead to a more conservative queueing performance), for the case of unconstrained traffic, and to zero for the constrained traffic.

Future work includes the integration of the proposed models in dynamic traffic policy schemes in real diffserv IP environments. Careful analysis and modelling of cases of semi-constrained traffic, although their counterpart "worst-case" full-constrained cases cover their traffic trends, is of particular interest, too.

Acknowledgements

The authors would like to thank Dr. K. Kontovasilis for valuable suggestions and for providing the "genflow" program. Finally, we wish to thank the anonymous reviewers for their helpful comments.

References

[1] P. A. Jacosb and P. A. W. Lewis, Time series generated by mixtures, *J. Time Series Anal.*, vol. 4, no. 1, pp. 19–36, 1983.

[2] B. Maglaris, D. Anastassiou, P. Sen, G. Karlsson and J. D. Robbins, Performance models of statistical multiplexing in packet video communications, *IEEE Trans. Commun.*, vol. 36, pp. 834–843, 1988.

[3] D. Mitra, Stochastic theory of a fluid model of producers and consumers coupled by a buffer, *Adv. Appl. Prob.*, vol. 20, pp. 646–676, 1988.

[4] R. Kishimoto, Y. Ogata and F. Inumaru, Generation interval distribution characteristics of packetized variable rate video coding data streams in an ATM network, *IEEE JSAC*, vol. 7, pp. 833–841, 1989.

[5] H. S. Chin, J. W. Goodge, R. Griffiths and D. J. Parish, Statistics of video signals for viewphone-type pictures, *IEEE JSAC*, vol. 7, pp. 826–832, 1989.

[6] M. Nomura, T. Fujii and N. Ohta, Basic characteristics of variable rate video coding in ATM environment, *IEEE JSAC*, vol. 7, pp. 752–760, 1989.

[7] D. P. Heyman, A. Tabatabai and T. V. Lakshman, Statistical analysis and simulation study of video teleconference traffic in ATM networks, *IEEE Trans. Circuits Syst. Video Technol.*, vol. 2, pp. 49–59, 1992.

[8] D. M. Cohen and D. P. Heyman, Performance modelling of video teleconferencing in ATM networks, *IEEE Trans. Circuits Syst. Video Technol.*, vol. 3, pp. 408–422, 1993.

[9] D. P. Heyman and T. V. Lakshman, *Modelling Teleconference Traffic from VBR Video Coders*, IEEE ICC, pp. 1744–1748, 1994.

[10] D. M. Lucantoni and M. F. Neuts, Methods for performance evaluation of VBR video traffic models, *IEEE/ACM Trans. Networking*, vol. 2, no. 2, pp. 176–180, 1994.

[11] R. Frederick, *Experiences with Real-Time Software Video Compression*, Xerox Parc, 1994.

[12] A. Elwalid, D. Heyman, T. V. Lakshman, D. Mitra and A. Weiss, Fundamental bounds and approximations for ATM multiplexers with applications to video teleconferencing, *IEEE JSAC*, vol. 13, pp. 1004–1016, 1995.

[13] S. R. McCanne, Scalable compression and transmission of internet multicast video, Report No. UCB/CSD-96-928, Computer Science Division (EECS), University of California, Berkeley, California 94720, 1996.

[14] A. Erramilli, O. Narayan and W. Willinger, Experimental queueing analysis with long-range dependent packet traffic, *IEEE/ACM Trans. Networking*, vol. 4, pp. 209–223, 1996.

[15] D. P. Heyman, The GBAR source model for VBR videoconferences, *IEEE/ACM Trans. Networking*, vol. 5, pp. 554–560, 1997.

[16] L. D. McMahan, Video conferencing over an ATM network, Thesis, California State University, Northridge, 1997.

[17] M. Law and W. D. Kelton, *Simulation Modelling and Analysis*, 3nd Edition, McGraw-Hill Higher Education, 1999.

[18] B. Ryu, Modelling and simulation of broadband satellite networks: Part II – Traffic modelling, *IEEE Communications Magazine*, 1999.

[19] S. Xu, Z. Huang and Y. Yao, An analytically tractable model for video conference traffic, *IEEE Trans. Circuits Syst. Video Technol.*, vol. 10, pp. 63–67, 2000.

[20] G. Sisodia, L. Guan, M. Hedley and S. De, A new modelling approach of H.263+ VBR coded video sources in ATM networks, *RealTimeImg*, no. 5, pp. 347–357, 2000.

[21] F. Fitzek and M. Reisslein, MPEG-4 and H.263 video traces for network performance evaluation, *IEEE Network*, vol. 15, no. 6, pp. 40–54, 2001.

[22] C. Skianis, K. Kontovasilis, A. Drigas and M. Moatsos, Measurement and statistical analysis of asymmetric multipoint videoconference traffic in IP networks, *Telecommun. Systems*, vol. 23, no. 1. pp. 95–122, 2003.

[23] F. Fitzek, P. Seeling and M. Reisslein, Using network simulators with video traces, web site.

[24] S. Domoxoudis, S. Kouremenos, V. Loumos and A. Drigas, Measurement, modelling and simulation of videoconference traffic from VBR video encoders, in *Second International Working Conference, HET-NETs '04, Performance Modelling and Evaluation of Heterogeneous Networks*, Ilkley, West Yorkshire, UK, 26–28 July 2004.

[25] ITU Recommendation, H.261: Video codec for audiovisual services at 64 kbit/s, 1993.

[26] ITU Recommendation, H.263: Video coding for low bit rate communication, 2005.

[27] Ç.263 Standard, Overview and TMS320C6x implementation, White Paper, www.ubvideo.com.

[28] T. Lakshman, A. Ortega and A. Reibman, Variable-Bit-Rate (VBR) Video: Tradeoffs and potentials, in *Proceedings of the IEEE*, 1998.

[29] D. Anick, D. Mitra and M. M. Sondhi, Stochastic theory of a data handling system with multiple sources, *Bell Systems Techn. J.*, vol. 61, no. 8, pp. 10–18, 1974.

[30] T. E. Stern and A. I. Elwalid, Analysis of separable Markov-modulated rate models for information-handling systems, *Adv. Appl. Probability*, vol. 23, pp. 105–139, 1991.

[31] N. Mitrou, S. D. Vamvakos and K. P. Kontovasilis, Modelling, parameter assessment and multiplexing analysis of bursty sources with hyper-exponentially distributed bursts, *Comput. Networks ISDN Systems*, vol. 27, pp. 1175–1192, 1995

[32] The ViC Tool, http://www-mice.cs.ucl.ac.uk/multimedia/software/vic/

[33] VCON Vpoint HD, http://www.vcon.com

[34] France Telecom eConf, http://www.rd.francetelecom.com

[35] Sorenson EnVision, http://www.sorensonvrs.com

[36] OpenH323 Project, http://openh323.org

[37] MorningSound, http://www.soundmorning.com/

PART TWO

QUEUEING AND INTERCONNECTION NETWORKS

6

Transient and Steady State Analysis in a Discrete-Time Queueing Network

Sebastian Kempken[1], Gerhard Hasslinger[2] and Wolfram Luther[1]

[1]*Abteilung Informatik, Universität Duisburg-Essen, Duisburg, Germany; e-mail: {kempken, luther}@inf.uni-due.de*
[2]*T-Systems ENPS, Deutsche Telekom, Darmstadt, Germany; e-mail: gerhard.hasslinger@t-systems.com*

Abstract

This paper describes a newly developed software toolkit that provides a means to analyze the transient and steady state behavior of a discrete-time open queueing network. Currently, the network under consideration consists of time-slotted queues with renewal processes describing the arrival of data. We consider the buffer occupancy at the network nodes and the resulting loss probabilities due to overflow events. The network nodes handle several data streams using one of two strategies, such as prioritization of each stream or bandwidth sharing, that is, multiple streams in one priority class. Additionally, probability distributions are computed from verified workload analysis. We describe sample settings and compare the resulting probabilities for transient and steady state analysis.

Keywords: Queueing systems, transient analysis, verified analysis.

Abbreviations Used

SMP   Semi-Markov process
GI   General independent distribution
G   General distribution

D. D. Kouvatsos (ed.), Traffic and Performance Engineering for Heterogeneous Networks, 137–154.

6.1 Introduction

Transient analysis of queueing systems is a method related to both steady state analysis and simulation. For single buffered network elements, matrix analytic methods [1–3] and factorization approaches [4–7] provide efficient and reliable computation techniques for the steady state solutions of systems with regular structures in the transition equations. Transient analysis is an alternate way to achieve these results in the long run and also shows the evolution of the system from a predefined starting situation over a limited period of time or in long-term development.

If the system considered is ergodic, a convergence to the steady-state solution is observed over time. Hence, steady state analysis is included in the usual transient behavior, but the computational effort is often much higher than for direct steady state solutions depending on the convergence behavior and on the relevant state space of the system.

Transient analysis is more flexible in dealing with exceptional cases and non-regular structures in the transition matrix. Furthermore, it allows consideration of a given system's development over a limited time period.

In comparison to simulation, transient analysis provides complete distribution functions for the states of the system at embedded points in time. In contrast, simulation follows randomly chosen paths of system evolution, thus yielding results that are subject to statistical deviations at a confidence level. For these reasons, transient analysis is gaining more attention for a variety of applications, especially in the field of telecommunications [8–11].

We considered the transient analysis of single renewal process queues in [12] and for semi-Markovian queues in [13].

In this paper, we present a means to compute the development of state probabilities of a discrete-time open queueing network by transient analysis. We relate this approach to the verified workload analysis of single-server queues introduced in previous work and express how the verified techniques available are applicable in the analysis of queueing networks. Since we apply interval arithmetic, we are able to give a number of iterations required in the transient case to approximate the steady state solution.

Data streams and link capacities are acquired using stochastic processes defined by their respective distribution functions. Multiple data streams arriving at a network element can be handled according to a priority scheme – either a strict prioritization of all streams or an assignment of multiple streams to a single priority class such that they share the available bandwidth equally. A separate buffer is assigned to each outgoing link at a node. In the case of limited buffers, we are able to compute the loss probability resulting from buffer overflow events.

In the following section, we give an introduction to stochastic traffic modeling and open queueing networks as a common model with focus on telecommunication networks. We consider the transient analysis of events for a single data stream arriving at a buffered network element in Section 6.3. Section 6.4 explains how multiple streams are handled according to a prioritization scheme or in a bandwidth-sharing scenario. In Section 6.4.2, we show how the outcome of a verified workload analysis can be used to determine the characteristics of outgoing traffic in the stationary case.

The departure probability distribution for each traffic stream that is routed through the network is the basis for the analysis of an open queueing network (Section 6.4.3). We explain the implementation of these considerations in a software toolkit in Section 6.4.4. Afterwards, some examples are given to illustrate the abilities of the toolkit. A conclusion and some details about intended further work complete this paper.

6.2 Stochastic Traffic Modeling

The nature of traffic in telecommunication networks is unpredictable from a variety of viewpoints. On the one hand, service providers cannot know the exact transmission demands – that is, the transmission time and volume – that will be generated by the users and applications. On the other hand, senders and receivers cannot predict the number of network resources they will require at the start of their communications. Randomly changing workload, routes and system parameters are relevant under normal operating conditions; failure events are also unpredictable.

For this reason, stochastic traffic models represent an appropriate basis for describing traffic flow over time. The usual stochastic traffic models in telecommunications consider random variables for

- the interarrival times of events such as arrivals of packets, flows, connections or other units relevant to network elements, like the classical approach in queueing and service systems, and
- the counting function of the number of arrivals in predefined intervals, for instance in time-slotted multiplexer models.

Open queueing networks provide a typical approach for the analysis of telecommunication systems being decomposed into nodes representing switches

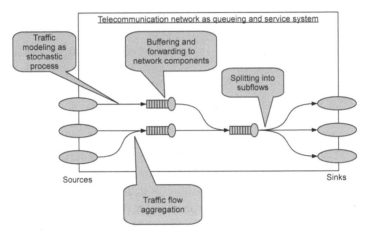

Figure 6.1 A queueing network with sources, sinks, buffering and forwarding components.

and routers and edges as transmission links (see Figure 6.1). The main items of this approach are

- traffic flows on the edges of a network topology being characterized by stochastic processes,
- server systems for switching and forwarding traffic with the help of buffers to store data in phases of temporary overload, and
- sources and destinations that generate and terminate traffic flows in the network.

In this paper, we consider discrete time-slotted queueing networks. Each network element accepts data packets that arrive over a number of input links and distributes them over a number of output links. The output line is chosen for each packet according to the particular routing directives. In most cases, there is also a buffer available for each outgoing link. The buffer stores all outgoing data packets in case of collision or temporary overload situations.

Thus, the usual router or switch model includes an output link with a given forwarding capacity S, a buffer of size B and an input process A describing all packets that arrive at the switch and are to be transmitted over the corresponding output link. The input process is determined by the distribution of the arriving data.

We compute the probability distribution of the amount of outgoing data at a network node to determine the renewal process for the amount of data arriving at the next node. Since we take buffers into consideration, the amount of departing data depends on the amount of data arriving at the node during the current and previous time slots. Hence, the departure process is autocorrelated. To reflect this characteristic, approximating semi-Markovian models can be derived [4]. In the approach presented below, however, we neglect this correlation.

6.3 Single Streams

We consider a single data stream arriving at a buffered network element and leaving over an exclusive output link in a time-slotted model. We assume that any arrivals are handled prior to the departure. The service discipline is nonpreemptive and the order of service is independent of the service time: for example, it may be first-come-first-served or random. The arrival process is given by the random variable A with a discrete distribution of the values $A_t \in \{0, \ldots, g\}, t \in \mathbb{N}_0$. A_t denotes the amount of data arriving in time slot t, measured e.g. in Byte. Correspondingly, the service capacity (also measured in Byte) is given by the variable S with the range $S_t \in \{0, \ldots, h\}, t \in \mathbb{N}_0$. In telecommunication networks, the value of S_t is often constant over time due to constant forwarding capacities of routers and switches. Different priorities for the data packets to be transmitted, however, lead to variable residual capacities in lower priority classes. The distributions of these variables are given by

$$a_t(k) := P(A_t = k), \quad k = 0, \ldots, g,$$

$$s_t(k) := P(S_t = k), \quad k = 0, \ldots, h.$$

To avoid overflows, they must satisfy $E(A) < E(S)$ as a stability constraint.

We assume a buffer of fixed size N available for the traffic flow. We consider the queueing system with reference to a one-dimensional state space describing the probability of the buffer occupancy immediately before the next arrival. We define $w_t(k)$ as the probability that the buffer will have an occupancy of k units, $0 \leq k \leq N$, at the beginning of time slot t. This is equivalent to the workload of the queueing system at time t and also to the waiting time of the current arrival.

Next, the data is stored in the buffer. We assume that the buffer capacity N is greater than the maximum amount of data arriving in a time slot $N > g$. The distribution of the occupancy after the arrival denoted by $\tilde{w}_t(k)$ is computed from the distributions of the buffer occupancy before the arrival and the current arrival distribution:

$$\tilde{w}_t(k) = \sum_{j=0}^{\min(g,k)} w_t(k-j)a_t(j), \quad k = 0, \ldots, N-1 \tag{6.1}$$

If the amount of data arriving is greater than the buffer capacity available, an overflow occurs. The probability of such overflows l_t – that is, the probability of data losses – is computed in a similar manner:

$$l_t = \sum_{i=0}^{g-1} \sum_{j=i+1}^{g} w_t(N-i)a_t(j) \tag{6.2}$$

In overflow situations, the buffer is filled to the maximum and the surplus is dropped. To reflect this, we add the overflow probability to $\tilde{w}_t(N)$:

$$\tilde{w}_t(N) = l_t + \sum_{j=0}^{g} w_t(N-j)a_t(j) \tag{6.3}$$

Afterwards, the buffer is emptied according to the service capacity available. The amount of data leaving during time slot t is given by the variable O_t. The probability distribution of this variable, $o_t(k) = P(O_t = k)$, $k = 0, \ldots, h$, is computed according to the following relation:

$$o_t(k) = \tilde{w}_t(k) \sum_{i=k}^{h} s_t(i) + s_t(k) \sum_{j=k+1}^{N} \tilde{w}_t(j), \quad k = 0, \ldots, h \tag{6.4}$$

This means that either the amount of data to be forwarded or the capacity available limits the actual amount of outgoing data. The buffer after the departure, that is, the buffer capacity at the beginning of the next time slot, can be computed from the buffer occupancy before the departure and the capacity available:

$$w_{t+1}(k) = \sum_{i=0}^{\min(h,N-k)} s_t(i)\tilde{w}_t(k+i), \quad k = 1, \ldots, N \tag{6.5}$$

The buffer is empty if there is more capacity available than data stored in the buffer:

$$w_{t+1}(0) = \sum_{i=0}^{h} s_t(i) \sum_{j=0}^{i} \tilde{w}_t(j) \tag{6.6}$$

Using these equations, we are able to iteratively compute the probabilities of the buffer occupancy at arrival instants as well as the probability of data loss in the long run depending on the buffer size. We denote these by

$$w(k) = \lim_{t \to \infty} w_t(k), \quad l = \lim_{t \to \infty} l_t \qquad (6.7)$$

Usually, we begin these computations with an empty system, that is,

$$w_0(0) = 1; \quad w_0(k) = 0, \quad k = 1, \ldots, N \qquad (6.8)$$

By introducing random variables W_t and $\tilde{W}_t$ according to their respective distributions $w_t(k)$ and $\tilde{w}_t(k)$, we relate the considerations to Lindley's equation [14]. It is

$$\tilde{W}_t = \min(W_{t-1} + A_t, N), \quad W_t = \max(\tilde{W}_t - S_t, 0).$$

6.4 Open Queueing Networks

6.4.1 Multiple Streams

The case of multiple data streams arriving at the same network element raises the question of how the bandwidth available is distributed among the particular data streams. The Internet Engineering Task Force (IETF) has introduced the concepts of integrated services (IntServ, [15]) and differentiated services (DiffServ, [16]).

The IntServ approach represents a reservation of the required bandwidth for each traffic flow. Hence, it requires signaling and state information to be maintained in each network element on the path in order to provide resource reservation for each flow. The differentiated services approach stands for a simpler concept based on classification and prioritization of traffic. For instance, each IP packet is marked to belong to a specific class using a 6-bit field in the header. This means that 64 priority classes can be distinguished. The actual classification in networks takes place at edge routers, whereupon core routers of a DiffServ network can set up class based scheduling. The main scheduling principles are strict priorities, class-based reservation and fair scheduling for flows within the same class. We enhance the concept presented in the previous section to model multiple data streams arriving at a data node and being forwarded over the same output link, thus competing for the bandwidth available. Since the internal handling capacity of routing elements is much higher than the sum of the outgoing link bandwidth capacities, we can consider the distributions on each link independently.

In the following, we assume that multiple data streams arrive at a node, and intend to compute the distribution of these streams after leaving the node. To do so, we have to consider the order in which the incoming packets are handled according to their original source, that is, the prioritization scheme. We consider the *Strict prioritization* and *Bandwidth sharing*, which are explained in detail in the following subsections.

As has been done for the single stream case, we consider the node as a time-slotted system. If the amount of data arriving from all sources in a time slot is higher than the capacity available, the surplus data packets are stored in a buffer up to a given limit. If the buffer is full, and further packets arrive at the node, these packets are dropped and lost. We assume that a separate buffer is available for each traffic flow. This way, we can consider the probability of remaining data packets from previous time slots needing to be forwarded per stream. In the following, we take a closer look at the prioritization techniques and describe how the corresponding output probabilities for each stream and their respective loss probabilities can be computed.

6.4.1.1 Strict Prioritization

In this case, all traffic sources that deliver packets being forwarded over the output link are ranked according to their priority. The capacity available is reserved for the traffic source with the highest priority. After the corresponding packets have been transmitted, the unused bandwidth is offered to the traffic source with the second highest priority. Thus, we have to compute the capacity available for each traffic source.

We use the same notation as in the previous section and consider the traffic flow with the highest priority. The remaining capacity can be computed from the original capacity and the probabilities for the amount of data forwarded of this traffic flow. We denote the remaining capacity distribution for the other traffic sources at time t by $s_t^{(2)}(k), k = 0, \ldots, h$. Correspondingly, the original capacity that is available for the stream with the highest priority is given by $s_t^{(1)}(k), k = 0 \ldots, h$ and the probability distribution of the output of this stream by $o_t^{(1)}(k), k = 0, \ldots, h$, which can be computed according to equation (6.4). The capacity available for the stream with the second highest priority can be computed from the departure distribution of the first stream, since the original capacity decreases accordingly:

$$s_t^{(2)}(k) = \sum_{j=0}^{h-k} s_t^{(1)}(k+j) o_t^{(1)}(j) \tag{6.9}$$

We may extend this consideration for the remaining data streams so that the capacity available for the stream with the n-th highest priority can be iteratively computed by

$$s_t^{(n)}(k) = \sum_{j=0}^{h-k} s_t^{(n-1)}(k+j) o_t^{(n-1)}(j) \tag{6.10}$$

The probability distribution of outgoing packets for each stream can be computed for their respective bandwidth available according to equation (6.4). In the case of an overflow – that is, when more data arrives than can be forwarded – the surplus data is stored in the buffer. We assume that a buffer of fixed size N is available for each traffic source to consider the buffer occupancy and loss probabilities on a per-stream basis. Analogous to Section 6.3, buffer occupancy and the resulting loss probability for any data stream are given by equations (6.1) to (6.6).

6.4.1.2 Bandwidth Sharing

In the case of bandwidth sharing, all traffic sources have equal status. An identical share of the bandwidth available is reserved for each traffic source. If some of the sources do not fully utilize their share, the remaining capacity is made available to the other traffic sources. We derive a means to compute the resulting output distributions and the loss probability for each data stream.

To do so, we first compute the bandwidth available for each stream. The total number of relevant traffic sources is denoted by n. If the distribution of the capacity available is $s_t(k)$, the amount reserved for any stream i is given by

$$s_t^{(i)}(k) = \sum_{j=kn}^{(k+1)n-1} s_t(j), \quad k = 0, \ldots, \left\lfloor \frac{h}{n} \right\rfloor \tag{6.11}$$

The values $s_t^{(i)}(k)$ are equal for all indices i, since the capacity is distributed uniformly. Please note that, due to the discrete nature of the approach, the discretization should be taking the number of traffic sources into account. To do so, it is often required to change the discretization for each node. This is straightforward if the number of discretization steps is increased or decreased by a power of two. In every case, the mean value of the distribution must be preserved.

If we look at the traffic originating from a source i, the bandwidth available to this stream is given by the equation above. Next, we compute the probability of bandwidth becoming available if the other streams do not fully utilize their share. The capacity available for all streams except i, which we denote by $\bar{s}_t^{(i)}(k)$, is given as the sum of their respective reservations. This can be computed with a series of aggregations:

$$\bar{s}_t^{(i)} = \sum_{j=0, i \neq j}^{n-1} s_t^{(j)} \tag{6.12}$$

To compute the probability of additional capacity becoming available, we apply the scheme derived in the previous section. We apply equation (6.10) for all other streams, starting with the capacity distribution given by $\bar{s}_t^{(i)}$ and the respective buffer occupations for each traffic flow. Please note that, since we intend to compute the probability of unused capacity only, these buffer occupancy probabilities are not changed during this step.

This unused bandwidth is added to the capacity for stream i by a convolution of the distributions. The actual output, buffer occupancy and loss probabilities for stream i are then computed using the usual equations (6.1) to (6.6). The procedure described is carried out for each data stream on the output link.

6.4.1.3 Combinations

The approaches presented may also be combined to reflect a broad class of forwarding schemes according to the DiffServ concept. In most cases, traffic is prioritized through assignment to one of several priority classes. These classes are processed according to their respective priorities, multiple data streams in one class are handled equally (fair share).

To reflect this, the service capacity available for each stream is computed according to a combination of both approaches. First, the whole capacity is reserved for the priority class with the highest priority. This bandwidth is shared among the respective traffic flows equally. By using convolutions to sum up the output probability distributions $o_t^{(i)}$, we can compute the probability of remaining bandwidth becoming available in accordance with equation (6.10). This capacity is now given to the next priority class, and so on.

6.4.2 Steady State Analysis

Traczinski [18] presents a means for verified steady state workload analysis. The workload distribution of a GI/G/1 queue, which is equivalent to the buffer occupation at arrival instants as investigated above, is determined either by factorization of the characteristic polynomial of the queueing system or by Wiener–Hopf factorization. The initial approximations of the factorization results are verified using interval arithmetic. The outcome of this technique is a verified enclosure of the workload distribution in tight intervals (about 10^{-13} for double precision arithmetic), denoted as $[W]$.

Given these workload enclosures, we are able to compute the distribution of the outgoing process $[O]$. As described in the previous sections, we assume a first-come-first-served discipline, which implies that the buffer resulting from previous arrivals is emptied first. If capacity still remains, some newly arriving data can be transmitted directly. The amount of outgoing data $[O]$ at a time t is therefore given by the actual buffer (workloads) plus the arriving data, limited by the available capacity. Hence, we can apply equations (6.1) to (6.4) analogously to efficiently determine verified enclosures for the distributions in the steady state. Similar considerations apply in the case of multiple streams and prioritization schemes: The relations derived allow the outcome of the verified workload analysis to be used directly for a verified steady state analysis of queueing networks.

6.4.3 Open Queueing Networks

Using the methods presented previously, we are able to compute the distribution of the amount of outgoing traffic for a network element for each particular incoming data stream. This distribution can be considered as the distibution of the traffic incoming at the next network node for the particular traffic source. This way, we are able to compute the buffer occupancy and loss probabilities for each traffic source at each node on the route to the particular final destination.

6.4.4 Implementation

We implemented the techniques presented so far as a C++ class library. This library is a new part of our software toolkit *InterVerdiKom* [5]. The main functionality of this toolkit is the verified computation of workload distributions at single GI/G/1 and SMP/G/1 queues using efficient techniques, such as the factorization of the characteristic polynomial of the queue or the Wiener–Hopf factorization. A verification of the results is included by means of interval arithmetic.

Currently, the proposed techniques are available as a software library only. A graphical user interface with the ability to display and modify the queueing network investigated is in development. Nevertheless, we give some sample results computed with these routines.

When computing these results, we have to face the problem of round-off errors, especially in the case of shared bandwidth. These errors cumulate for every iteration step, so that without error handling, the results would be unreliable. Hence, we apply interval arithmetic as a means of round-off error control in every iteration step. We yield upper and lower bounds for each point of the probability distribution. Since all computations are carried out using outward rounding, the correct result must be contained within the resulting intervals. These intervals are compared with the outcome of the steady state analysis in the subsequent section.

6.5 Examples

In this section, we provide examples of small queueing networks to illustrate the abilities of the methods presented and their implementation. We also take a look at convergence behavior in these examples; that is, we determine the number of iterations required to approximate the steady state probability distributions derived from the corresponding analysis.

6.5.1 Tandem Network

In our first example, we consider a two-node tandem queueing network. The traffic source A is determined as a renewal process with distribution

$$P(A = 10) = 0.8, \quad P(A = 50) = 0.2,$$

$$P(A = k) = 0, \quad k \in 0, \dots, 49, \quad k \neq 10.$$

The time-constant service capacity $S1$ is given by

$$P(S1 = 15) = 0.4, \quad P(S1 = 30) = 0.6$$

$$P(S1 = k) = 0, \quad k \in 0, \dots, 29, \quad k \neq 15,$$

and the constant capacity $S2$ by

$$P(S2 = 15) = 0.4, \quad P(S2 = 25) = 0.6$$

$$P(S2 = k) = 0, \quad k \in 0, \dots, 24, \quad k \neq 15.$$

Table 6.1 Example 1: Tandem queue.

Iterations	$n = 1$	$n = 10$	$n = 100$	$n = 1000$	Steady
P(O1=10)	0.800000	0.429697	0.395529	0.395508	0.395508
P(O1=15)	0.080000	0.238338	0.252520	0.252528	0.252528
P(O1=20)	0.000000	0.016772	0.016790	0.016789	0.016789
P(O1=25)	0.000000	0.028437	0.026807	0.026806	0.026806
P(O1=30)	0.120000	0.286756	0.308355	0.308369	0.308369
P(O2=10)	0.800000	0.225638	0.164625	0.164547	0.164547
P(O2=15)	0.128000	0.424777	0.430891	0.430885	0.430885
P(O2=20)	0.000000	0.046818	0.044601	0.044588	0.044588
P(O2=25)	0.072000	0.302766	0.359882	0.359979	0.359979

We compute the output probabilities $O1$ and $O2$ for a buffer size of 2000 units. The results after $n = 1, 10, 100, 1000$ iterations are displayed in Table 6.1. We also give the results derived from steady state analysis.

6.5.2 Multiple Streams at a Node

In our second example, we consider three streams arriving at a single node. The streams are all given by the identical distribution

$$P(A = 1) = 0.5, \quad P(A = 5) = 0.5,$$

$$P(A = k) = 0, \quad k \in 0, \ldots, 4, \quad k \neq 1.$$

The service capacity is assumed constant:

$$P(S = 10) = 1, \quad P(S \neq 10) = 0.$$

The network node is considered for each of the two prioritization schemes. We assume a buffer capacity of 2000 units. The numerical results are given in Tables 6.2 and 6.3. In the case of strict prioritization, the capacity available is sufficient for the two streams with the highest priority in every case. The results for the high priority streams are therefore omitted.

As examples for the computation of the loss probability, the evolution of this value for the considered node under the two prioritizaton schemes and an assumed buffer size of just 20 units is given in Figures 6.2 and 6.3.

Table 6.2 Example 2: Multiple streams, strict prioritization.

Iterations	$n = 1$	$n = 10$	$n = 100$	$n = 1000$	Steady
P(O3=0)	0.250000	0.250000	0.250000	0.250000	0.250000
P(O3=1)	0.375000	0.134012	0.121170	0.121146	0.121146
P(O3=2)	0.000000	0.060947	0.052180	0.052170	0.052170
P(O3=3)	0.000000	0.031388	0.030264	0.030259	0.030259
P(O3=4)	0.250000	0.355717	0.370920	0.370944	0.370944
P(O3=5)	0.125000	0.049269	0.045102	0.045094	0.045094
P(O3=6)	0.000000	0.030586	0.025915	0.025911	0.025911
P(O3=7)	0.000000	0.016645	0.016785	0.016782	0.016782
P(O3=8)	0.000000	0.071437	0.087664	0.087694	0.087694

Table 6.3 Example 2: Multiple streams, shared bandwidth.

Iterations	$n = 1$	$n = 10$	$n = 100$	$n = 1000$	Steady
P(O=1)	0.500000	0.275653	0.265999	0.265999	0.265999
P(O=2)	0.000000	0.028555	0.029798	0.029798	0.029798
P(O=3)	0.312500	0.443874	0.449457	0.449457	0.449457
P(O=4)	0.000000	0.067099	0.071159	0.071159	0.071159
P(O=5)	0.187500	0.120769	0.115938	0.115938	0.115938
P(O=6)	0.000000	0.019687	0.021340	0.021340	0.021340
P(O=7)	0.000000	0.036486	0.036811	0.036811	0.036811
P(O=8)	0.000000	0.007878	0.009500	0.009500	0.009500

6.5.3 Convergence Behavior

In the two examples presented, buffer occupancy and loss probabilities converge to steady state distributions. The number of iterations required to reach the steady state distribution up to a given approximation accuracy is an important indicator for the significance of steady state models: If, for instance, the steady state distribution is reached after a time period longer than the transmission time of the traffic sources, the distributions computed are never actually attained. The application of interval arithmetic allows to determine a worst-case number of iterations required to reach a given accuracy ϵ, that is,

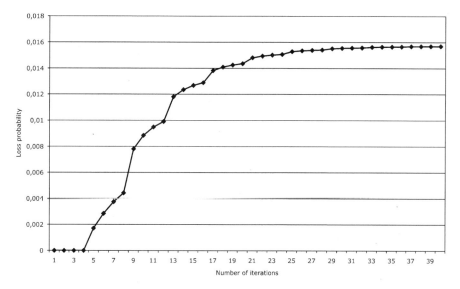

Figure 6.2 Example 2: Three streams at a node, strict prioritization, buffer limited to 20 units, loss probability of stream with least priority ($L3$).

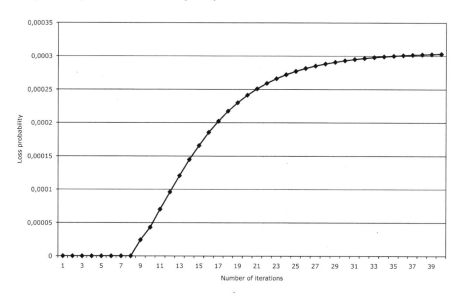

Figure 6.3 Example 2: Three streams at a node, shared bandwidth, buffer limited to 20 units, loss probability of each stream (L).

Table 6.4 Required number of iterations for different ϵ.

$\epsilon = 10^{...}$	−1	−2	−3	−4	−5	−6	−7	−8	−9
Iterations	5	21	48	80	115	151	190	229	269

the distance between the respective outer bounds of the intervals computed in the transient and the steady state analysis is smaller than ϵ. We consider a single-server GI/G/1 example taken from [20, example 4], however, this technique is applicable to all queueing networks modeled as described in the previous sections. The arrival and service distributions are given by

$$P(A = 15) = 2/5 \quad P(A = 30) = 3/5,$$

$$P(S = 10) = 4/5 \quad P(S = 50) = 1/5.$$

The required numbers of iterations are given in Table 6.4, the intervals begin to intersect after 444 iterations. Please note that the results depend on the actual implementation, in our case, we applied IEEE 754 double precision intervals.

6.6 Conclusion and Further Work

In this paper, we have presented a means to compute the traffic distributions in time-slotted open queueing networks in transient states. Traffic sources are modeled by discrete distributions; the data packets are routed through a network of queues with given buffer capacity for each flow. The methods presented in this work allow the probability distributions to be computed for outgoing traffic, buffer occupancy and data loss at each node for each data flow. Multiple data streams arriving at a network element can be handled according to a strict prioritization scheme or by sharing the available bandwidth equally. Both approaches can be combined to provide a more realistic model of the traffic in modern communication networks by assigning the data flows to priority classes. The relations derived for the transient case are applied to derive the corresponding steady state probability distributions from a verified workload analysis of a single GI/G/1 queue. We investigated some examples of small queueing networks that are analyzed by the techniques proposed in this paper.

The techniques are implemented as part of the software toolkit *InterVerdiKom*. Its main focus is on the verified computation of the transient and steady state workload distributions of GI/G/1 and SMP/G/1 queues, but the authors intend to enhance it to allow for the verified analysis of queueing networks as presented in this work.

We also analyzed some small queueing networks using the techniques proposed in this paper and examined the convergence behavior of the probability distributions by considering the number of iterations needed to approximate the steady state distribution.

We intend to further explore the possibilities of the approach described in this paper. We will consider and implement additional prioritization schemes like packet-drop policy, which means that packets arriving at a node are dropped if the buffer is full, regardless of the packet's source. In related work, we focused on modeling video traffic sources using semi-Markov processes [4, 19]. We intend to include these types of traffic sources as well.

Regarding implementation aspects, we want to provide a convienient graphical interface for the methods presented. The user should be able to look at the queueing network being analyzed and to modify the prioritization schemes of nodes, the capacity of links and the routing of traffic flows.

Acknowledgements

This research and software development was carried out by the authors in the project *Verdikom* funded by the German Research Council (DFG).

References

[1] A. Alfa, Combined elapsed time and matrix-analytic method for the discrete time GI/GI/1 and GI^X/G/1 system, *Queueing Systems*, vol. 45, pp. 5–25, 2003.

[2] A. Alfa and W. Li, Matrix-geometric analysis of the discrete time GI/G/1 system, *Stochastic Models*, vol. 17, pp. 541–554, 2001.

[3] G. Latouche and V. Ramasvami *Introduction to Matrix Analytic Methods in Stochastic Modeling*, ASA-SIAM, 1999.

[4] G. Haßlinger, Waiting times, busy periods and output models of a server analyzed via Wiener–Hopf factorization, *Performance Evaluation*, vol. 40, pp. 3–26, 2000.

[5] S. Kempken, W. Luther and G. Haßlinger, A tool for verified analysis of transient and steady states of queues, in *Proceedings of the First International Conference on Performance Evaluation Methodologies and Tools (VALUETOOLS'06)*, 2006.

[6] S. Q. Li, A general solution technique for discrete queueing analysis of multi-media traffic on ATM, *IEEE Transactions on Communication*, vol. 39, pp. 1115–1132, 1991.

[7] D. Traczinski, W. Luther and G. Haßlinger, Polynomial factorization for servers with semi-markovian workload: Performance and numerical aspects of a verified solution technique, *Stochastic Models*, vol. 21, pp. 643–668, 2005.

[8] S. Ahn and V. Ramasvami, Efficient algorithms for transient analysis of stochastic fluid flow models, *Journal of Applied Probability*, vol. 42, pp. 531–549, 2005.

[9] L. Breuer, Numerical results for the transient distribution of the GI/G/1 queue in discrete time, in *Proceedings of the 13th GI/ITG Conference on Measuring, Modelling and Evaluation of Computer and Communication Systems (MMB)*, pp. 209–218, 2006.

[10] A. Pantazi and T. Antonakopoulos, Equilibrium point analysis fot the binary exponential backoff algorithm, *Computer Communications*, vol. 24, pp. 1759–1768, 2001.

[11] S. Rank and H. P. Schwefel, Transient analysis of buffer-occupancy fluctuations and relevant time scales, *Performance Evaluation*, pp. 725–742, 2006.

[12] G. Haßlinger and S. Kempken, Transient analysis of a single server system in a compact state space, in *Proceedings of the 13th International Conference on Analytical and Stochastic Modelling Techniques and Applications*, 2006.

[13] G. Haßlinger, Transient analysis of semi-markovian switching systems in telecommunication networks, in *Proceedings of The 20th European Simulation and Modelling Conference (ESM)*, 2006.

[14] D. Lindley, The theory of queues with a single server. *Proceedings of Cambridge Philosophical Society*, vol. 48, pp. 277–289, 1952.

[15] J. Wroclawski, The use of RSVP with IETF integrated services (proposed IETF standard, RFC 2210), 1997.

[16] B. Carpenter and K. Nichols, Differentiated services in the Internet, in *Proceedings of the IEEE*, pp. 1479–1494, 2002.

[17] S. Floyd and V. Jacobsen, Random early detection gateways for congestion avoidance. *IEEE/ACM Transactions on Networking*, vol. 1, pp. 397–413, 1993.

[18] D. Traczinski, Faktorisierungslösungen für die Workload in Bedsystemen mit Ergebnisverifikation, PhD Thesis, Universität Duisburg-Essen, 2005.

[19] S. Kempken and W. Luther, Modeling of H.264 high definition video traffic using discrete-time semi-Markov processes, in *Proceedings of the 20th International Teletraffic Congress*, LNCS, Springer, 2007.

[20] M. Chaudhry, Alternative numerical solutions of stationary queueing-time distributions in discrete-time queues: GI/G/1, *Journal of the Operational Research Society*, vol. 44, no. 10, pp. 1035–1051, 1993.

7

An Analytical Model for Hypercubic Networks under Correlated Traffic with Non-Uniform Destination Distributions

Yulei Wu[1], Geyong Min[1], Mohamed Ould-Khaoua[2] and
Demetres D. Kouvatsos[1]

[1]*Department of Computing, School of Informatics, University of Bradford,
Bradford BD7 1DP, U.K.; e-mail: {y.l.wu, g.min, d.d.kouvatsos}@brad.ac.uk*
[2]*Department of Computing Science, University of Glasgow, Glasgow G12 8RZ,
U.K.; e-mail: mohamed@dcs.gla.ac.uk*

Abstract

Non-uniformly distributed message destinations can cause catastrophic congestion and loss of throughput. A number of recent studies have shown that the network traffic in parallel computation environments exhibits a high degree of burstiness and possesses strong correlations in the number of message arrivals between adjacent time intervals, and message destinations are non-uniformly distributed over the network nodes. Therefore, it is important to gain a clear understanding of the performance behaviours of interconnection networks under such traffic patterns. However, simulation-based approaches may be time-consuming and costly. This study develops an analytical model to compute the message latency in hypercubic interconnection networks under the correlated traffic with non-uniformly distributed message destinations. The accuracy of the model is validated through extensive simulation experiments.

Keywords: Interconnection networks, adaptive routing, wormhole switching, MMPP/G/1, performance modelling.

*D. D. Kouvatsos (ed.), Traffic and Performance Engineering for Heterogeneous
Networks,* 155–171.

7.1 Introduction

The interconnection network is one of the most critical architectural components in large-scale multi-computers as any interaction between the processors ultimately depends on its effectiveness [4, 28]. Hypercube has become one of the most important message-passing architectures for multi-computers owing to its attractive properties, such as regularity, symmetry, low diameter, and high connectivity to deal with fault-tolerance [2]. An n-dimensional hypercubic network is composed of $N (N = 2^n)$ nodes with 2 nodes in each dimension [2]. Each node contains a processing element (PE) and a router, as depicted in Figure 7.1. The PE consists of a processor and some local memory. The router has $(n + 1)$ input and $(n + 1)$ output channels. Each node is connected to its n neighboring nodes through n input and n output channels. The remaining channels are used by the PE to inject/eject messages to/from the network respectively. The router contains buffers for input virtual channels, where a virtual channel shares the bandwidth of the physical channel with other virtual channels in a time-multiplexed fashion [3]. The input and output channels are connected by an $(n + 1)V$-way crossbar switch, V being the number of virtual channels, which can simultaneously connect multiple input to multiple output channels. Wormhole switching technique and adaptive routing algorithm have been extensively used in routers in the current-generation of high-end parallel systems to reduce message latency and support efficient communication [1, 2, 4, 5, 7, 9, 13, 23, 26].

The message destinations have been shown to exhibit the non-uniform distribution [15–17, 22, 24, 29], rather than only the simplified uniform-distribution. Uniform destination distribution means that messages generated by each source node are sent to other network nodes with equal probability. On the other hand, hot-spot destination distribution is a typical example of the non-uniform message destination distributions [16]. The hot-spot node often receives a larger amount of traffic than the other network nodes, which causes the higher traffic loads on network channels located closer to the hot-spot node. Pfister and Norton [16] have found that non-uniformly distributed message destinations, e.g., hot-spots, can cause catastrophic congestion and loss of throughput. There are many reasons for the existence of hot-spot nodes. For example, global synchronization where each node sends a synchronization message to a distinguished node which coordinates the synchronization activity of the system, is a typical situation that can produce hot-spots [15, 17]. Recent measurement studies [6, 20, 21] have shown that the traffic arrival process in high-performance computing systems reveals bursty nature (i.e.,

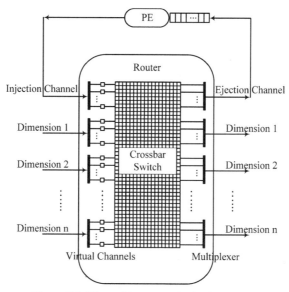

Figure 7.1 The node structure in the hypercube.

time-varying arrival rates) and possesses strong correlations in the number of message arrivals between adjacent time intervals, which have different theoretical properties from the conventional Poisson process and can significantly affect queueing performance [14, 19, 27]. However, most of the performance models [1, 3, 5, 7, 9, 13, 26] for interconnection networks have been widely reported based on the simplified assumptions that messages follow the Poisson arrival process and the destinations are uniformly distributed. Recently, several analytical models have been reported to handle the correlated traffic [14] or non-uniform message destination distribution [15, 17, 29], separately.

It is important to gain a clear understanding of the performance behaviours of interconnection networks under the correlated traffic pattern and non-uniformly distributed destinations simultaneously. However, simulation-based approaches may be time-consuming and costly. To this end, this study develops an analytical model to calculate the message latency in adaptive-routed wormhole-switched hypercubic interconnection networks in the presence of correlated traffic with hot-spot destination distribution. The message arrivals from each source node follow the well-known Markov-Modulated Poisson Process (MMPP), which is a doubly stochastic process with the arrival rate varying according to a multi-state ergodic continuous time Markov chain [8, 10, 25] and is able to model the bursty and correlated nature of the message arrival process [10, 14]. The average degree of virtual channel multiplexing is computed by employing the MMPP/G/1 queueing system with infinite buffer capacity [1]. The validity of the model is demonstrated by comparing analytical results to those obtained through simulation experiments of the actual system.

The rest of the study is organised as follows. Section 7.2 derives the analytical model. Simulation experiments are used to validate the model in Section 7.3. Finally, Section 7.4 concludes this study.

7.2 The Analytical Model

The model is based on the following assumptions. Assumptions (a) and (b) are used to generate correlated traffic with hot-spot destination distribution; other assumptions are widely used in the related studies [1, 3, 5, 7, 9, 13–17, 19, 26, 27, 29].

(a) The arrivals of messages generated by each source node are bursty and correlated and follow an independent two-state $MMPP_s$ (the subscript s denoting the traffic of source nodes). The $MMPP_s$ is characterised by the infinitesimal generator $\mathbf{Q}_s$ of the underlying Markov chain and the rate matrix, $\mathbf{\Lambda}_s$, as

$$\mathbf{Q}_s = \begin{bmatrix} -\varphi_{s1} & \varphi_{s1} \\ \varphi_{s2} & -\varphi_{s2} \end{bmatrix} \quad \text{and} \quad \mathbf{\Lambda}_s = \text{diag}(\lambda_{s1}, \lambda_{s2}), \qquad (7.1)$$

where the element φ_{s1} is the transition rate from state 1 to 2 and φ_{s2} is the rate out of state 2 to 1. λ_{s1} and λ_{s2} are the traffic rate when the Markov chain is in state 1 and 2, respectively.

(b) The message destinations are non-uniformly distributed across the network. Each generated message has the probability, δ, to be directed to the hot-spot node, and the probability, $(1 - \delta)$, of being evenly directed to all network nodes [16]. Let us refer to these two types of messages as hot-spot and regular messages, respectively.

(c) Message length is m flits, where m is a random variable. The Laplace–Stieltjes transform of m is given by $F_m^*(s)$. Each flit requires one-cycle transmission time to cross a physical channel.

(d) The local queue in the source node has infinite buffer capacity.

(e) Messages are routed adaptively through the network according to Duato's routing algorithm [2]. In this algorithm, V ($V \geq 2$) virtual channels are used per physical channel and are divided into two classes.

The first class contains $(V - 1)$ virtual channels, which are crossed adaptively, and the second class contains one virtual channel, which is crossed in an increasing order of dimensions.

7.2.1 Network Latency for Regular and Hot-spot Messages

A regular message generated by a given node is equi-probably destined to any of $(N - 1)$ nodes in the hypercubic networks. The destination of an i-hop message can be any of the $\binom{n}{i}$ nodes that are i hops away from its source node [9]. Thus, the probability, P_i $(1 \leq i \leq n)$, that a source node generates an i-hop message can therefore be written as

$$P_i = \frac{\binom{n}{i}}{N - 1}. \tag{7.2}$$

The network latency of a message consists of two parts: one is the actual time to transmit a message from its source to destination, and another is the delay due to blocking along its network path. Since the adaptive routing distributes the regular messages evenly across the network and the symmetry of the hypercube topology, the network latency for a regular message, t_r, can be written as

$$t_r = m + \bar{d}_r + b_r, \tag{7.3}$$

where $\bar{d}_r$ is the mean message distance of a regular message, and b_r is the blocking time experienced by a regular message to cross the network. $\bar{d}_r$ and b_r can be given by [13]

$$\bar{d}_r = \sum_{i=1}^{n} i P_i = \frac{\sum_{i=1}^{n} i \binom{n}{i}}{N - 1} = \frac{n}{2} \frac{N}{N - 1}, \tag{7.4}$$

$$b_r = \sum_{i=1}^{\bar{d}_r} B_{r_i}, \tag{7.5}$$

where B_{r_i} is the blocking time seen by the regular message at its ith-hop channel $(1 \leq i \leq \bar{d}_r)$.

Let $F_{t_r}(x) = \text{Pr}(m + \bar{d}_r + b_r \leq x)$ denote the distribution function of the network latency for a regular message, t_r. Since the Laplace–Stieltjes transform of the sum of independent random variables is equal to the product of their transforms [11, 12], the Laplace–Stieltjes transform of t_r can be given by

$$F_{t_r}^*(s) = \int_0^\infty e^{-sx} \, dF_{t_r}(x) = F_{b_r}^*(s) e^{-s\bar{d}_r} F_m^*(s), \tag{7.6}$$

where $F_{b_r}^*(s)$ is the Laplace–Stieltjes transform of the blocking time for a regular message.

Under the hot-spot traffic, the message destinations are non-uniformly distributed over the network. As a consequence, the channels located closer to the hot-spot node receive higher traffic loads than those farther away. Let t_{h_j} be a random variable that denotes the network latency seen by a hot-spot message located j ($1 \le j \le n$) hops away from the hot-spot node. According to Equation (7.3), we can obtain the network latency for a j-hop hot-spot message, t_{h_j}, as

$$t_{h_j} = m + j + b_{h_j}, \tag{7.7}$$

where b_{h_j} is a random variable denoting the blocking time experienced by a hot-spot message that is j ($1 \le j \le n$) hops away from the hot-spot node. b_{h_j} can be calculated as

$$b_{h_j} = \sum_{k=1}^{j} B_{h_{j,k}}, \tag{7.8}$$

where $B_{h_{j,k}}$ is the blocking time for a j-hop hot-spot message at its kth-hop channel along its network path towards the hot-spot node ($1 \le k \le j$). The Laplace–Stieltjes transform of t_{h_j} can be written as

$$F_{t_{h_j}}^*(s) = F_{b_{h_j}}^*(s)\, e^{-sj}\, F_m^*(s), \tag{7.9}$$

where $F_{b_{h_j}}^*(s)$ is the Laplace–Stieltjes transform of the blocking time for a hot-spot message located j ($1 \le j \le n$) hops away from the hot-spot node.

The probability that a message needs to make j hops to reach its destination is

$$P_{h_j} = \frac{\binom{n}{j}}{N - 1}. \tag{7.10}$$

Averaging over all the possible hops j ($1 \le j \le n$) yields the network latency, t_h, for hot-spot messages as

$$t_h = \sum_{j=1}^{n} P_{h_j} t_{h_j}. \tag{7.11}$$

The regular and hot-spot messages experience different network latencies due to the non-uniform traffic loads and varying blocking time over different network channels, depending on their locations with respect to the hot-spot node.

Therefore, taking both the regular and hot-spot messages with appropriate probabilities into consideration yields the mean network latency as

$$\bar{T} = (1 - \delta)t_r + \delta t_h. \qquad (7.12)$$

7.2.2 Modelling of the Traffic Characteristics on Network Channels

The traffic arrival rate of the MMPP_s and its covariance function, which relates to how dependent the rate at one instant of time is to that at another instant of time, play a major role in the method for determining the traffic arrival process at network channels. We can re-collect the mean, variance, third central moment, covariance function, and the integral of covariance function of the traffic rate from [10].

To identify the traffic characteristics on network channels, let us refer to the loads formed by the regular messages and hot-spot messages as the regular traffic and hot-spot traffic, respectively. The loads at network channels located at j ($1 \leq j \leq n$) hops away from the hot-spot node are composed of both the regular and hot-spot traffic.

As the regular messages are evenly directed to all network nodes, the use of adaptive routing, which allows messages to cross any channel to advance towards its destination, results in the balanced loads of regular messages on all channels. Equation (7.4) reveals that the average number of channels visited by regular messages, $\bar{d}_r$, is always less than the network dimension, n. Therefore, the regular traffic at a given network channel is a fraction of the traffic generated by a source node. This fraction, f_r, is given by

$$f_r = \frac{N\bar{d}_r(1 - \delta)}{Nn} = \frac{(1 - \delta)N}{2(N - 1)}. \qquad (7.13)$$

Let MMPP_{cr} model the regular traffic on the network channel, which is the result from the splitting of MMPP_s with the splitting probability, f_r. As the splitting of a traffic flow with MMPP_s gives rise to a new MMPP, the corresponding infinitesimal generator $\mathbf{Q}_{cr}$ and the rate matrix $\mathbf{\Lambda}_{cr}$ of the MMPP_{cr} can be given by [8]

$$\mathbf{Q}_{cr} = \mathbf{Q}_s \quad \text{and} \quad \mathbf{\Lambda}_{cr} = f_r \mathbf{\Lambda}_s. \qquad (7.14)$$

In the case of hot-spot traffic, being different from the regular traffic, messages are non-uniformly distributed over the network channels. Since there are

$$(n - j + 1) \binom{n}{j - 1}$$

channels located j hops away from the hot-spot node [15], the hot-spot traffic arriving at one of these channels from a given source node, which is located more than j hops away from the hot-spot node, is a fraction of the traffic generated by the source node. This fraction, f_{h_j}, can be expressed as

$$f_{h_j} = \frac{\delta}{(n - j + 1)\binom{n}{j-1}}. \tag{7.15}$$

Let MMPP_{h_j} model the hot-spot traffic at a network channel located j hops away from the hot-spot node arriving from a source node which is located more than j hops away from the hot-spot node. Following equation (7.14), MMPP_{ch_j} can be derived by splitting MMPP_s, with the splitting probability, f_{h_j}.

Given that there are

$$N - \sum_{k=0}^{j-1} \binom{n}{k} = \sum_{k=j}^{n} \binom{n}{k}$$

source nodes located more than j $(1 \leq j \leq n)$ hops away from the hot-spot node, the hot-spot traffic on a network channel located j hops away from the hot-spot node, modelled by MMPP_{ch_j}, is the superposition of $\sum_{k=j}^{n} \binom{n}{k}$ traffic flows modelled by MMPP_{h_j}. Using the method presented in [10] to derive the parameter matrices of the two-state MMPP resulting from the superposition of multiple MMPPs, we can obtain the infinitesimal generator $\mathbf{Q}_{ch_j}$ and the rate matrix $\mathbf{\Lambda}_{ch_j}$ of MMPP_{ch_j}. Since the network channels located j $(1 \leq j \leq n)$ hops away from the hot-spot node receive both regular and hot-spot traffic, the superposition of these two types of traffic yields the traffic loads at network channels. Let MMPP_{c_j} model the traffic arriving at the network channel that is j $(1 \leq j \leq n)$ hops away from the hot-spot node. Adopting the method used to compute MMPP_{ch_j}, we can easily obtain the parameter matrices, $\mathbf{Q}_{c_j}$ and $\mathbf{\Lambda}_{c_j}$, of MMPP_{c_j}.

As adaptive routing distributes regular traffic evenly across the network channels, and due to the symmetry of the hypercube topology, the mean service time seen by a regular message on each channel is identical and is equal to its network latency, t_r. However, the presence of hot-spot traffic causes the service time to vary from one network channel to another. Taking both regular and hot-spot traffic with their appropriate weights into account yields the service time at network channels located j ($1 \leq j \leq n$) hops away from the hot-spot node, t_{c_j} as [29]

$$t_{c_j} = \frac{\lambda_{cr}}{\lambda_{c_j}} t_r + \frac{\lambda_{ch_j}}{\lambda_{c_j}} t_{h_j}, \tag{7.16}$$

where λ_{cr}, λ_{ch_j}, and λ_{c_j} represent the mean arrival rate of regular traffic, hot-spot traffic, and the superposed traffic on network channels located j ($1 \leq j \leq n$) hops away from the hot-spot node, respectively. The Laplace–Stieltjes transform, $F_{t_{c_j}}^*(s)$, of the service time at network channels is required to calculate the mean waiting time, W_{c_j}, in equation (7.21). Following formulae (7.6) and (7.9), $F_{t_{c_j}}^*$ can be expressed as

$$F_{t_{c_j}}^*(s) = \frac{\lambda_{cr}}{\lambda_{c_j}} F_{t_r}^*(s) \frac{\lambda_{ch_j}}{\lambda_{c_j}} F_{t_{h_j}}^*(s). \tag{7.17}$$

7.2.3 Calculation of the Laplace–Stieltjes Transforms of the Blocking Time for Regular Messages and Hot-spot Messages

In order to calculate the Laplace–Stieltjes transform, $F_{b_r}^*(s)$ and $F_{b_{h_j}}^*(s)$, let us first compute the blocking time, b_r and b_{h_j}, at network channels. A regular or hot-spot message is blocked at a network channel when all the adaptive virtual channels of the remaining dimensions to be visited, and also the deterministic virtual channel of the lowest dimension still to be crossed are busy [2]. The blocking time can be determined by the probability of blocking at a given network channel and the waiting time to acquire a deterministic virtual channel when blocking occurs.

When the regular message has reached its ith-hop channel along its network path, this channel can be from 1 to n hops away from the hot-spot node. Let $Pb_{r_{i,j}}$ represent the probability that this message is blocked at its ith-hop channel located j ($1 \leq j \leq n$) hops away from the hot-spot node. Let ξ_j be the ratio of the number of channels that are j ($1 \leq j \leq n$) hops away from the hot-spot node out of the total number of channels in the network. ξ_j is given by [15]

$$\xi_j = \frac{(n-j+1)\binom{n}{j-1}}{nN}. \tag{7.18}$$

Therefore, the blocking time, b_{r_i}, for the typical regular message can be written as

$$B_{r_i} = \sum_{j=1}^{n} \xi_j P b_{r_{i,j}} W_{c_j}, \qquad (7.19)$$

where W_{c_j} denotes the waiting time experienced by a message to acquire a deterministic virtual channel at the lowest dimension when blocking occurs at the channel located j $(1 \leq j \leq n)$ hops away from the hot-spot node. To compute W_{c_j}, we model the network channel as an MMPP/G/1 queueing system. The virtual waiting time, W_j, and the actual waiting time, W_{c_j}, can be expressed as [8]

$$W_j = \frac{2\rho_j + \lambda_{c_j} t_{c_j}^{(2)} - 2t_{c_j}((1 - \rho_j)\mathbf{g} + t_{c_j}\boldsymbol{\pi}_j \Lambda_{c_j})(\mathbf{Q}_{c_j} + \mathbf{e}\boldsymbol{\pi}_j)^{-1}\hat{\lambda}}{2(1 - \rho_j)}, \qquad (7.20)$$

$$W_{c_j} = \frac{1}{\rho_j}\left(W_j - \frac{1}{2}\lambda_v t_{c_j}^{(2)}\right), \qquad (7.21)$$

where t_{c_j} and $t_{c_j}^{(2)}$ represent the first two moments of the service time on network channels located j $(1 \leq j \leq n)$ hops away from the hot-spot node; they can be computed by differentiating $F_{t_{c_j}}^*(s)$ and setting $s = 0$ [11]. The traffic intensity, $\rho_j = t_{c_j} \lambda_{c_j}$. $\boldsymbol{\pi}_j$ is the steady-state vector of the MMPP$_{c_j}$ and $\hat{\lambda} = \Lambda_{c_j} \mathbf{e}$. $\mathbf{e}$ is the unit column vector. The algorithm for computing the matrix $\mathbf{g}$ can be found in [8].

Let P_{a_j} denote the probability that all adaptive virtual channels at a physical network channel located j $(1 \leq j \leq n)$ hops away from the hot-spot node are busy and $P_{a\&d_j}$ be the probability that all adaptive and also deterministic virtual channels of the physical channel are busy. The regular message reaching its ith-hop channel can use $(\bar{d}_r - i)(V - 1)$ adaptive virtual channels at the remaining $(\bar{d}_r - i + 1)$ dimensions to be crossed. It can also use any virtual channel at the lowest dimension still to be visited. Thus, the probability, $P b_{r_{i,j}}$, can be computed by [9]

$$P b_{r_{i,j}} = P_{a_j}^{\bar{d}_r - i} P_{a\&d_j}, \qquad (7.22)$$

where P_{a_j} and $P_{a\&d_j}$ are given by [13]

$$P_{a_j} = P_{V_j} + \frac{P_{(V-1)_j}}{\binom{V}{V-1}} \quad \text{and} \quad P_{a\&d_j} = P_{V_j}, \qquad (7.23)$$

where P_{i_j} is the probability that i $(0 \leq i \leq V)$ virtual channels at a given physical channel located j $(1 \leq j \leq n)$ hops away from the hot-spot node are busy. P_{i_j} will be calculated in Section 7.2.4.

By examining the expression of b_r given by formulae (7.5) and (7.19), it is infeasible to find the exact expression for its Laplace–Stieltjes transform, $F_{b_r}^*(s)$. Given that any distribution function can be approximated arbitrarily closely by a series-parallel stage-type of exponential distributions [11]. Accordingly, the distribution of the blocking time can be reasonably approximated by an exponential distribution. Thus, the probability density function of the blocking time for a regular message, $f_{b_r}(x)$, and its Laplace–Stieltjes transform, $F_{b_r}^*(s)$, can be written as

$$f_{b_r}(x) = \alpha e^{-\alpha x} \ (\alpha > 0) \quad \text{and} \quad F_{b_r}^*(s) = \frac{\alpha}{\alpha + s}, \tag{7.24}$$

where α is selected to match the mean blocking time, b_r, for a regular message. α is found to be

$$\alpha = \frac{1}{b_r}. \tag{7.25}$$

In the case of hot-spot messages, let us consider a j-hop hot-spot message reaching its kth-hop channel which is $(j - k + 1)$ hops away from the hot-spot node. Let $Pbh_{j,j-k+1,k}$ represent the probability that the j-hop hot-spot message is blocked at its kth-hop channel and $W_{c_{j-k+1}}$ denote the waiting time experienced by a j-hop hot-spot message to acquire a deterministic virtual channel at the lowest dimension when blocking occurs. Therefore, the blocking time, $B_{h_{j,k}}$, for the j-hop hot-spot message at its kth-hop channel is given by

$$B_{h_{j,k}} = Pbh_{j,j-k+1,k} \ W_{c_{j-k+1}}, \tag{7.26}$$

where the blocking probability of the hot-spot message is $Pbh_{j,j-k+1,k} = P_{a_{j-k+1}}^{j-k} \cdot P_{a\&d_{j-k+1}}$. Similarly, the Laplace–Stieltjes transform, $F_{b_{h_j}}^*$, is given by

$$F_{b_{h_j}}^*(s) = \frac{\beta}{\beta + s} \tag{7.27}$$

with β matching the mean blocking time for a j-hop hot-spot message.

7.2.4 Calculation of the Probability of Virtual Channel Occupancy

The probability, P_{i_j}, that i $(0 \leq i \leq V)$ virtual channels at a given physical channel located j $(1 \leq j \leq n)$ hops away from the hot-spot node are busy can be determined using the probability that there are i packets in an MMPP/G/1 infinite queueing system [1] with the arrival process MMPP_{c_j} and the service time t_{c_j}. Specifically, the probability that there are i, $0 \le i \le V - 1$, virtual channel busy corresponds to the probability that there are i packets in the queueing system,

$$P_{i,j} = \pi_j (\mathbf{I} - \mathbf{R}_j) \mathbf{R}_j^i \, \mathbf{e}, \quad 0 \le i \le V - 1 \qquad (7.28)$$

and the probability that all virtual channels are busy is the summation of the probabilities of i, $V \le i \le \infty$, packets in the queueing system

$$P_{V,j} = \pi_j \mathbf{R}_j^V \mathbf{e}. \qquad (7.29)$$

In the above equation, the Matrix–Geometric Factor, $\mathbf{R}_j$, can be computed by solving the quadratic matrix equation [18]

$$\mathbf{A}_j + \mathbf{R}_j \mathbf{B}_j + \mathbf{R}_j^2 \mathbf{C}_j = 0 \qquad (7.30)$$

with $\mathbf{A}_j = \Lambda_{c_j}$, $\mathbf{B}_j = \mathbf{Q}_{c_j} - \Lambda_{c_j} - \mathbf{I}/t_{c_j}$, and $\mathbf{C}_j = \mathbf{I}/t_{c_j}$. A simple iterative method can be used to solve the above equation as follows [18]:

1. Initializing $\mathbf{R}_j = -\mathbf{A}_j / \mathbf{B}_j$;
2. At each iteration step, $\mathbf{R}_j = -(\mathbf{A}_j + \mathbf{R}_j^2 \mathbf{C}_j) \mathbf{B}_j^{-1}$, until $\|\mathbf{A}_j + \mathbf{R}_j \mathbf{B}_j + \mathbf{R}_j^2 \mathbf{C}_j\| \le 10^{-10}$.

π_j is the steady-state vector of the MMPP_{c_j}, $\mathbf{I}$ is the identity matrix of size 2, and $\mathbf{e} = (1, 1)^T$.

In virtual channel flow control, multiple virtual channels share the bandwidth of a physical channel in a time-multiplexed manner. The average degree of multiplexing of virtual channels at a given physical channel can be computed according to the method presented in [3] and [15].

7.2.5 Calculation of the Mean Waiting Time at the Source

Messages generated by source nodes follows an MMPP_s and can enter the network through one of the V injection virtual channels with equal probability $1/V$. Let the traffic arriving at an injection virtual channel in the source node located j ($1 \le j \le n$) hops away from the hot-spot node denoted by MMPP_v, resulting from the splitting of MMPP_s with the splitting probability $1/V$; the infinitesimal generator $\mathbf{Q}_v$ and the rate matrix Λ_v of MMPP_v can be derived according to equation (7.14).

To determine the waiting time, W_{s_j}, experienced by a message at the source node that is j ($1 \le j \le n$) hops away from the hot-spot node, the local queue is modelled as an MMPP/G/1 queueing system. The derivation of W_{s_j} is similar to that used above for the actual waiting time, W_{c_j}, in equation (7.21). Averaging over all possible values of j ($1 \leq j \leq n$) yields the mean waiting time at a given source node as

$$\bar{W}_s = \sum_{j=1}^{n} P_{h_j} W_{s_j}. \tag{7.31}$$

Examining formulae (7.20) and (7.21) reveals that the Laplace–Stieltjes transform of the service time for a message originating from a source node is required to compute W_{s_j}. A regular message injected from a source node located j ($1 \leq j \leq n$) hops away from the hot-spot node sees a network latency of t_r given by equation (7.3), whereas a hot-spot message sees the latency of t_{h_j} given by equation (7.7). Taking both the regular and hot-spot messages with their appropriate weights into consideration yields the service time for a message originating from a source node located j ($1 \leq j \leq n$) hops away from the hot-spot node as $t_{s_j} = (1 - \delta)t_r + \delta t_{h_j}$. The Laplace–Stieltjes transform of the service time, t_{s_j}, can be computed similarly to that used to derive $F_{t_{c_j}}^*(s)$ in equation (7.17).

   The mean message latency consists of the mean network latency, $\bar{T}$, that is the mean time to cross the network, and the mean waiting time at the source node, $\bar{W}_s$. In order to model the effect of virtual channel multiplexing, the mean message latency has to be scaled by a factor, $\bar{V}$, of average degree of virtual channel multiplexing. Thus, we can write

$$\text{Latency} = (\bar{T} + \bar{W}_s)\bar{V}. \tag{7.32}$$

7.3 Validation of the Model

A discrete-event simulator, operating at the flit level, has been developed to validate the accuracy of the derived analytical model. Numerous simulation experiments have been performed to validate the model for various combinations of network sizes, message lengths, number of virtual channels per physical channel, different $MMPP_s$ input traffic, and various hot-spot fractions. However, for the sake of specific illustration, latency results are presented for the following cases only: network size is $N = 2^6, 2^8$, and 2^{10} nodes; message length is $m = 32$ and 64 flits; number of virtual channels $V = 2, 3$, and 4 per physical channel; infinitesimal generator, $\mathbf{Q}_s$, of $MMPP_s$,

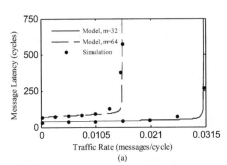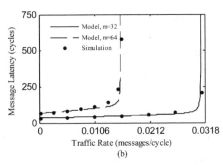

Figure 7.2 Latency predicted by the model against simulation experience in the 6-dimensional hypercube, $m = 32$ and 64 flits: (a) $V = 2$, $\varphi_{s1} = 0.7$, $\varphi_{s2} = 0.35$, $\delta = 0.25$ and (b) $V = 3$, $\varphi_{s1} = 0.08$, $\varphi_{s2} = 0.08$, $\delta = 0.15$.

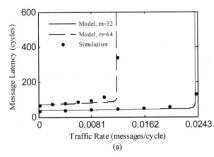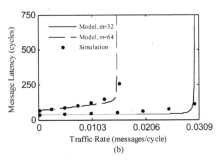

Figure 7.3 Latency predicted by the model against simulation experience in the 8-dimensional hypercube, $m = 32$ and 64 flits: (a) $V = 3$, $\varphi_{s1} = 0.9$, $\varphi_{s2} = 0.3$, $\delta = 0.15$ and (b) $V = 4$, $\varphi_{s1} = 0.005$, $\varphi_{s2} = 0.005$, $\delta = 0.05$.

denoting the different degrees of burstiness and correlations of traffic generated by source nodes, are set in the captions of the figures; hot spot fraction $\delta = 0.05, 0.15$, and 0.25, represents the probability of messages generated by source nodes being sent to the hot-spot node.

Figures 7.2–7.4 depict the performance results for the message latency predicted by the above analytical model plotted against those provided by the simulator as a function of the generated traffic into the 6-, 8-, and 10-dimensional hypercubes, respectively. In these figures, the horizontal axis represents the traffic rate, λ_{s1}, at which a node injects messages into the network when the MMPP$_s$ is at state 1, while the vertical axis denotes the mean message latency obtained from the above model. For the sake of clarity of the figures, we have deliberately set the arrival rate, λ_{s2}, at state 2 at zero; otherwise we need to use three-dimensional graphs to represent the results.

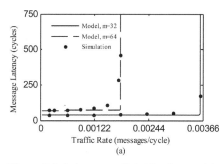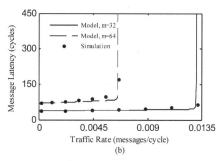

Figure 7.4 Latency predicted by the model against simulation experience in the 10-dimensional hypercube, $m = 32$ and 64 flits: (a) $V = 2$, $\varphi_{s1} = 0.9$, $\varphi_{s2} = 0.45$, $\delta = 0.25$ and (b) $V = 4$, $\varphi_{s1} = 0.8$, $\varphi_{s2} = 0.6$, $\delta = 0.05$.

These figures reveal that the mean message latency obtained from the above model closely match those obtained from the simulation in the steady-state region, and its tractability makes it a practical and cost-effective evaluation tool to gain insight into the behaviour of interconnection networks under the correlated traffic with hot-spot destination distribution.

7.4 Conclusion

Recent studies have shown that the network traffic exhibits a high degree of burstiness and possesses strong correlations in the number of message arrivals between adjacent time intervals in parallel computation environments and message destinations are non-uniformly distributed over the network nodes. This study has developed an analytical model in the hypercubic interconnection networks in the presence of the correlated traffic with non-uniformly distributed message destinations. The traffic generated from each source node follows the well-known MMPP in order to capture the properties of bursty and correlated traffic. The message destinations are non-uniformly distributed across the network nodes according to the characteristics of the hot-spot distribution. The average degree of virtual channel multiplexing has been computed by employing the MMPP/G/1 queueing system with infinite buffer capacity. Results obtained from simulation experiments have demonstrated that the proposed model exhibits a good degree of accuracy for various combinations of operating conditions. The accuracy and tractability of the model make it a practical and cost-effective evaluation tool to gain insight into the behaviour of hypercubic networks under the correlated traffic with hot-spot destination distribution.

Acknowledgements

This work is supported in part by the EU Network of Excellence (NoE) Euro-FGI (NoE 028022), UK EPSRC research grant (EP/C525027/1), and the EU IST STREP Project VITAL (IST-034284 STREP).

References

[1] N. Alzeidi, M. Ould-Khaoua and A. Khonsari, A new general method to compute virtual channels occupancy probabilities in wormhole networks, *Journal of Computer and System Sciences*, vol. 74, no. 6, pp. 1033–1042, 2008.

[2] J. Duato, S. Yalamanchili and L. Ni, *Interconnection Networks: An Engineering Approach*, Morgan Kaufmann, 2003.

[3] W. J. Dally, Virtual channel flow control, *IEEE Trans. Parallel and Distributed Systems*, vol. 3, no. 2, pp. 194–205, 1992.

[4] W. J. Dally and B. P. Towles, *Principles and Practices of Interconnection Network*, Morgan Kaufmann, 2004.

[5] Y. M. Boura and C. R. Das, Performance analysis of buffering schemes in wormhole routers, *IEEE Trans. Computers*, vol. 46, no. 5, pp. 580–587, 1997.

[6] P. A. Dinda, B. Garcia and K. S. Leung, The measured network traffic of compiler parallelized programs, in *Proceedings of the 30th IEEE International Conference on Parallel Processing (ICPP'01)*, pp. 175–184, 2001.

[7] J. T. Draper and J. Ghosh, A comprehensive analytical model for wormhole routing in multicomputer systems, *Journal of Parallel & Distributed Computing*, vol. 32, no. 2, pp. 202–214, 1994.

[8] W. Fischer and K. Meier-Hellstern, The Markov-modulated Poisson process (MMPP) cookbook, *Performance Evaluation*, vol. 18, no. 2, pp. 149–171, 1993.

[9] J. Kim and C. R. Das, Hypercube communication delay with wormhole routing, *IEEE Trans. Computers*, vol. 43, no. 7, pp. 806–814, 1994.

[10] H. Heffes, A class of data traffic processes-covariance function characterization and related queueing results, *Bell System Technical Journal*, vol. 59, no. 6, pp. 897–929, 1980.

[11] L. Kleinrock, *Queueing Systems*, vol. 1, New York, John Wiley, 1975.

[12] R. Nelson, *Probability, Stochastic Processes, and Queuing Theory: The Mathematics of Computer Performance Modeling*, Springer, 1995.

[13] S. Loucif, M. Ould-Khaoua and L. M. Mackenzie, Modelling fully-adaptive routing in hypercubes, *Telecommunication Systems*, vol. 13, no. 1, pp. 111–118, 2000.

[14] G. Min and M. Ould-Khaoua, On the performance of cubic networks under correlated traffic pattern, in *Proceedings of the 16th International Parallel and Distributed Processing Symposium (IPDPS'02)*, pp. 257–264, 2002.

[15] M. Ould-Khaoua and H. Sarbazi-Azad, An analytical model of adaptive wormhole routing in hypercubes in the presence of hot spot traffic, *IEEE Trans. Parallel and Distributed Systems*, vol. 12, no. 3, pp. 283–292, 2001.

[16] G. J. Pfister and V. A. Norton, Hot-spot contention and combining in multistage interconnection networks, *IEEE Trans. Computers*, vol. 34, no. 10, pp. 943–948, 1985.

[17] H. Sarbazi-Azad, M. Ould-Khaoua and L. M. Mackenzie, Analytical modeling of wormhole-routed k-ary n-cubes in the presence of hot-spot traffic, *IEEE Trans. Computers*, vol. 50, no. 7, pp. 623–634, 2001.

[18] H.-P. Schwefel, Performance analysis of intermediate systems serving aggregated ON/OFF traffic with long-range dependent properties, PhD Dissertation, Technische Universität München, 2000.

[19] Y. Wu, G. Min and L. Wang, Performance analysis of interconnection networks under bursty and batch arrival traffic, in *Algorithms and Architectures for Parallel Processing*, H. Jin, O. F. Rana, Y. Pan and V. K. Prasanna (Eds.), Lecture Notes in Computer Science, vol. 4494, Springer, pp. 25–36, 2007.

[20] F. Silla, M. P. Malumbres, J. Duato, D. Dai and D. K. Panda, Impact of adaptivity on the behavior of networks of workstations under bursty traffic, in *Proceedings of International Conference on Parallel Processing (ICPP'98)*, pp. 88–95, 1998.

[21] M. S. Squillante, D. D. Yao and L. Zhang, Analysis of job arrival patterns and parallel scheduling performance, *Performance Evaluation*, vols. 36–37, no. 1, pp. 137–163, 1999.

[22] C. Izu, J. Miguel-Alonso and J.A. Gregorio, Evaluation of interconnection network performance under heavy non-uniform loads, in *Distributed and Parallel Computing*, M. Hobbs, A. Goscinski and W. Zhou (Eds.), Lecture Notes in Computer Science, vol. 3719, Springer, pp. 396–405, 2005.

[23] E. Baydal, P. Lopez and J. Duato, A family of mechanisms for congestion control in wormhole networks, *IEEE Trans. Parallel and Distributed Systems*, vol. 16, no. 9, pp. 772–784, 2005.

[24] B. Matthews and I. Elhanany, On the performance of output-queued cell switches with non-uniformly distributed bursty arrivals, *IEEE Proceedings on Communications*, vol. 153, no. 2, pp. 201–204, 2006.

[25] A. Heindl, Decomposition of general queueing networks with MMPP inputs and customer losses, *Performance Evaluation*, vol. 51, nos. 2–4, pp. 117–136, 2003.

[26] N. Alzeidi, M. Ould-Khaoua, L. M. Mackenzie and A. Khonsari, Performance analysis of adaptively-routed wormhole-switched networks with finite buffers, in *Proceedings of IEEE International Conference on Communications (ICC'07)*, pp. 38–43, June 2007.

[27] D. D. Kouvatsos, S. Assi and M. Ould-khaoua, Performance modelling of wormhole-routed hypercubes with bursty traffic and finite buffers, *International Journal of Simulation, Practics, Systems, Science & Technology*, vol. 6, nos. 3–4, pp. 69–81, 2005.

[28] J. Miguel-Alonso, C. Izu and J. A. Gregorio, Improving the performance of large interconnection networks using congestion-control mechanisms, *Performance Evaluation*, vol. 65, nos. 3–4, pp. 203–211, March 2008.

[29] F. Safaei, A. Khonsari, M. Fathy and M. Ould-Khaoua, Performance modelling of pipelined circuit switching in hypercubes with hot spot traffic, *Microprocessors and Microsystems*, vol. 32, no. 1, pp. 10–22, 2008.

PART THREE
PERFORMANCE EVALUATION STUDIES

8

Service Differentiation through Priority Jumps

Tom Maertens, Joris Walraevens and Herwig Bruneel

SMACS Research Group, Department of Telecommunications and Information Processing (IR07), Ghent University – UGent, Sint-Pietersnieuwstraat 41, B-9000 Gent, Belgium; e-mail: {tmaerten,jw,hb}@telin.ugent.be

Abstract

In this paper, we consider a queueing system with a priority scheduling scheme with priority jumps. Expressions for the probability generating functions of the queue contents and the packet delays have been derived in a previous paper. In the current paper, we determine expressions for some performance measures, i.e., mean values, and approximate tail distributions. These performance measures are furthermore used to illustrate the impact of the priority scheme on the performance of an output queue in a packet switch. We thereby compare the dynamic priority scheme with the FIFO scheme. At the end, we show that the results of this paper can be used in the performance study of more complicated models.

Keywords: Discrete-time, queueing theory, dynamic priority scheduling, performance analysis.

8.1 Introduction

One of the main keys to a succesful telecommunication network nowadays is the ability to efficiently support different services. Different services generate different types of traffic, and different types of traffic have extremely diverse *Quality-of-Service* (QoS) requirements. For *real-time* traffic for example, it is important that mean delay and delay jitter are not too large, while some

D. D. Kouvatsos (ed.), Traffic and Performance Engineering for Heterogeneous Networks, 175–198.

loss is allowed. For *non-real-time traffic* on the other hand, the packet loss ratio is the restrictive quantity, and not so much the delay. In this paper, we further focus on delay as QoS measure. Taking into account their different delay requirements, we then categorize real-time traffic as *delay-sensitive*, and non-real-time traffic as *delay-tolerant*.

To support both types of traffic in a telecommunication network, they are scheduled according to a *priority* scheme. Two priority levels are provided, i.e., the *high- and low-priority* level, and some scheduling rules are introduced between both levels. In the Head-of-Line (HOL) priority scheme for instance, the priority is always given to the delay-sensitive traffic. This priority scheme is not very efficient. Indeed, the HOL scheme provides relatively low delays for the delay-sensitive traffic, but when the system is highly loaded and a large portion of the system traffic consists of delay-sensitive traffic, it can cause excessive delays for the delay-tolerant traffic (see e.g., [1, 5, 13, 15, 20, 31]). Although this type of traffic tolerates a certain amount of delay, extreme values have to be avoided as much as possible. The Transmission Control Protocol (TCP) for example, could consider a delay-tolerant packet with a too big delay as lost, and would consequently decrease its transmission rate. This would decrease the throughput – which is particularly detrimental to data services – but is thus unnecessary since the packet is not lost. The service differentiation between both types of traffic may thus be too drastic in some cases. This is because the HOL scheme is *static*: the priority levels never change in time, and packets of the low-priority level are only served when there are no high-priority packets present in the system.

Dynamic priority schemes aim for a more *gradual* service differentiation. The priority levels of both types of traffic can for instance be *varied dynamically* with time. When there is too much delay-tolerant traffic in the system, this type of traffic gets service priority for a while (see e.g., [4, 6, 7, 10–12, 16, 19]). Another way to reduce performance degradation for the delay-tolerant traffic, is to serve the priority levels in a *weighted* order. The priority levels do not change during time here, but packets of the low-priority level are with a certain regularity scheduled for service before the high-priority packets (see e.g., [9, 17, 25, 26, 28–30, 33]). In a third class of dynamic priority schemes, packets of the low-priority level can in the course of time *jump* to the high-priority level. From the server's point of view, such a scheme is then similar to the static HOL scheme: the server always chooses the packet for service at the head of the highest non-empty priority level. Many criteria can be used to decide when low-priority packets jump to the high-priority level: a maximum queueing delay in the low-priority queue [21], a queue-length-threshold of the high- or low-priority queue [14,23], a random jumping probability per time unit [22], the arrival characteristics of one type of traffic [24], etc.

In this paper, we consider a system that adopts a scheduling scheme with priority jumps. Particularly, we introduce a parameter β, which gives the probability that the total content of the low-priority queue jumps to the end of the high-priority queue. We opt for this straightforward model, so that we can analytically study the effect of priority jumps, and the influence of the system parameters on the performance of the system. The introduction of a jumping parameter β makes the model moreover very efficient. Indeed, the value of β can be chosen in such a way that the delay-tolerant traffic stays within its delay requirements: e.g., the more stringent the delay requirement, the larger the value of β. The service differentiation between both types of traffic can thus be adjusted by changing the value of the jumping parameter.

In a previous paper [22], the authors have tackled the problem of obtaining analytical results for the probability generating functions (pgfs) of the contents of the high- and low-priority queue, and the pgfs of the delays of both types of packets. In the present paper, we concentrate on the derivation of expressions for some performance measures, such as the mean values and (approximate) tail probabilities. It is thereby shown that tail behavior of a performance characteristic is not necessarily geometric. We further use these performance measures to illustrate the impact of the priority scheme with priority jumps on the performance of a specific queueing system. The results of this paper can moreover be applied to predict the performance of more complicated queueing systems.

The outline of the paper is as follows. In the following section, we summarize the results of [22]. In Sections 8.3 and 8.4, we calculate the moments of the queue contents and packet delays and study the tail behavior of these quantities. We apply the obtained results to an output-queueing switch, and discuss the impact of the scheduling scheme in Section 8.5. Some conclusions are finally formulated in Section 8.6.

8.2 Previous Results

We consider a *discrete-time* queueing system with *one server* and *two queues* of *infinite capacity*. The service time of a packet is one slot. We assume that two types of traffic arrive at the system: packets of type 1, representing the delay-sensitive traffic, and packets of type 2, which are delay-tolerant. Both types of traffic enter the system in separate queues. The numbers of arrivals are independent and identically distributed (i.i.d.) from slot-to-slot, but can be correlated in one slot. This dependence is expressed in their joint pgf $A(z_1, z_2)$. We further define the marginal pgfs $A_1(z)$, $A_2(z)$, and $A_T(z)$, of the number of type-1 arrivals, of the number of type-2 arrivals, and of the total number of arrivals, in one slot. The corresponding arrival rates are then $\lambda_j = A'_j(1)$ ($j = 1, 2$), and $\lambda_T = A'_T(1)$ (with $\lambda_T = \lambda_1 + \lambda_2$).

Newly arriving packets can enter service at the beginning of the slot following their arrival slot at the earliest. At the end of each slot, the content of the queue in which type-2 packets are originally stored, jumps with probability β to the other queue, where they join type-1 packets and previously jumped type-2 packets. The packets in the latter queue have service priority, and only when there are no packets present in this queue, type-2 packets in the other queue can be served. Within a queue, the service discipline is FIFO. For convenience, we further denote the queues by the high- and low-priority queue respectively. Note that the jumps occur at the end of a slot, and that the jumping packets are thus stored after the content of the high-priority queue at the end of the slot.

In [22], the authors have derived expressions for the pgf of the total system content at the beginning of a slot, for the pgfs of the contents of both queues at the beginning of a slot, and for the pgfs of the delays of both types of packets. For the total system content and the queue contents, they have found

$$U_T(z) = \frac{(1 - \lambda_T)A_T(z)(z - 1)}{z - A_T(z)}, \tag{8.1}$$

$$U_H(z) = \frac{(1 - \lambda_T)\beta A_T(z)z(z - 1)}{(z - A_T(z))(z - (1 - \beta)A_1(z))}$$
$$- \frac{(1 - \lambda_T)(1 - \beta)\beta A_T(Y(1))A_1(z)(z - 1)}{(z - (1 - \beta)A_1(z))(Y(1) - A_T(Y(1)))}, \tag{8.2}$$

$$U_L(z) = \frac{(1 - \lambda_T)(1 - \beta)A_2(z)(z - 1)(1 - Y(z))}{(1 - (1 - \beta)A_2(z))(z - Y(z))}$$
$$\times \frac{Y(z) - (1 - \beta)A_T(Y(z))}{Y(z) - A_T(Y(z))} + \frac{\beta}{1 - (1 - \beta)A_2(z)}, \tag{8.3}$$

with $Y(z)$ implicitly defined as $(1 - \beta)A(Y(z), z)$. For the packet delays, following pgfs have been determined:

$$D_1(z) = \frac{\beta(1 - \lambda_T)z(A_1(z) - 1)(Y(1)A_T(z) - A_T(Y(1))z)}{\lambda_1(Y(1) - A_T(Y(1)))(z - A_T(z))(z - (1 - \beta)A_1(z))}, \qquad (8.4)$$

$$D_2(z) = \frac{\beta(1 - \lambda_T)}{\lambda_2} \frac{z(A_T(z) - A_1(z))}{(z - A_T(z))(1 - (1 - \beta)A_1(z))}$$
$$+ \frac{(1 - \beta)(1 - \lambda_T)}{\lambda_2} \frac{z(A_T(V_0(z)) - A_1(V_0(z)))(1 - A_1(z))}{(V_0(z) - A_T(V_0(z)))(1 - (1 - \beta)A_1(z))}, \qquad (8.5)$$

with $V_0(z)$ implicitly given by $(1 - \beta)zA_1(V_0(z))$. When we choose $\beta = 0$ in all these expressions, we obtain the same pgfs of the studied quantities as in [31]. This is expected, because when $\beta = 0$, type-2 packets never jump to the high-priority queue, and we thus have the same situation as in the static HOL priority scheme.

8.3 Calculation of the Moments

In this section, we give expressions for the mean values of the studied quantities. Expressions for higher moments can be obtained in a similar way, but are omitted because of their size. We however illustrate them in figures in Section 8.5. To make the expressions more readable, we define λ_{11} and λ_{TT} as

$$\lambda_{11} \triangleq \left. \frac{\partial^2 A(z_1, z_2)}{\partial z_1 \partial z_1} \right|_{z_1 = z_2 = 1} \quad \text{and} \quad \lambda_{TT} \triangleq \left. \frac{\partial^2 A_T(z)}{\partial z^2} \right|_{z=1}$$

respectively. By taking the first derivative of the respective pgfs for $z = 1$, we obtain

$$E[u_T] = \lambda_T + \frac{\lambda_{TT}}{2(1 - \lambda_T)},$$

$$E[u_H] = 1 + \lambda_2 - \frac{1 - \lambda_1}{\beta} + \frac{\lambda_{TT}}{2(1 - \lambda_T)} - \frac{(1 - \lambda_T)(1 - \beta)A_T(Y(1))}{Y(1) - A_T(Y(1))},$$

$$E[u_L] = \frac{(1 - \beta)\lambda_2}{\beta} + \frac{(1 - \lambda_T)(1 - \beta)(Y(1) - (1 - \beta)A_T(Y(1)))}{\beta(Y(1) - A_T(Y(1)))}.$$

It is easily verified that these expressions satisfy $E[u_T] = E[u_H] + E[u_L]$, as expected. For the mean values of the packet delays, we find

$$E[d_1] = 1 + \lambda_2 - \frac{1 - \lambda_1}{\beta} + \frac{\lambda_{TT}\lambda_1 + \lambda_{11} - \lambda_{11}\lambda_T}{2(1 - \lambda_T)\lambda_1}$$

$$- \frac{(1 - \lambda_T)A_T(Y(1))}{Y(1) - A_T(Y(1))},$$

$$E[d_2] = 1 - \lambda_1 + \frac{\lambda_1}{\beta} + \frac{(1 - \lambda_1)\lambda_{TT} - (1 - \lambda_T)\lambda_{11}}{2(1 - \lambda_T)\lambda_2}$$

$$- \frac{(1 - \beta)(1 - \lambda_T)(A_T(V_0(1)) - A_1(V_0(1)))\lambda_1}{\lambda_2(V_0(1) - A_T(V_0(1)))\beta}.$$

Notice that $E[u_H] \neq \lambda_1 E[d_1]$ and that $E[u_L] \neq \lambda_2 E[d_2]$, as one would – at first – expect according to Little's law. The reason for this is that in the calculation of the system content, packets of the low-priority queue jump to the high-priority queue and from that moment on, they are treated as part of the content of the high-priority queue. This is of course not the case in the calculation of the delay of a type-2 packet. So basically, the system content is analyed on a "queue"-basis, while the packet delays are analyzed on "packet"-basis.

8.4 Calculation of the Tail Probabilities

Another important performance characteristic, besides the moments, is the (tail) distribution of the studied quantities. The tail probabilities, i.e., the probability mass function (pmf) for large values, typically represent the 'exceptional' situations in a queueing system. The probability that the delay is larger than a given value N, or the packet loss, are examples of interesting performance measures for which the calculation of the tail probability is usually sufficient. The tail distribution is thus often used to impose statistical bounds on the guaranteed QoS for both types of traffic.

Exact theoretical solutions for this inversion problem make use of the probability generating property of pgfs, or of residue theory. However, since these solution methods need a lot of derivations, they are often quite unpractical. We will therefore use an *approximate* solution technique, which is known to be quite popular: the dominant-singularity method. In [3] for example, it has been shown that the pmf $x(n)$ of a discrete variable X is –

for high n – dominated by the contribution of the singularity of the corresponding pgf $X(z)$ with the smallest absolute value. Because of a property of pgfs, this *dominant* singularity is necessarily positive real and larger than 1. In this section, we derive expressions for the tail probabilities of the total system content, of the contents of the high- and low-priority queue separately, and of the delay of a type-1 packet, by using this dominant-singularity approximation method and Darboux's theorem (see Appendix A).

It is assumed in the remainder that the pgfs $A_T(z)$, $A_1(z)$, and $A_2(z)$, and their derivatives go to infinity for z equal to their radii of convergence or for $z \to \infty$. This includes all "usual" arrival processes, and is thus not a restrictive assumption. We furthermore suppose that $\beta > 0$. For $\beta = 0$ (i.e., the static HOL scheme), we refer to [31].

8.4.1 Content of the Total System

The tail behavior of the total system content has also been investigated in [31]. The following approximation is there found:

$$\text{Prob}\,[u_T = n] \approx \frac{(1 - \lambda_T)(s_T - 1)s_T^{-n}}{A_T'(s_T) - 1}, \tag{8.6}$$

with s_T respresenting the dominant singularity of the pgf $U_T(z)$. It is the (dominant) positive real zero larger than 1 of $z - A_T(z)$, i.e., the denominator of $U_T(z)$.

8.4.2 Content of the High-Priority Queue

Let us further concentrate on the tail behavior of the content of the high-priority queue. Two singularities may play a role here, namely the dominant positive real zeros larger than 1 of $z - A_T(z)$ and $z - (1 - \beta)A_1(z)$. We denote them by s_T and s_1 respectively. For z positive real, larger than 1, and in the mutual regions of convergence of $A_T(z)$ and $A_1(z)$, we can however easily verify that $A_T(z) > (1 - \beta)A_1(z)$. So s_T is always smaller than s_1, and as a consequence, s_T is the dominant singularity of $U_H(z)$.

Since the first derivative of $U_H(z)$ stays finite for $z = s_T$, this singularity is a pole with multiplicity one. In the neighborhood of this pole, we can approximate $U_H(z)$ by $K_{U_H}/(s_T - z)$. The constant K_{U_H} is obtained by calculating $\lim_{z \to s_T} U_H(z)(s_T - z)$. By using Darboux's theorem (see Appendix A), we subsequently find

$$\text{Prob}\,[u_H = n] = K_{U_H} s_T^{-n-1}. \tag{8.7}$$

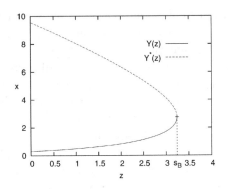

Figure 8.1 Solutions of $x - (1 - \beta)A(x, z) = 0$.

8.4.3 The Function $Y(z)$

The tail behavior of the content of the low-priority queue is not so straightforward. This is in the first place due to the appearance of $Y(z)$ in the expression of $U_L(z)$ (see (8.3)). We first take a closer look at this implicitly defined function on the positive real axis.

As z increases along the positive real axis, a branch point s_B is encountered where $Y(z)$ stays finite, but where $Y'(z) \to \infty$ (see, e.g., [18, 31] for similar cases). s_B is thus the solution of

$$\begin{cases} Y(s_B) = (1 - \beta)A(Y(s_B), s_B) \\ Y'(s_B) \to \infty \end{cases} \Rightarrow \begin{cases} Y(s_B) - (1 - \beta)A(Y(s_B), s_B) = 0 \\ (1 - \beta)A^{(1)}(Y(s_B), s_B) = 1 \end{cases}.$$

(8.8)

For values of z beyond s_B, $Y(z)$ is no longer properly defined. Note that $Y(z)$ is a solution of the functional equation $x - (1 - \beta)A(x, z) = 0$. This equation has another positive real solution $Y^*(z)$, which decreases as z increases (see Figure 8.1). Both solutions then coincide for $z = s_B$. By applying the results of [8], one can show that in the neighborhood of s_B, $Y(z)$ is approximately given by

$$Y(z) \approx Y(s_B) - K_Y(s_B - z)^{1/2},$$

(8.9)

with

$$K_Y = \sqrt{\frac{2A^{(2)}(Y(s_B), s_B)}{A^{(11)}(Y(s_B), s_B)}}.$$

(8.10)

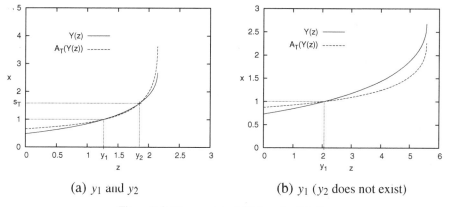

(a) y_1 and y_2 (b) y_1 (y_2 does not exist)

Figure 8.2 The zero(s) of $Y(z) - A_T(Y(z))$.

The constant K_Y is found by substituting $z = s_B$ in (8.9). Since $Y(z)$ appears in the expression of $U_L(z)$, s_B is also a singularity of this pgf, and may thus play a role in the tail behavior of the content of the low-priority queue.

8.4.4 Content of the Low-Priority Queue

Secondly, we show that none of the (dominant) positive real solutions larger than 1 of $1 - (1 - \beta)A_2(z)$ – denoted by s_2 – and $z - Y(z)$ – denoted by s_Y – are potential singularities of $U_L(z)$.

Since s_2 is a zero of $1 - (1 - \beta)A(1, z)$, it is easily seen that $(x, z) = (1, s_2)$ is a solution of $x - (1 - \beta)A(x, z) = 0$. s_2 is thus smaller than or equal to s_B, because the latter equation has no solution for $z > s_B$ (see previous subsection). We have also shown that this equation has two positive real solutions for $z \leq s_B$, namely $(x, z) = (Y(z), z)$ and $(x, z) = (Y^*(z), z)$. So when $z = s_2$, there are two possibilities for x, i.e., $x = Y(s_2)$ and $x = Y^*(s_2)$. One of them must equal 1. Since $Y(s_B) > 1$ and $Y^*(z) > Y(s_B)$, $Y(s_2) = 1$. The numerator of $U_L(z)$ however also vanishes for $z = s_2$ when $Y(s_2) = 1$, which means that s_2 is not a singularity of $U_L(z)$. Likewise, choosing $z = s_Y$ in the numerator of $U_L(z)$ and using the definition of $Y(z)$ (see Section 8.2), also leads to zero. s_Y is thus not a singularity of $U_L(z)$ as well.

Finally, we look at the zeros of $Y(z) - A_T(Y(z))$, i.e., the last factor of the denominator which may possibly yield a potential singularity. Note that the zero(s) of this factor are smaller than or equal to s_B, since $Y(z)$ appears in the factor and since $Y(z)$ does not exist for $z > s_B$. The equation $x - A_T(x) = 0$ has $x = 1$ and $x = s_T$ as solutions on the positive real axis (with $s_T > 1$). We can thus easily verify that the smallest positive real zero of $Y(z) - A_T(Y(z))$ satisfies $Y(z) = 1$ (see Figure 8.2a). This zero, denoted by y_1, coincides with s_2, and we have already shown that this is not a singularity of $U_L(z)$. On the other hand, the second positive real zero y_2 of $Y(z) - A_T(Y(z))$, which satisfies $Y(z) = s_T$ (see Figure 8.2a), seems to be a potential singularity, since y_2 is not a zero of the numerator. y_2 does however not always exist. When $s_T > A_T(Y(s_B))$, the functions $Y(z)$ and $A_T(Y(z))$ cease to exist before they intersect once more (see Figure 8.2b). In this case, they have only point of intersection, namely y_1.

In summary, the tail behavior of the content of the low-priority queue is dominated by the singularities s_B or y_2, depending on whether y_2 exists or not. Three cases can occur: y_2 exists and $y_2 < s_B$, y_2 exists and $y_2 = s_B$, or y_2 does not exist. In the first case, the singularity y_2 is dominant. This singularity is a pole with multiplicity one. Consequently,

$$U_L(z) \approx \frac{K_{U_L}^{(1)}}{y_2 - z}$$

for $z \to y_2$. The constant $K_{U_L}^{(1)}$ can be obtained by determining $\lim_{z \to s_T} U_L(z)(y_2 - z)$:

$$K_{U_L}^{(1)} = \frac{(1 - \lambda_T)\beta(1 - \beta)A_2(y_2)(y_2 - 1)(s_T - 1)s_T}{Y'(y_2)(A_T'(s_T) - 1)(1 - (1 - \beta)A_2(y_2))(y_2 - s_T)}, \qquad (8.11)$$

where we have used the fact that $Y(y_2) = s_T$. In the second case, y_2 and s_B coincide, and are so-called *co-dominant*. Using expression (8.9) in (8.3), and taking into account the fact that $Y(s_B) = s_T$, yields

$$U_L(z) \approx$$

$$\frac{\left\{ \begin{array}{l} (1 - \lambda_T)(1 - \beta)A_2(z)(z - 1)\left(1 - s_T + K_Y(s_B - z)^{1/2}\right) \\ \times \left(s_T - K_Y(s_B - z)^{1/2} - (1 - \beta)s_T + (1 - \beta)K_Y A_T'(s_T)(s_B - z)^{1/2}\right) \end{array} \right\}}{(s_B - z)^{1/2}K_Y(A_T'(s_T) - 1)(1 - (1 - \beta)A_2(z))\left(z - s_T + K_Y(s_B - z)^{1/2}\right)}.$$

$$(8.12)$$

The pgf $U_L(z)$ can thus be approximated by $K_{U_L}^{(2)}/(s_B - z)^{1/2}$ in the neighborhood of $y_2 = s_B$, with

$$K_{U_L}^{(2)} = \frac{(1 - \lambda_T)(1 - \beta)\beta s_T A_2(s_B)(s_B - 1)(1 - s_T)}{K_Y(A_T'(s_T) - 1)(1 - (1 - \beta)A_2(s_B))(s_B - s_T)}. \qquad (8.13)$$

In the third case, when y_2 does not exist, the branch point s_B is dominant. By substituting expression (8.9) in (8.3), we obtain

$$U_L(z) \approx$$

$$\frac{\left\{ \begin{array}{l} (1 - \lambda_T)(1 - \beta)A_2(z)(z - 1)\left(1 - Y(s_B) + K_Y(s_B - z)^{1/2}\right) \\ \times \left(Y(s_B) - (1 - \beta)A_T(Y(s_B)) + K_Y(s_B - z)^{1/2}((1 - \beta)A'_T(Y(s_B)) - 1)\right) \\ \times \left(z - Y(s_B) - K_Y(s_B - z)^{1/2}\right) \\ \times \left(Y(s_B) - A_T(Y(s_B)) - K_Y(s_B - z)^{1/2}(A'_T(Y(s_B)) - 1)\right) \end{array} \right\}}{\left\{ \begin{array}{l} (1 - (1 - \beta)A_2(z))\left((z - Y(s_B))^2 - K_Y^2(s_B - z)\right) \\ \times \left((Y(s_B) - A_T(Y(s_B)))^2 - K_Y^2(s_B - z)(A'_T(Y(s_B)) - 1)^2\right) \end{array} \right\}}.$$

$$(8.14)$$

This expression leads to $U_L(z) \approx U_L(s_B) - K_{U_L}^{(3)}(s_B - z)^{1/2}$ in the neighborhood of s_B, with

$$K_{U_L}^{(3)} =$$

$$\frac{\left\{ \begin{array}{l} (1 - \lambda_T)(1 - \beta)\left(\beta(1 - Y(s_B))(s_B - Y(s_B))(Y(s_B)A'_T(Y(s_B)) - A_T(Y(s_B))) \\ -(s_B - 1)(Y(s_B) - A_T(Y(s_B)))(Y(s_B) - (1 - \beta)A_T(Y(s_B)))\right)(s_B - 1)A_2(s_B) \end{array} \right\}}{(s_B - Y(s_B))^2 (Y(s_B) - A_T(Y(s_B)))^2 (1 - (1 - \beta)A_2(s_B))}.$$

$$(8.15)$$

We now have approximate expressions for $U_L(z)$ in the neighborhood of its dominant singularity, for the three possible cases. By using Darboux's theorem with these approximations, we find the corresponding tail probabilities:

$$\text{Prob}[u_L = n] = \begin{cases} K_{U_L}^{(1)} y_2^{-n-1} & \text{if } y_2 < s_B, \\ \dfrac{K_{U_L}^{(2)} n^{-1/2} s_B^{-n}}{\sqrt{\pi s_B}} & \text{if } y_2 = s_B, \\ \dfrac{K_{U_L}^{(3)} n^{-3/2} s_B^{-n}}{2\sqrt{\pi / s_B}} & \text{if } y_2 \text{ does not exist,} \end{cases} \qquad (8.16)$$

where the constants $K_{U_L}^{(i)}$ ($i = 1, 2, 3$) are given by (8.11), (8.13) and (8.15) respectively. The first expression constitutes a typical geometric tail behavior, while the others are of a non-geometric nature.

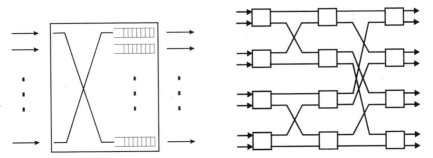

(a) An $N \times N$ output-queueing switch   (b) An 8×8 self-routing 3-stage switching network

Figure 8.3 The switching environment.

8.4.5 Delay of a Type-1 Packet

The dominant singularity of $D_1(z)$ is the same as the dominant singularity of $U_H(z)$ (also with multiplicity one). In the neighborhood of s_T, $D_1(z)$ is approximated by

$$D_1(z) \approx \frac{\beta(1 - \lambda_T)s_T^2(A_1(s_T) - 1)}{\lambda_1(s_T - (1 - \beta)A_1(z))(A_T'(s_T) - 1)(s_T - z)}. \tag{8.17}$$

For the tail probabilities of the delay of a type-1 packet, we obtain

$$\text{Prob}\,[d_1 = n] = \frac{\beta(1 - \lambda_T)s_T^{1-n}(A_1(s_T) - 1)}{\lambda_1(s_T - (1 - \beta)A_1(s_T))(A_T'(s_T) - 1)}. \tag{8.18}$$

8.4.6 Delay of a Type-2 Packet

The tail behavior of the delay of type-2 packet is again more complicated. It can namely be characterized by four singularities: the positive real zeros larger than 1 of $z - A_T(z)$, $V_0(z) - A_T(V_0(z))$, and $1 - (1 - \beta)A_1(z)$, plus the branch point of $V_0(z)$. We may thus have quite a lot of different cases with respect to the dominant singularity of $D_2(z)$. A special paper is therefore devoted to this [23].

8.5 Application

The results obtained in the former sections are now applied to an output-queueing switch (see Figure 8.3a). This output-queueing switch has N inlets

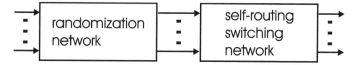

Figure 8.4 A switching network with traffic randomization.

and N outlets and we assume that two types of traffic arrive at the switch: traffic of type 1, which is delay-sensitive, and traffic of type 2, respresenting delay-tolerant traffic. The packet arrivals on the inlets are generated by independent and identically distributed (i.i.d.) Bernoulli processes with arrival rate λ_T. An arriving packet is assumed to be of type j with probability λ_j/λ_T ($j = 1, 2$). So $\lambda_1 + \lambda_2 = \lambda_T$. The incoming packets are routed to the output queue corresponding to their destination, in an independent and uniform way. The output queues thus all behave identically, and we can concentrate on the study of one. The numbers of arriving packets to an output queue in one slot are generated according to a two-dimensional binomial process, which is fully characterized by the joint pgf

$$A(z_1, z_2) = \left(1 - \frac{\lambda_1}{N}(1 - z_1) - \frac{\lambda_2}{N}(1 - z_2) \right)^N . \qquad (8.19)$$

Obviously, the numbers of packets entering an output queue are correlated within one slot. When m type-1 packets arrive during a slot ($0 \leq m \leq N$), the maximum number of type-2 arrivals during the same slot is limited by $N - m$ (because there are only N inlets). An output queue is furthermore assumed to consist of two logical queues. Type-1 packets arrive to the first queue, and type-2 packets arrive to the second queue. The packets of the first queue have service priority over the packets of the second, and in each slot, the contents of the second jumps to the first with probability β. The results obtained in the former sections can thus be used to study the performance of an output queue in a switch.

The choice for Bernoulli arrivals on the inlets of the switch is motivated as follows. An $N \times N$ switching element, as described above, is the smallest building block of a $P \times P$ self-routing switching network (see e.g., [27, 32]). The number of stages in the switching network, denoted by K, is then equal to $log_N P$. In Figure 8.3b, we for example see a 8×8 self-routing 3-stage switching network, consisting of 12 2×2 switching elements. In [27], the authors state that in the case of random (i.e., uncorrelated in time) input traffic, a Bernoulli proces is a reasonably good candidate to represent the arrival

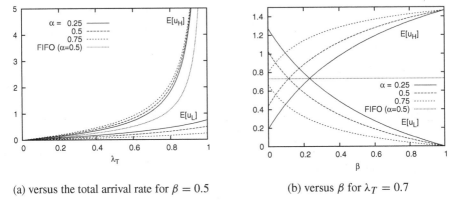

(a) versus the total arrival rate for $\beta = 0.5$ (b) versus β for $\lambda_T = 0.7$

Figure 8.5 Mean values of the queue contents.

process on the inlets of the switching elements. However, due to the integrated traffic (voice and data) in real networks, the input traffic is rather bursty (i.e., correlated). It is known that the traffic burstiness (or correlation) adversely affects the performance. By adding a randomization network in front of the switching network (see Figure 8.4), the performance of the switching network can be significantly improved. A randomazation network distributes the bursty input traffic among all the inlets of the self-routing network. As a consequence, the performance of the switching network is made less sensitive to the bursty input traffic (see e.g. [32]). For the ideally randomization network, its output traffic is assumed to be random traffic, independent of the input traffic. This brings us back to a Bernoulli arrival proces.

Let us now study the impact of the priority scheme with priority jumps on the performance of an output queue in a switch. We therefore consider some performance measures, such as the mean values and the variances of the queue contents and the packet delays. The performance study is focused on the comparison between queues with the dynamic priority scheme and the FIFO scheme. Note that we assume a 16×16-switch ($N = 16$). We finally define α as the fraction of type-1 arrivals in the overall traffic mix (i.e., $\alpha \triangleq \lambda_1/\lambda_T$).

In Figure 8.5a, the mean values of the contents of the high- and low-priority queue are shown as functions of the total arrival rate λ_T, for $\beta = 0.5$ and $\alpha = 0.25, 0.5$ and 0.75 respectively. In order to compare the dynamic priority scheme with the FIFO scheme, we have applied a FIFO scheduling on a joint queue in which the packets of both types of traffic are mixed up

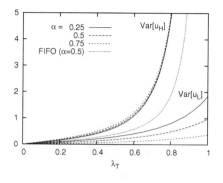

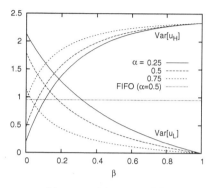

(a) versus the total arrival rate for $\beta = 0.5$ (b) versus β for $\lambda_T = 0.7$

Figure 8.6 Variances of the queue contents.

(according to their arrivals). We have plotted the mean number of packets of one type of traffic present in the system, since for $\alpha = 0.5$ the mean number of type-1 packets in the system equals the mean number of type-2 packets. We can easily see that $E[u_H]$ is larger for the dynamic priority scheme than for the FIFO scheme. For $E[u_L]$, the opposite holds. This can be explained as follows: packets of the high-priority queue have priority over packets of the low-priority queue. So without priority jumps, the low-priority queue would build up as long as there are packets in the high-priority queue. Because of the priority jumps, the content of the low-priority queue however jumps once every two slots to the high-priority queue, thereby leaving the low-priority queue totally empty. As a consequence, $E[u_H]$ is larger than $E[u_L]$. The figure also shows that $E[u_H]$ increases when α increases. This is expected, since a higher value of α means a larger fraction of type-1 packets in the arrival stream. The opposite again holds for $E[u_L]$.

Figure 8.5b shows the mean values of the queue contents as functions of β, for $\lambda_T = 0.7$ and $\alpha = 0.25$, 0.5 and 0.75 respectively. The influence of β is quite obvious: larger β means more jumps (on average), resulting into a higher $E[u_H]$ and a lower $E[u_L]$. The figure also shows that the two curves for $\alpha = 0.25$ and the two curves for $\alpha = 0.5$ intersect each other – and the FIFO curve – for certain values of β. This means that from those β-values on, $E[u_H]$ is larger than $E[u_L]$ for the respective values of α. For $\alpha = 0.75$, $E[u_H]$ is always larger than $E[u_L]$ (when $\lambda_T = 0.7$). When $\beta = 1$, we can easily see that $E[u_L] = 0$. In each slot, the newly arriving type-2 packets immediately jump to the high-priority queue. The low-priority queue is thus

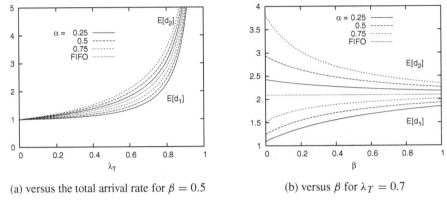

(a) versus the total arrival rate for $\beta = 0.5$ (b) versus β for $\lambda_T = 0.7$

Figure 8.7 Mean values of the packet delays.

always empty then. For the variances of the queue contents (see Figure 8.6), the same conclusions can be drawn as for the mean values.

In Figure 8.7a, we depict the mean values of the packet delays of both types of traffic as functions of the total arrival rate, for $\beta = 0.5$ and $\alpha = 0.25$, 0.5 and 0.75 respectively. We also show the mean value of the packet delay for the FIFO scheme. The packet delays are then the same for type-1 and type-2 traffic (independent of α), and can thus be calculated as if there is only one type of traffic arriving at the system, according to an arrival process with pgf $A(z, z)$ (see [2]). The influence of the priority scheduling is quite obvious: $E[d_1]$ is smaller for the dynamic priority scheme than for the FIFO scheme. For $E[d_2]$, the opposite holds. The reason is clear: the type-1 packets have priority over the type-2 packets. The influence of the dynamic priority scheme is however limited. The mean delay of a type-1 packet reduces only moderately in comparison with the mean delay for the FIFO scheme, while the price to pay, a higher mean delay of a type-2 packet, is also rather small. Note further that it follows from this figure that increasing the fraction of type-1 packets in the overall traffic mix (i.e., increasing α), increases the mean delay of both types of packets. Indeed, the smaller amount of type-2 packets suffer from larger delays, and thus give cause for a larger $E[d_2]$ as well.

In Figure 8.7b, the mean values of the packet delays are shown as functions of β, for $\lambda_T = 0.7$ and $\alpha = 0.25$, 0.5 and 0.75. A larger value of β implies more jumps, and as a consequence, a lower negative effect from the priority scheduling on $E[d_2]$. The price to pay is a higher $E[d_1]$. We can derive similar conclusions with respect to the delay jitter (see Figure 8.8). We

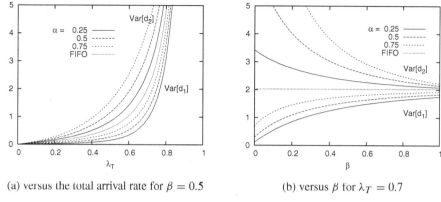

(a) versus the total arrival rate for $\beta = 0.5$ (b) versus β for $\lambda_T = 0.7$

Figure 8.8 Variances of the packet delays.

can here conclude that the dynamic priority scheme does what it is designed for: lowering the delay of the type-1 packets (which are delay-sensitive), but in contrast with the static HOL priority scheme (see [31]), taking into consideration the delay of the type-2 packets (being delay-tolerant). The parameter β can be chosen depending on the delay guarantees for both types of traffic. A low β will highly favor the delay-sensitive traffic, while choosing β higher will give the delay-sensitive traffic only a small delay reduction (compared to the FIFO scheme).

All numerical examples until now, assume that the number of arrivals of one type of traffic is correlated with the number of arrivals of the other type of traffic in one slot (as described in Section 8.2). Our model however does not include correlation amongst consecutive slots. The results of the current paper are therefore not directly useful to describe a queueing system with correlation in time. In the following, we propose an approximation for a system with correlated arrivals based on our results of the system with no correlation in time.

Before we propose our approximation, we describe the modelling of the correlation that is used to validate the approximation. The queueing system is modelled with a single server and two priority queues of infinite capacity. The queues are fed by m_j number of two-state (i.e., an idle and an active state) independent Markov sources, with $j = 1, 2$ denoting the high- and low-priority queue respectively. The arrival process to each queue is thus correlated within itself, but the two arrival processes are independent in this paragraph. For type-j sources, a transition from idle state to active state oc-

Table 8.1   Validation of simulation results versus approximate results for $E[d_1]$.

λ_T	$K = 5$		$K = 10$		$K = 20$	
	$E[d_1]_{\text{sim}}$	$E[d_1]_{\text{approx}}$	$E[d_1]_{\text{sim}}$	$E[d_1]_{\text{approx}}$	$E[d_1]_{\text{sim}}$	$E[d_1]_{\text{approx}}$
0.1	1.394	1.402	1.906	1.888	2.973	2.860
0.2	1.892	1.909	3.509	3.003	5.415	5.190
0.3	2.558	2.556	4.555	4.441	8.675	8.191
0.4	3.453	3.450	6.575	6.367	12.870	12.200
0.5	4.754	4.698	9.457	9.073	18.906	17.823
0.6	6.751	6.590	13.785	13.153	26.810	26.278
0.7	10.099	9.778	20.920	19.986	43.001	40.403
0.8	16.967	16.232	35.778	33.732	72.657	68.732
0.9	37.399	35.836	79.012	75.211	163.005	153.961

curs with probability $1 - \gamma_j$, while the probability of a transition from active to idle state occurs with probability $1 - \alpha_j$. We assume that an active source generates one packet per slot, whereas an idle source generates no packets during a slot. σ_j represents the fraction of time a type-j user is in the active state, and α_j and γ_j are selected in accordance with

$$\text{mean active period of a type-}j\text{ user} = \frac{1}{1 - \alpha_j} = \frac{K}{1 - \sigma_j}, \qquad (8.20)$$

and

$$\text{mean passive period of a type-}j\text{ user} = \frac{1}{1 - \gamma_j} = \frac{K}{\sigma_j}. \qquad (8.21)$$

Note that K is hereby defined as the burstiness factor of both types of traffic.

Now, the basic idea of our approximation is the fact that the influence of time-correlation on the performance measures is similar in case of the FIFO scheme as in the case of the HOL-PJ (Head-of-Line with Priority Jumps) scheme. More precisely, we calculate the difference between $E[d_1]$ for the HOL-PJ scheme and $E[d]$ for FIFO scheme, and we assume that this difference is independent from the burstiness factor K. Or, said in a more readable manner:

$$E[d]_{\text{FIFO}} - E[d_1]_{\text{HOL-PJ}} \approx \text{independent from } K. \qquad (8.22)$$

Table 8.2 Validation of simulation results versus approximate results for $E[d_2]$.

λ_T	$K = 5$		$K = 10$		$K = 20$	
	$E[d_2]_{\text{sim}}$	$E[d_2]_{\text{approx}}$	$E[d_2]_{\text{sim}}$	$E[d_2]_{\text{approx}}$	$E[d_2]_{\text{sim}}$	$E[d_2]_{\text{approx}}$
0.1	1.482	1.452	1.988	1.938	2.968	2.910
0.2	2.085	2.017	3.226	3.110	5.501	5.298
0.3	2.853	2.741	4.811	4.616	8.706	8.366
0.4	3.868	3.704	6.909	6.620	12.996	12.454
0.5	5.445	5.048	9.850	9.423	18.994	18.173
0.6	7.407	7.058	14.261	13.620	28.147	26.745
0.7	10.883	10.396	21.541	20.605	43.239	41.021
0.8	17.864	17.047	36.302	34.547	72.760	69.547
0.9	38.412	36.919	79.851	76.294	162.865	155.044

This independency then leads to

$$E[d_1]_{\text{HOL-PJ, general } K} \approx E[d]_{\text{FIFO, general } K}$$

$$- (E[d]_{\text{FIFO, } K = 1} - E[d_1]_{\text{HOL-PJ, } K = 1}). \qquad (8.23)$$

All quantities in the right-hand side of (8.23) can be explicitly calculated. $E[d]_{\text{FIFO, general } K}$ and $E[d]_{\text{FIFO, } K = 1}$ can be easily obtained from [2], where a single-class FIFO queue with the arrival process as described in this paragraph is analyzed. Further, the quantity $E[d_1]_{\text{HOL-PJ, } K = 1}$ of the left-hand side of the equation, is determined in the current paper. To validate our approximation, we have compared the approximate results with simulation results, for $(m_1, m_2) = (8, 8)$ and various β. Table 8.1 shows the simulation results and the approximate results, for $\beta = 0.5$, and $K = 5$, 10 and 20 respectively. The table shows that our approximations are very good. We can thus predict the behavior of a queueing system with the HOL-PJ scheduling scheme and with correlated arrivals, by combining the results obtained for a queue with the HOL-PJ scheme and uncorrelated arrivals, and the results for a FIFO queue and correlated arrivals. Simulations for other values of β learn that for lower β these approximations are slightly worse, but still very reasonable, while for higher β even better approximations are found. Note finally that a similar discussion can be followed with respect to the mean packet delay of type-2 packets (see Table 8.2).

Let us finally take a look at the tail behavior of the content of the low-priority queue. We have shown that three types of behavior are encountered,

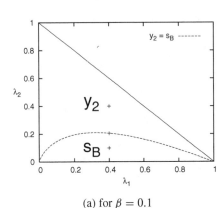
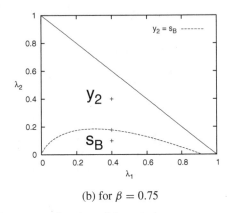

(a) for $\beta = 0.1$ (b) for $\beta = 0.75$

Figure 8.9 Regions for tail behavior of $U_2(z)$ as a function of the arrival rates.

depending on which singularity is dominant. In case of the output-queueing switch considered in this section, Figure 8.9 shows for which combinations of type-1 and type-2 arrival rates, $y_2 = s_B$ (i.e., the singularities y_2 and s_B coincide). The curve splits the (λ_1, λ_2)-space in 2 regions: a region in which y_2 does not exist, making s_B dominant (below the curve), and a region in which y_2 is dominant (above the curve) – since y_2, when it exists, is smaller than s_B. Note that in the area above the linear line (defined by $\lambda_1 + \lambda_2 = 1$), the total arrival rate is larger than 1, resulting in an unstable system. When we compare Figure 8.9a ($\beta = 0.1$) with Figure 8.9b ($\beta = 0.75$), we notice the role of β: the region below the curve, where y_2 does not exist, becomes smaller for increasing β. Although the role of β is limited, we can conclude that the values of *all* system parameters have an influence on the tail behavior of $U_2(z)$.

Figures 8.10a and 8.10b then show the tail probabilities of the content of the low-priority queue, for the (λ_1, λ_2)-combinations indicated by the marks in Figures 8.9a and 8.9b, respectively. We have compared our approximations with simulation results (marks in Figures 8.10a and 8.10b). The figures show that all approximations are excellent.

8.6 Conclusions

In this paper, we have considered a queueing system with a priority scheme with priority jumps. We have derived explicit expressions for the mean values of the queue contents and the packet delays, and determined approximate ex-

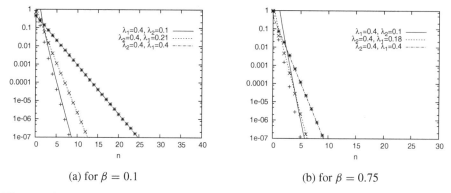

(a) for $\beta = 0.1$ (b) for $\beta = 0.75$

Figure 8.10   Tail behavior of the queue-2 contents for some combinations of type-1 and type-2 arrival rates.

pressions for the tail probabilities of the studied quantities. It is thereby shown that non-geometric tails can occur for the content of the low-priority queue. In the numerical examples, we have furthermore illustrated the impact of the priority scheme on the performance of an output queue in a packet switch. The results of this paper can moreover be used to predict the performance of a queueing system with the HOL-PJ scheme and with time-correlation in the arrival process.

Appendix A: Darboux's Theorem

Theorem 1.1 *Suppose* $X(z) = \sum_{n=0}^{\infty} x(n)z^n$ *with positive real coefficients* $x(n)$ *is analytic near* 0 *and has only algebraic singularities* α_k *on its circle of convergence* $|z| = R$, *in other words, in a neighborhood of* α_k *we have*

$$X(z) \sim \left(1 - \frac{z}{\alpha_k}\right)^{-\omega_k} G_k(z),\qquad (8.24)$$

where $\omega_k \neq 0, -1, -2, \ldots$ *and* $G_k(z)$ *denotes a nonzero analytic function near* α_k. *Let* $\omega = \max_k Re(\omega_k)$ *denote the maximum of the real parts of the* ω_k. *Then we have*

$$x(n) = \sum_{j} \frac{G_j(\alpha_j)}{\Gamma(\omega_j)} n^{\omega_j-1} \alpha_j^{-n} + o(n^{\omega-1} R^{-n}),\qquad (8.25)$$

with the sum taken over all j *with* $Re(\omega_j) = \omega$ *and* $\Gamma(\omega)$ *the Gamma-function of* ω *(with* $\Gamma(n) = (n-1)!$ *for* n *discrete).*

Acknowledgements

We wish to thank the referees for their comments which lead to a considerable improvement of this paper. Note also that the second author is a Postdoctoral Fellow with the Fund for Scientific Research, Flanders (F.W.O.-Vlaanderen), Belgium.

References

[1] J. J. Bae and T. Suda, Survey of traffic control schemes and protocols in ATM networks, *ACM Transactions in Networking*, vol. 2, no. 5, pp. 508–519, 1994.

[2] H. Bruneel, B. Steyaert, E. Desmet and G. H. Petit, An analytical technique for the derivation of the delay performance of ATM switches with multiserver output queues, *International Journal of Digital and Analog Cummunication Systems*, vol. 5, pp. 193–201, 1992.

[3] H. Bruneel, B. Steyaert, E. Desmet and G. H. Petit, Analytic derivation of tail probabilities for queue lengths and waiting times in ATM multiserver queues, *European Journal of Operational Research*, vol. 76, pp. 563–572, 1994.

[4] R. Chipalkatti, J. F. Kurose and D. Towsley, Scheduling policies for real-time and nonreal-time traffic packet switching node, in *Proceedings IEEE INFOCOM '89*, pp. 774–783, 1989.

[5] B. D. Choi, B. I. Choi, Y. Lee and D. K. Sung, Priority queueing system with fixed length packet-train arrivals, *IEE Proceedings-Communications*, vol. 145, no. 2, pp. 331–341, 1998.

[6] B. I. Choi and Y. Lee, Performance analysis of a dynamic priority queue for traffic control of bursty traffics in ATM networks, *IEE Proceedings-Communications*, vol. 148, no. 3, pp. 181–187, 2001.

[7] D. I. Choi, B. D. Choi and D. K. Sung, Performance analysis of priority leaky bucket scheme with queue-length-treshold scheduling policy, *IEE Proceedings-Communications*, vol. 145, no. 6, 1998.

[8] M. Drmota, Systems of functional equations, *Random Structures & Algorithms*, vol. 10, nos. 1–2, pp. 103–124, 1997.

[9] A. Francini, F. M. Chiussi, R. T. Clancy, K. D. Drucker and N. E. Idirene, Enhanced weighted round robin schedulers for accurate bandwidth distribution in packet networks, *Computer Networks*, vol. 37, no. 5, pp. 561–578, 2001.

[10] S. Fratini, Analysis of a dynamic priority queue, *Commun. Statist.-Stochastic Models*, vol. 6, no. 3, pp. 415–444, 1990.

[11] P. A. Ganos, M. N. Koukias and G. K. Kokkinakis, ATM switch with multimedia traffic priority control, *European Transactions on Telecommunications*, vol. 7, no. 6, pp. 527–540, 1996.

[12] E. Gelenbe, Approximate analysis of coupled queueing in ATM networks, *IEEE Communications Letters*, vol. 3, no. 2, pp. 31–33, 1999.

[13] A. Gravey and G. Hebuterne, Mixing time and loss priorities in a single server queue, in *Proceedings of 13th International Teletraffic Congress*, pp. 147–152, 1991.

[14] J. S. Jang, S. H. Schim and B. C. Shin, Analysis of DQLT scheduling policy for an ATM multiplexer, *IEEE Communications Letters*, vol. 1, no. 6, pp. 175–177, 1997.

[15] L. Kleinrock, *Queueing Systems*, Wiley, New York, 1975.

[16] C. Knessl, D. I. Choi and C. Tier, A dynamic priority queue model for simultaneous service of voice and data traffic, *SIAM Journal on Applied Mathematics*, vol. 63, no. 2, pp. 398–422, 2002.

[17] A. Kuurne and A. P. Miettinen, Weighted round robin scheduling strategies in (e)gprs radio interface, in *Proceedings of the Vehicular Technology Conference (VTC2006-Fall)*, vol. 5, pp. 3155–3159, 2004.

[18] K. Laevens and H. Bruneel, Discrete-time multiserver queues with priorities, *Performance Evaluation*, vol. 33, no. 4, pp. 249–275, 1998.

[19] J. T. Lee and Y. H. Kim, Performance analysis of a hybrid priority control scheme for input and output queueing ATM switches, in *Proceedings INFOCOM '98*, 1998.

[20] Y. Lee and B. D. Choi, Queueing system with multiple delay and loss priorities for ATM networks, *Information Sciences*, vol. 138, nos. 1–4, pp. 7–29, 2001.

[21] Y. Lim and J. E. Kobza, Analysis of a delay-dependent priority discipline in an integrated multiclass traffic fast packet switch, *IEEE Transactions on Communications*, vol. 38, no. 5, pp. 659–685, 1990.

[22] T. Maertens, J. Walraevens and H. Bruneel, On priority queues with priority jumps, *Performance Evaluation*, vol. 63, no. 12, pp. 1235–1252, 2006.

[23] T. Maertens, J. Walraevens and H. Bruneel, Priority queueing systems: From probability generating functions to tail probabilities, *Queueing Systems*, vol. 55, no. 1, pp. 27–39, 2007.

[24] T. Maertens, J. Walraevens, M. Moeneclaey and H. Bruneel, Performance analysis of a discrete-time queueing system with priority jumps, *International Journal of Electronics and Communications (AEÜ)*, in press, available online.

[25] M. B. Mamoun, J.-M. Fourneau and N. Pekergin, Analyzing weighted round robin policies with a stochastic comparison approach, *Computers and Operations Research*, vol. 35, no. 8, pp. 2420–2431, 2008.

[26] A. K. Parekh and R. G. Gallager, A generalized processor sharing approach to flow control in integrated services network: the single node case, *IEEE/ACM Transactions on Networking*, vol. 1, no. 3, pp. 344–357, 1993.

[27] G. H. Petit, A. Buchheister, A. Guerrero and P. Parmentier, Performance evaluation methods applicable to an ATM multi-path self-routing switching network, in *Proceedings of the Thirteenth International Teletraffic Congress (ITC 13)*, vol. 14, pp. 917–922, 1991.

[28] H. Shimonishi, M. Yoshida, F. Ruixue and H. Suzuki, An improvement of weighted round robin cell scheduling in ATM networks, in *Proceedings of the 1997 Global Telecommunications Conference (GLOBECOM 1997)*, vol. 2, pp. 1119–1123, 1997.

[29] M. Shreedhar and G. Varghese, Efficient fair queueing using deficit round robin, *ACM SIGCOMM Computer Communication Review*, vol. 25, no. 4, p. 231, 1995.

[30] D. Stiliadis and A. Varma, Latency-rate servers: a general model for analysis of traffic scheduling algorithms, *IEEE/ACM Transactions on Networking*, vol. 6, no. 5, pp. 611–624, 1998.

[31] J. Walraevens, B. Steyaert and H. Bruneel, Performance analysis of a single-server ATM queue with a priority scheduling, *Computers and Operations Research*, vol. 30, no. 12, pp. 1807–1829, 2003.

[32] Y. Xiong, H. Bruneel and G. H. Petit, Performance study of an ATM self-routing multistage switch with bursty traffic: simulation and analytic approximation, *European Transactions on Telecommunications*, vol. 4, no. 4, pp. 443–453, 1993.

[33] Y.-S. Yen, W. Chen, J.-C. Zhhuang and H.-H. Chao, A novel sliding weighted fair queueing scheme for multimedia transmission, in *Proceedings of the 19th International Conference on Advanced Information Networking and Applications (AINA 2005)*, vol. 1, pp. 15–20, 2005.

9

A Simulation Study of "Retreat FIFOs" for the HoL Blocking Problem

Fotios K. Liotopoulos

Senior Member IEEE; partly affiliated with Virtual Trip, Ltd. and with the Hellenic Open University, e-mail: liotop@gmail.com.

Abstract

The Head-of-Line (HoL) blocking problem, which occurs in input-buffered service systems, degrades their performance, especially in "hot-spot" traffic conditions. In this paper, we study a variation of the FIFO queues, called "Retreat FIFOs" or "FIFOs with Retreat" (RFIFOs), which significantly improves the performance of plain FIFOs, with respect to the HoL blocking problem. We simulate and compare three FIFO types; a typical FIFO, an RFIFO and a "FIFO with Overflow" (VFIFO). RFIFOs are simulated with a variable number of "retreat steps" (a variation referred to as "RFIFO-S"). Our simulation results indicate that the performance of the RFIFO is significantly better compared to plain FIFOs, and VFIFOs, especially for hot-spot traffic. Furthermore, the RFIFO-S has improved throughput over the RFIFO in congested conditions.

Keywords: HoL blocking, packet switching, packet queuing, VOQ, FIFO, RFIFO.

9.1 Introduction

Input buffered switching systems suffer from the well known "Head of Line (HoL) blocking" problem. Packets arriving at an input port are buffered

D. D. Kouvatsos (ed.), Traffic and Performance Engineering for Heterogeneous Networks, 199–215.

in a FIFO queue waiting for service. If the first in line packet is blocked, then all subsequent packets are denied service, even if their corresponding destinations are not blocked. This phenomenon contributes significantly to congestion, by increasing the average queue lengths, packet latency and packet loss probability and by decreasing the useful throughput of the system.

The negative effects of the HoL blocking problem are not only exhibited in input-buffered switching systems, but in all input-buffered service systems, in general. Moreover, even output FIFO queues with multiple destinations suffer from this problem. The performance degradation of these systems is more apparent in service conditions with hot-spot traffic, (i.e. traffic targeted to a specific destination with increased intensity). Some earlier switching fabrics used a centralized memory with random access to implement the switching function, thus avoiding the HoL blocking problem. However, in high performance switching systems, the centralized memory becomes a bottleneck and nowadays this approach tends to be abandoned.

In an effort to solve or at least alleviate the problem, several researchers [1–9] focus on improving the throughput of input-queued switches, by applying various packet scheduling algorithms and heuristics. The poor performance of the maximum matching (63.2%) seems to be an important factor for the poor performance (58.6%) of general input-queued (FIFO) switches [10]. Some proposed algorithms have managed to improve this performance (e.g., increase the maximum switch throughput to 98.1%) under certain conditions [11].

Virtual Output Queuing (VOQ), originally invented by Tamir and Frazier [12] in 1988 and then developed by Anderson et al. [13] in the early 1990s, has been proposed as a solution to the HoL blocking problem. In an $M \times N$ switching fabric with M inputs and N outputs, each input queue is replaced by a set of N queues, each of which is destined to a specific output. Therefore, packets with different destinations are queued in different queues and no HoL blocking occurs. One key factor in achieving high performance using VOQ is the scheduling algorithm, which is responsible for the selection of packets to be transmitted in each time unit from the input queues (or VOQs) to the output lines. Several algorithms, such as parallel iterative matching (PIM), iSLIP, Oldest Queue First (OQF), Longest Queue First (LQF), have been proposed in the literature [14, 15]. It has been shown that, with as few as four iterations of the above iterative scheduling algorithms, the throughput of the switch exceeds 99%. Several current high capacity routers have adopted architectures combining VOQ and output queuing with a non-blocking switch-fabric. For example, the Cisco GSR routers (12000 series) use a Modified Deficit Round

Robin scheme for servicing the per-CoS (Class of Service) VOQs, as well as the per-CoS output queues [16, 17]. A review of existing multi-stage queuing and scheduling schemes can be found in [18].

However, every VOQ architecture requires N times more resources (i.e., NM input queues, N schedulers/multiplexers, more fabric area for control, etc.) and worse average utilization of the queues due to the fragmentation of the buffering resources. Moreover, for a centralized "intelligent" scheduler the overhead of scheduling NM queues may be significant and too expensive to perform for large N and M (e.g. for $N = M = 1024$ and 10 Gbps line-rates). Our approach alleviates the HoL problem by merely modifying the existing input FIFOs transparently, thus eliminating the global scheduling bottlenecks and large implementation complexities that characterize the afore-mentioned VOQ-based approaches.

In this paper, we study a variation of the typical FIFO queue design that remedies the HoL blocking problem, without affecting significantly the time-space complexity of the FIFO design. The new FIFO, named "RFIFO" (or "Retreat FIFO") [19], enables a "packet retreat mechanism" when the packet at the top of the FIFO is blocked. In such a case, the packet is repeatedly swapped with its successors, until either the end of the queue or a packet with the same destination is reached. In effect, the blocked packet retreats towards the end of the queue, thus unblocking its successors, which may have better chances for service. This simple mechanism can be easily implemented in hardware with a small space-time overhead and may result in significant performance improvements, as it is demonstrated by our simulation results presented below. In the scope of this paper and for the reasons mentioned above, we do not compare the RFIFO's performance with other VOQ-based approaches.

9.2 RFIFO Operation and Design

For simplicity, we assume that each FIFO line consists of a single packet, or an atomic element that requires service, in general. Therefore, each line of a typical FIFO consists of: (a) a destination address and (b) the packet data. In addition to those, an RFIFO also consists of a status TAG associated with each line (see Figure 9.1). The status TAG is used to indicate a retreat process in progress and can take three possible values: "S"wap, "L"ast, or "–" (blank), therefore only two bits are enough to encode it.

Given an RFIFO, let L_i denote its i-th line (element or packet), D_i the destination of L_i and T_i the TAG of L_i. Line numbering begins at 1, with

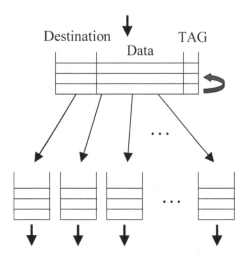

Figure 9.1 The architecture of an RFIFO-based queuing system.

L_1 being the top of the queue. Initially, all tags are set to "E" (empty). As packets are inserted in the queue, they are tagged "–" (blank), except for the last (occupied) line, which is tagged "L" (last). As packets are serviced and lines are emptied, they are tagged "–" (blank) again.

When the top of the RFIFO, (HoL), is blocked, then T_1 is set to "S" (i.e., the line is to be swapped), indicating that a retreat process is about to begin. While T_i = "S" and $D_i \neq D_{i+1}$, the L_i and L_{i+1} lines are swapped, until T_{i+1} = "L". If a packet with the same destination (i.e., $D_k = D_{k+1}$) is reached, then the retreat process is paused. However, in order to handle the case of "multiple HoL blocking" (e.g. $D_1 = D_2 = \ldots$, all blocked), we tag the next packet as "S" (D_k = "S"), thus relaying the retreat process to L_{k+1} and enabling it to begin a retreat process itself, too. As the L_{k+1} packet retreats, it allows its predecessor, L_k, to resume its retreat process. Figure 9.2 and Table 9.1 show these transitions in more detail.

To further illustrate the Retreat Process, let us consider the example of Figure 9.3. Packets labeled "3a" and "3b" have the same destination (output no. 3), which is blocked. This implies that "3a" must be tagged "S" and swapped with "4" and "1". Subsequently, packet "3a" relays the retreat to packet "3b", which is also tagged "S". At this point, packets "4" and "1" are asynchronously serviced and removed from the queue. "3a" becomes HoL again, but it is still blocked. Therefore, both "3a" and "3b" still remain in "retreat" mode and they are successively swapped with packets "6", "2" and

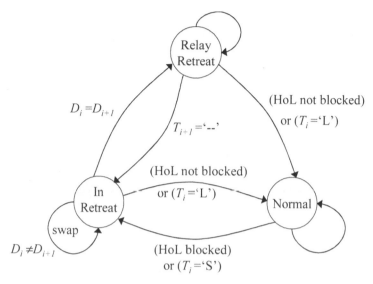

Figure 9.2 The state transition diagram of a packet in an RFIFO-based queuing system.

"5", which are eventually serviced and removed from the queue. Meanwhile, packet "3b" reaches the end of the queue and is tagged "L" (thus, ending its retreat). Packet "3a" then becomes HoL and is not blocked any more. Therefore, both "3a" and "3b" are tagged "–" (blank) and both packets are serviced.

9.3 Bounded-Retreat FIFO

Assuming that the average queue length is usually small, the retreat process is not limited to a fixed number of retreat-steps. If such a variation is desired, then the 2-bit tag may be replaced by an n-bit counter, indicating the "remaining swaps until the end of the retreat". This variation is called "*Bounded-Retreat FIFO with S steps*", or simply "*RFIFO-S*", where n is the maximum number of retreat steps that are allowed for each retreat process. Note that the retreat process of a packet is considered complete only when its tag changes from "S" to "–", and not when it relays the retreat to a subsequent packet. In situations with intense traffic conditions and high blocking, the queue lengths grow significantly and an unbounded retreat process contributes to higher packet latency. This effect is clearly demonstrated by our simulation analysis, below.

Table 9.1 State transitions during a packet's (unbounded) retreat process.

No.	Condition	TAGs (before the Action)		TAGs (after the Action)		ACTION
		T_i	T_{i+1}	T_i	T_{i+1}	
1	If not HoL or HoL_blocked	S	S	S	S	/* In Relay */
2	If HoL and not_blocked	S	S, –	–	–	/* End the Retreat */
3	If HoL and not_blocked	S	L	–	L	/* End the Retreat */
4	If same destination	S	L	–	L	/* End the Retreat */
5	If different destination	S	L	–	L	Swap the two packets
6	If same destination	S	–	S	S	/* Relay the Retreat */
7	If different destination	S	–	–	S	Swap the two packets
8	–	L	S	L	–	*** Impossible ***
9	–	L	L	L	–	*** Impossible ***
10	–	L	–	L	–	–
11	–	–	S	–	S	–
12	If HoL and blocked	–	L	S	L	/* Start the Retreat */
13	If HoL and blocked	–	–	S	–	/* Start the Retreat */

An RFIFO-S requires a tag with $S+2$ distinct values ("–", "1", ..., "S-1", "S", "L"). Therefore, an RFIFO-S $t$ag can be implemented with $\log_2(S+2)$ state bits. For example, an RFIFO-6, i.e. an RFIFO with a maximum of 6 retreat steps is implemented with a 3-bit tag, taking the values ("–", "1", "2", "3", "4", "5", "6", "L").

Whenever a retreat process is initiated, the tag is set to the value "S". Every time a packet in retreat is swapped, its tag value is decremented by one. If the tag is "1", then a subsequent swapping changes it to "–" and the packet's retreat is terminated. If a packet in retreat becomes the last packet of the RFIFO, then its tag becomes "L" and its retreat is also terminated, irrespective of the number of retreat steps already preformed. If the packet enters a "Relay Retreat" state, then its tag value may either be reset to "–" or remain unchanged until the subsequent packet begins its own retreat, in which case the former packet's tag resumes its "count-down", (the latter option is shown in Table 9.2). Given these rules, the above state-transition table (Table 9.1) is modified accordingly, as shown in Table 9.2.

9.4 FIFO with Overflow

If the HoL packet of a FIFO queue is blocked then an alternative way to alleviate the problem is to redirect the blocked (and potentially blocking) packet to an overflow FIFO, associated with the former one. Therefore, each

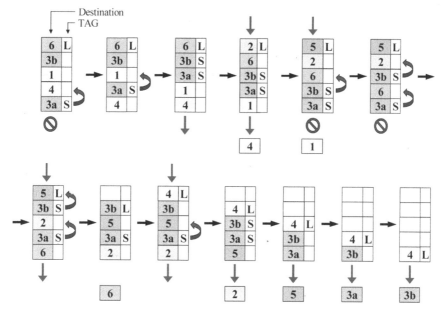

Figure 9.3 Example of the retreat process in an RFIFO queue.

input (regular) FIFO is replaced by a pair of FIFOs ("Main FIFO" and "Overflow FIFO"). From this point forward, we will refer to this combination as "VFIFO". The VFIFO variation alleviates the HoL blocking problem by removing potentially blocking packets from the Main FIFOs, thus allowing subsequent packets to be delivered. In addition to this improvement, it further improves the packet throughput, compared to a regular FIFO, since for each input port two FIFOs are now contributing packets to the overall throughput.

Figure 9.4 depicts the architecture of the "VFIFO". It consists of two FIFO queues, the "Main FIFO" and the "Overflow FIFO" and a 2-to-1 multiplexer. Packets enter the MAIN FIFO and, if the HoL packet is blocked, then it is re-directed to the Overflow FIFO. If the HoL packet of the Overflow FIFO is blocked, then it is "recycled", i.e., it is de-queued and immediately re-inserted in the Overflow FIFO in order to allow subsequent non-blocked packets to be serviced. The output of the VFIFO is produced by multiplexing the outputs of the two FIFOs, with the Overflow FIFO having the highest priority between the two. In other words, if both FIFOs have a non-blocked packet available for delivery, then preference is given to the Overflow FIFO. This choice is justified by the observation that packets found in the Overflow FIFO are already delayed and blocked and they should not be further

Table 9.2 State transitions during a packet's (bounded) retreat process for an RFIFO-S.

No.	Condition	TAGs (before the Action)		TAGs (after the Action)		ACTION
		T_i	T_{i+1}	T_i	T_{i+1}	
1	If not HoL or HoL_blocked	P	Q	P	Q	/* In Relay */
2	If HoL and not_blocked	P	Q, –	–	–	/* End the Retreat */
3	If HoL and not_blocked	P	L	–	L	/* End the Retreat */
4	If same destination	P	L	–	L	/* End the Retreat */
5	If different destination	P	L	–	L	Swap the two packets
6	If same destination	P	–	P	S	/* Relay the Retreat */
7	If different destination	P	–	–	P-1	Swap the two packets
8	–	L	P	L	–	*** Impossible ***
9	–	L	L	L	–	*** Impossible ***
10	–	L	–	L	–	–
11	–	–	P	–	P	–
12	If HoL and blocked	–	L	S	L	/* Start the Retreat */
13	If HoL and blocked	–	–	S	–	/* Start the Retreat */

Note: P, Q are tag values   {1,2,...,S}.

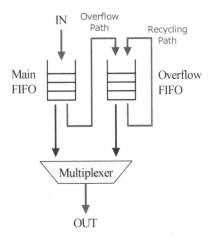

Figure 9.4 The architecture of the VFIFO.

delayed by later arrivals residing in the Main FIFO. The VFIFO is considered "blocked", iff at least one of the two FIFOs is non-empty and the HoL packet(s) is (are) blocked.

One disadvantage of the VFIFO is that it does not guarantee in-order packet delivery. To illustrate this case, simply consider a case where a packet

A packet is *generated* during each
timeslot, with probability IGP

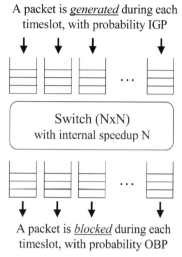

Figure 9.5 The RFIFO simulation model ($N = 8$).

destined for output A is blocked and is re-directed to the Overflow FIFO, where it is blocked again by a predecessor packet. At this point, A becomes unblocked and a subsequent packet with the same destination (A) may then be delivered through the Main FIFO before its predecessor.

9.5 Simulation Results

In order to evaluate the effectiveness of our RFIFO design, we developed a packet-level simulator with time resolution (timeslot) equal to the transmission time of a single "packet" (as defined above). We simulated an 8×8 switching system with speedup=8, and combined input-output queuing (see Figure 9.5). We assumed that normal FIFOs and RFIFOs have queue sizes equal to 16, while VFIFOs consist of two FIFO queues of size 8 each. All output queues are plain FIFO queues and all input queues can either be FIFOs, or RFIFOs, or RFIFO-Ss, or VFIFOs, user selectable. Input traffic for input port i is generated (one packet arrival per timeslot) with probability IGP_i, (Input Generation Probability). Every output FIFO j delivers one packet in every timeslot, with blocking probability OBP_j (Output Blocking Probability). The packet destinations are uniformly distributed.

The simulator reports (among other) the following performance metrics:

- *Blocked packets*: The number of arriving packets that are blocked because their corresponding input queues are full.
- *Delivered packets*: The number of packets delivered by the output queues.
- *Packet Latency*: The average packet delay from generation to delivery.

All simulation experiments were run for 100,000 cycles (timeslots), which is enough for the switching system to overcome any start-up effects and exhibit steady-state behaviour. We simulated both with uniform input traffic (Case I) and with hot-spot input traffic (Case II), to evaluate the behaviour of RFIFO under average and worst input traffic distribution conditions, respectively.

Case I: Uniform Traffic [$\text{IGP}_i = \text{IGP}, \forall i$, and $\text{OBP}_j = \text{OBP}, \forall j$]

In all graphs of Figure 9.6, we present the throughput, blocking and latency properties of a switching system, based on four different types of queues; a normal FIFO ("NORMAL"), an RFIFO ("RFIFO-MAX"), an RFIFO-3 ("RFIFO-3") and a VFIFO ("OVERFLOW"). In particular, we study how throughput, blocking and latency are affected, as we vary the Input Generating Probability (IGP) and the Output Blocking Probability (OBP) of the switch. In all graphs of Figure 9.6, we assume a uniform traffic, that is, all IGPs and OBPs being equal.

In Figures 9.6a, 9.6b and 9.6c we varied OBP between 5 and 40% with $\text{IGP} = 80\%$. For $\text{OBP} < 20\%$, the input flow is less than the maximum output capability of the system and the system is stable. In these conditions, RFIFOs exhibit near constant throughput, less blocking and less latency, compared to all other queue types. RFIFOs perform slightly better that RFIFO-s and better than VFIFOs, but for $\text{OBP} > 24\%$ RFIFO-s's are outperformed by VFIFOs in all graphs. For $\text{OBP} > 20\%$, although the performance degrades for all types of queues, RFIFOs still outperform all the others, in terms of packet blocking and throughput. Only in terms of average latency and for $\text{OBP} > 20\%$ is the situation completely reversed. The reason for this inefficiency is due to the fact that, in congested situations, RFIFOs perform unnecessary swaps, which increase the average residency time of a packet in each queue that it goes through.

In Figures 9.6d, 9.6e and 9.6f we varied IGP between 55 and 90% with $\text{OBP} = 30\%$. For $\text{IGP} < 70\%$, the input flow is less than the maximum output capability of the system and the system is stable. In these conditions, RFIFOs exhibit constantly higher throughput, less blocking and less latency, compared to all other FIFOs. For $\text{IGP} > 70\%$, although the throughput

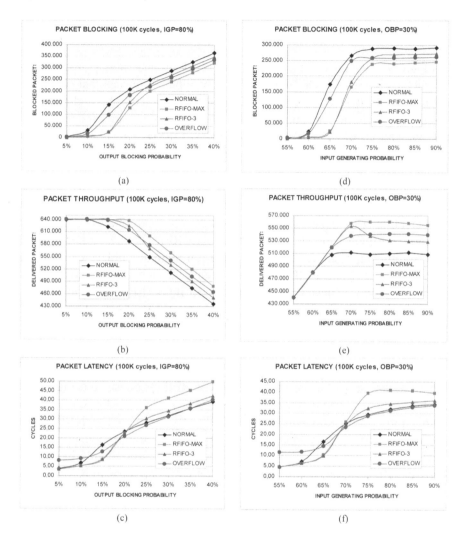

Figure 9.6 Simulation results for uniform (symmetric) traffic.

tends to saturate for all types of queues, RFIFOs clearly outperform all other types of FIFOs in terms of both packet blocking and throughput. Again, just as before, in terms of average latency and for IGP > 70% the situation is completely reversed.

Case II: Hot-spot traffic [IGP_i = IGP, $\forall i$, and OBP_j = OBP, $\forall j \neq k$, $OBP_k > OBP_j$]

In Figure 9.7, we present the throughput, blocking and latency properties of a switching system, based on the same four types of queues that were considered earlier, that is, a normal FIFO, an RFIFO, an RFIFO-3 and a VFIFO. Again, we study how throughput, blocking and latency are affected, as we vary the Output Blocking Probability (OBP) of the switch. The difference is that, in these graphs, we assume hot-spot traffic, that is, one queue having higher OBP than all the others.

For the experiments of Figure 9.7, we set IGP = 50% and OBP = 20%. Hot-spot traffic conditions were simulated by ranging OBP5 from 40 to 75%, thus creating a bottleneck at output port 5. The performance degradation of normal FIFOs and VFIFOs compared to RFIFOs and RFIFO-s's was quite significant, in all measures; blocking, throughput and latency. With the introduction of "Retreat FIFOs", near constant throughput, low blocking and low latency was achieved, indicating that RFIFOs (RFIFO-s's, in general) is a very effective solution for the "HoL blocking" problem, especially with hot-spot traffic conditions.

Case III: Simulating various RFIFO types

From the definitions of the various types of RFIFOs considered thus far, it can be easily observed that there should be a "continuous behaviour" starting from a normal FIFO (equivalent to an RFIFO-0), continuing with RFIFO-s's (for $s = 1$ to $M - 1$), up to an RFIFO (which is by definition identical to an RFIFO-M, for some integer M). For this reason, we were highly inclined to study the effect of varying the number of retreat steps, s, of an RFIFO-s and compare these cases against a normal FIFO, an RFIFO and a VFIFO.

Figures 9.8a, 9.8b and 9.8c depict the behaviour of the switch in terms of packet blocking, throughput and latency for all the different types of queues considered in our experiments. We simulated two types of traffic conditions; one with IGP = 80% and OBP = 50% (high traffic load) and one with IGP = 60% and OBP = 35% (lower traffic load). Each graph is a bar graph for 12 different x-values: (from left to right) a VFIFO, an RFIFO, nine RFIFO-s types (for $s = (14, 12, 10, 8, 6, 4, 3, 2, 1)$) and finally a normal FIFO. The experimental results verified our intuition about the "continuous behaviour" of the RFIFOs, as explained above.

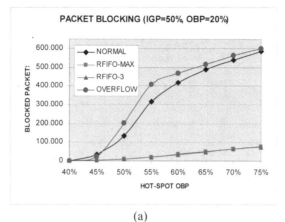

(a)

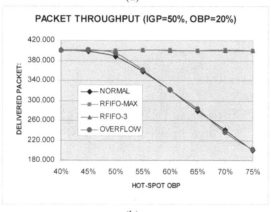

(b)

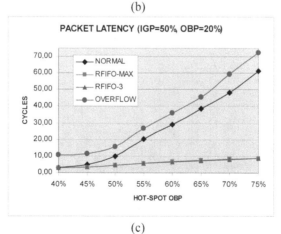

(c)

Figure 9.7 Simulation results for hot-spot traffic.

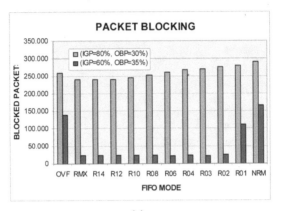

(a)

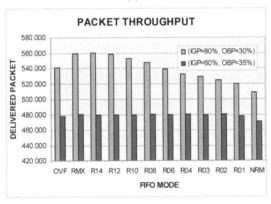

(b)

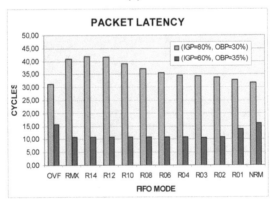

(c)

Figure 9.8 Comparison of various RFIFO types.

9.6 Conclusions

In this paper, we proposed an effective variation of the FIFO queue, called RFIFO (or "Retreat FIFO"), which can be used to alleviate the "HoL blocking" problem; a problem that occurs in all input-buffered service systems, including network switches and routers.

RFIFOs were evaluated by simulation for both uniform and hot-spot traffic conditions and they were compared against three types of FIFOs: normal FIFOs, FIFOs with overflow (termed VFIFOs) and RFIFOs with a variable number of retreat steps (termed RFIFO-s). The performance of RFIFOs proved significantly superior to both VFIFOs and plain FIFOs, especially for hot-spot traffic. Under uniform traffic, the throughput and blocking improvement against normal FIFOs was approximately 10%, with a relative increase in latency under congestion. With hot-spot traffic the RFIFOs managed to maintain near constant throughput, as well as lower blocking and lower latency, compared to the severe performance degradation exhibited by plain FIFOs and VFIFOs. If latency is of concern, then the standard RFIFO is apparently not a good choice, since its performance is the worst of all. However, an RFIFO-s (for small values of s) performs better than an RFIFO under congestion. Therefore, a hybrid design combining "the best of both worlds" would be to use an RFIFO under non-congested conditions and simply limit the number of retreat steps when congestion is detected. Such a design would perform very well under all traffic conditions.

Considering their implementation complexity, RFIFOs are relatively easy to implement, not only in software, but in hardware as well. A 2-bit tag per queue line and a swapping mechanism of consecutive queue lines are required for a fast hardware implementation.

The replacement of plain FIFO buffers with hardware RFIFOs is expected to significantly improve the performance of network switches and routers, thus resulting in faster, more stable and more efficient data communication networks.

Out future work plans include the modification of the RFIFO mechanism to support more advanced traffic models with QoS and CoS priorities. We also plan to validate our simulation results against a hardware prototype and real application traces.

References

[1] G. Thomas, Improved windowing rule for input buffered packet switches, *Electron. Lett.*, vol. 29, pp. 393–395, 1993.

[2] R. Y. Awdeh and H. T. Mouftah, Survey of ATM switch architectures, *Computer Networks ISDN Syst.*, vol. 27, pp. 1567–1613, 1995.

[3] J.-J. Li, Improving the input-queueing switch under bursty traffic, *Electron. Lett.*, vol. 31, pp. 854–855, 1995.

[4] N. McKeown, V. Anantharam and J. Walrand, Achieving 100% throughput in an input-queued switch, in *Proc. IEEE Infocom*, San Francisco, CA, March 1996, vol. 1, pp. 3A.4.1–3A.4.6, 1996.

[5] G. Nong, J. K. Muppala and M. Hamdi, A performance model for ATM switches with multiple input queues, in *Proc. IEEE ICCCN*, Las Vegas, NV, pp. 222–227, 1997.

[6] N.K. Sharma and M.R. Pinnu, An efficient implementation of bypass queue under bursty traffic, *Parallel Computing*, vol. 23, pp. 777–781, 1997.

[7] G. Thomas, Bifurcated queuing for throughput enhancement in input queued switches, *IEEE Commun. Lett.*, vol. 1, pp. 56–57, 1997.

[8] J. S.-C. Wu and Y.-D. Lin, An efficient and orderly implementation of bypass queue under bursty traffic, *Parallel Computing*, vol. 24, pp. 2143–2148, 1998.

[9] K. Yoshigoe and K. J. Christensen, An evolution to crossbar switches with virtual output queuing and buffered cross points, *IEEE Network*, vol. 17, no. 5, pp. 48–56, 2003.

[10] M. Karol and M. Hluchyj, Input versus output queuing in a space division switch, *IEEE Trans. Commun.*, vol. 35, no. 12, pp. 1347–1356, 1987.

[11] J. S.-C. Wu and Y.-D. Lin, A novel pairing algorithm for high-speed large-scale switches, *IEEE Comm. Lett.*, vol. 4, no. 1, pp. 23–25, 2000.

[12] Y. Tamir and G. Frazier, High performance multi-queue buffers for VLSI communications switches, in *Proc. Computer Architecture*, pp. 343–354, June 1988.

[13] T. Anderson et al., High-speed switch scheduling for local-area networks, *ACM Trans. Computer Systems*, vol. 11, no. 4, pp. 319–352, 1993.

[14] G. Nong and M. Hamdi, On the provision of quality-of-service guarantees for input queued switches, *IEEE Commun. Mag.*, vol. 38, no. 12, pp. 62–69, 2000.

[15] C. Minkenberg and T. Engbersen, A combined input and output queued packet-switched system based on PRIZMA switch-on-a-chip technology, *IEEE Commun. Mag.*, vol. 38, no. 12, pp. 70–77, 2000.

[16] Cisco Systems Inc., Basic scalability and performance considerations for evaluating large-scale router designs, Cisco 12000 Series Routers, White Paper, 2004. http://www.cisco.com/en/US/products/hw/routers/ps167/products_white_paper09186a 0080091fdf.shtml.

[17] Cisco Systems Inc., Designing service provider core networks to deliver real-time services, Cisco 12000 Series Routers, White Paper, 2004. http://www.cisco.com/en/US/products/hw/routers/ps167/products_white_paper09186a0080091fdc.shtml

[18] J. Soldatos, E. Vayias and G. Kormentzas, On the building blocks of quality of service in heterogeneous IP networks, *IEEE Commun. Surveys & Tutorials*, vol.7 no. 1, pp. 70–89, 2005.

[19] F. K. Liotopoulos, RFIFO: Retreat FIFOs for the head-of-line blocking problem, in *Proc. 2nd Int. Conf. on the Performance Modelling and Evaluation of Heterogeneous Networks (HET-NETs '04)*, Ilkley, UK, pp. P6/1–7, July 2004.

10

Modeling of the HTTP Protocol in Control Systems Environment

Marek Fiuk[1] and Tadeusz Czachórski[2]

[1]*Pelco, 10 Corporate Drive, Orangeburg, NY, U.S.A.; e-mail: mfiuk@speakeasy.net;*
[2]*IITIS-PAN, ul. Bałtycka 5, 44-100 Gliwice, Poland; e-mail: tadek@iitis.gliwice.pl*

Abstract

Several modern communication protocols used in networked control and monitoring applications employ combination of XML and SOAP standards, which in turn use the HTTP protocol to provide the actual network communication services. Consequently, we postulate that analysis of the network related performance issues in such applications can be reduced to the performance analysis of the underlying HTTP protocol operating in those environments. In this paper we present a model of a networked control system based on the UPnP communication standard (which is built on top of SOAP, XML and HTTP) together with an efficient algorithm for the analytical examination of that model.

Keywords: HTTP, UPnP, modeling, simulation, CTMC, tandem queue.

10.1 Introduction

The rapid growth of the Internet created new opportunities for network-based Automation, Control and Monitoring systems. Attracted by the broad range of open standards and protocols (TCP/IP, HTTP, XML. SOAP) numerous teams – from both research and industrial communities, have made attempts to use Internet technologies in Networked Control Systems. One of the first

D. D. Kouvatsos (ed.), Traffic and Performance Engineering for Heterogeneous Networks, 217–232.

was the 1994 Mercury Project at USC which enabled control of a robotic manipulator using the HTTP protocol [8]. It was followed be an Internet-based mobile robot developed at Carnegie-Mellon University in 1995 [9]. During the past decade, research teams from around the world have internet-enabled dozens of control applications, including industrial process supervision and control, robotics, home automation, control of telephony switches, operating planetary rovers and spacecraft and many others. In that process several standards and protocols were created, addressing specific needs of particular classes of control and monitoring applications. Included among those are:

- OPC XML DA – OPC Foundation's software interface used in automation industry that adopts the XML set of technologies to facilitate an exchange of plant data across the internet, and upwards into the enterprise domain. It is based on DCOM and Web Services, it follows the Client/Server approach [5].
- BACnet/WS – an addition to a communication protocol (BACnet) commonly used in the Building Automation Control industry, provides a set of generic Web Services that can be used to implement an interface to any other building automation protocol [6].
- VoiceXML – a markup language derived from XML for writing telephone call handling applications [7].
- UPnP (Universal Plug and Play) – an architecture and protocols for peer-to-peer network connectivity of intelligent appliances, wireless devices, and PCs.
- Web Services – set of interfaces describing operations (services) that are network-accessible through the standardized XML messaging. Interfaces are defined using a formal XML notation (service description) that provides details necessary to interact with the service, including message structure, transport protocols, and location.

More often than not, these protocols employ combination of XML and SOAP to define necessary data structure and implement the messaging scheme. In turn, these standards use the HTTP protocol to provide the actual network communication services.

Thus, a common pattern can be observed in the architecture of several new communication protocols and standards, used in networked control and monitoring applications (see Figure 10.1). The high(er) level control/communication protocol (e.g. BACnet/WS, VoiceXML, UPnP) resides on top of the XML/SOAP combination which in turn resides on top of HTTP. As a result of that, all the networked control applications build around such

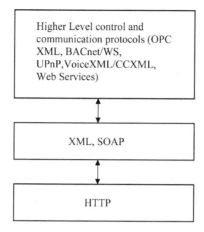

Figure 10.1 Typical stack of modern communication protocols.

high level protocols use HTTP as the only network communication platform. Consequently, analysis of the network related performance issues in networked control applications can be reduced to the performance analysis of the HTTP protocol operating in those environments.

Several different techniques can be employed to perform that analysis. One is to collect and examine performance measurements of a real networked control system, another is to develop analytical apparatus to capture such system's fundamental properties, yet another is to construct a simulation model and then use it to perform simulation runs to collect performance data.

We have employed both the analytical method and the simulation to study performance of the HTTP protocol in control applications. Furthermore, we have concentrated on systems using UPnP as the higher level control/communication protocol. To that end, we have constructed a model of a networked control system employing combination of UPnP, SOAP, XML and HTTP. In this paper we discuss an efficient algorithm we have developed for the analytical examination of that model. Currently, we are also implementing a simulator that we are going to use to validate the approach taken in the algorithm.

10.2 UPnP – Example of an HTTP Based Control Protocol

Universal Plug and Play (UPnP) is architecture for peer-to-peer network connectivity of intelligent appliances, wireless devices, and PCs. It leverages

TCP/IP and the Web technologies (IP, TCP, UDP, HTTP, SOAP and XML) to enable seamless proximity networking in the home, office, and public spaces. UPnP allows automatic discovery and control of services available on the network without user intervention. Devices that act as servers can advertise their services to clients. Clients, known as Control Points, can search for specific services on the network. When they find the Devices with the desired services, the Control Points can retrieve detailed descriptions of these services and interact with them from that point on.

UPnP operations can be divided into five basic phases:

- *Discovery*. In this first phase, Control Points search for devices and services and Devices multicast announcements of services they offer to Control Points using the Simple Service Discovery Protocol (SSDP).
- *Description*. Once a Control Point finds an interesting service, it requests from the corresponding Device its complete description. The description is an XML document, it contains (among others) manufacturer information, version, and a list of services supported by the Device.
- *Control*. This phase allows Control Points to control one or more of the services contained in a Device by enacting changes in the state of the device. The Simple Object Access Protocol (SOAP) allows a Control Point to query or change elements in a service's state table.
- *Eventing*. This phase allows Control Points to keep in sync with the state of services in which it is interested. Control Points subscribe to the event server for a particular service and receive event notifications when that service's state changes.
- *Presentation*. The presentation phase allows a Device to host a document, written in standard HTML, which can implement a user interface for that device. This document can be downloaded by Control Points and used to perform UI operations, since it provides means of control and status display.

10.3 UPnP Network Model

We present here a model of an UPnP network operating in the control application environment. It was constructed based on the UPnP specification document "Universal Plug and Play Device Architecture" [1] and an analysis of the source code provided in the "Intel SDK for UPnP™ Devices Version 1.2.1" package [2].

Two types of UPnP network operations can be distinguished – the "Power-up/Network Initialization" operations and the "Control Phase" operations. Since Control Points and Devices can be powered-up/attached to or detached from the network at any time, operation of both types can potentially overlap on the network. For example, a Control Point can receive SSDP NOTIFY command from a Device while sending SOAP request (Control Phase) to another Device. However, under typical conditions a Device will be involved in the Control Phase operations much more often than in the Power-up Phase operations (will send and receive many more Control Phase Commands and Event Notifications). Therefore, currently only the Control Phase operations are reflected in the model.

Figure 10.2 presents the UPnP Queuing Network (QN) Model for the Control Phase Operations. It consists of Control Point elements (numbered 1–P), Device elements (numbered 1–Q) and a single Network element. Control Points and Devices are all connected to the Network with connections symbolizing either the UPnP Request flow (solid line) or the UPnP Response flow (dashed line).

10.3.1 Control Phase Operations

A UPnP Control Point generates Control Requests at the rate G_r (requests per second) each destined to a single UPnP Device. Requests generated by a given Control Point may all go to the same Device or may be distributed among several Devices. After a Control Request is sent to its target Device, the Control Point continues to keep track of its status until that Device indicates that the request was received and fully processed (consumed). The Control Point is capable of maintaining (keeping track of) K outstanding Control Requests.

Once received, the Control Requests are serviced by the Device, with each request requiring processing time tpc. The Device is capable of processing M request simultaneously. Control Requests may be used to affect the Plant (controlled object) associated with the Device by changing value of its State Variables, or to query the status of that Plant. Changes to State Variables may in turn result in sending the Event Notification Request to all Control Points that registered for that particular event. There exist a correlation factor (coefficient) M_s between the number of received Control Requests and the number of resulting *State Variable Change* Event Notification Requests.

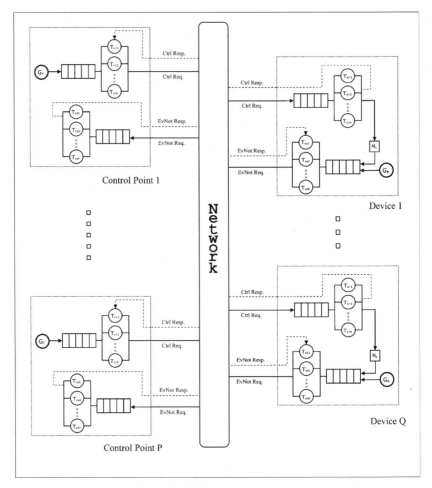

Figure 10.2 UPnP network model.

After processing is finished, the Device generates the Control Response which is sent back to the Control Point thus completing the handling of the Control Request there.

As the Plant's state changes (due to the processing of the Control Requests and the influence of the external environment) it generates (at the rate G_e) *Plant State Change* Event Notification Requests that are sent to Control Points that registered for them. After the Event Notification Request is sent to a Control Point, the Device continues to keep track of its status until that

Control Point indicates that the request was received and fully processed (consumed).

Once received, the Event Notification Requests are processed by the Control Point, with each Notification Request requiring processing time t_{pe}. The Control Point is capable of processing T Event Notification Requests simultaneously. After processing is finished, the Control Point generates the Event Notification Response which is sent back to the Device thus completing the handling of the corresponding Event Notification Request there.

10.3.1.1 The Control Point

Control Requests generated by the source G_r are queued before being sent to Devices by the concurrent transmission tasks $T_{cr1} - T_{crK}$. After sending a request, the transmission task waits for the Control Response to arrive from the Device and only then it can proceed to process next request from the queue. Event Notification requests received from Devices are queued before being processed by the concurrent event handling tasks $T_{ce1} - T_{ceT}$ (this processing includes sending an Event Notification Response back to the Device).

10.3.1.2 The Device

Control Requests received from Control Points are queued before being processed by the concurrent tasks $T_{sr1} - T_{srM}$. For each processed Control Request a corresponding Event Notification Request will be generated (with a probability of M_s) and queued. After processing of the request is completed, the task sends a Control Response back the Control Point.

Event Notification Requests (for the events corresponding to operations of the Plant) generated by the source G_e are queued before being sent to a Device by the concurrent transmission tasks $T_{se1} - T_{seN}$. After sending a request, the transmission task waits for the Event Notification Response to arrive from the Control Point and only then it can proceed to process next request from the queue.

10.3.1.3 The Network

Currently, a simplistic Constant Delay network model is used.

10.4 UPnP Network Analysis

The complexity of the UPnP network model presented in the previous paragraph makes its analysis rather difficult. Therefore, its further reduction is

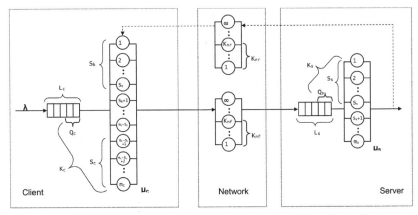

Figure 10.3 Tandem queue model of the HTTP sender/receiver pair.

needed before formal analysis can be attempted. It can be easily seen in the network model that any communication requires an HTTP connection between the HTTP sender (consisting of a queue and a number of sending tasks) and the HTTP receiver (consisting of a queue and a number of request/job processing tasks). Although a particular connection can be of one of the two types – the first involving the Control Point sending the Control Requests to the Device (with sending tasks $T_{cr1} - T_{crK}$ and request processing tasks $T_{sr1} - T_{srM}$), and the second involving the Device sending the Event Notification Requests to the Control Point (with sending tasks $T_{se1} - T_{seN}$ and request processing tasks $T_{ce1} - T_{ceT}$), the structure of the sender/receiver arrangement is identical in both cases. Since the outlined above HTTP sender/receiver relationship is of the client/server nature, we will use the later terms for the remaining of this paragraph. Thus, we reduce analysis of the UPnP network model to the analysis of an HTTP client/server pair which can be represented by a two-node tandem queue arrangement with an additional blocking mechanism, as shown in Figure 10.3.

There are several parameters defined for this new model, among them:

K_c – total number of jobs at the client $\qquad K_c = S_c + Q_c$

K_s – total number of jobs at the server $\qquad K_s = S_s + Q_s$

S_b – number of client service stations blocked by jobs waiting/serviced at the server and jobs in the network $\qquad S_b = K_s + K_{nf} + K_{nr}$

S_c – number of jobs currently serviced at the client

$$0 \leq S_c + S_b \leq m_c$$

S_s – number of jobs currently serviced
      at the server

We limit our considerations to the cases where the time needed to process a job at both the client and the server is significantly larger than time needed to deliver messages over the network, therefore we assume in our analysis that there are no "in flight" jobs in the system, e.g. K_{nf}, $K_{nr} = 0$.

The blocking mechanism present in the system makes the ratio between the number of service stations at the client (m_c) and the total capacity of the server ($L_s + m_s$) significant. We assume that the system is balanced in respect to that ratio, e.g. $m_c = L_s + m_s$. The rational here is that in a system with $m_c < L_s + m_s$ some of the server resources (queue, service stations) will simply never get fully utilized, therefore such system will effectively get reduced to a balanced one with smaller L_s and (possibly) m_s.

On the other hand, in a system with $mc > L_s + m_s$ some of the jobs leaving the client may get dropped, but since the space in the servers queue is usually inexpensive, it is unlikely that such (possibly harmful) imbalance would be allowed in a practical application. Therefore, we consider this case highly artificial.

We use Continuous Time Markov Chains (CTMC) to analyze the two-node tandem queue system that represents the reduced UPnP network model. Such a tandem queue system can be described by a phase/level random process with the phase (K_c in our case) giving the state of the client and the level (K_s in our case) giving the state of the server. The combination of the phase and the level defines the state of the whole system. The transitions between states can be represented by the state flow diagram which in turn is used to construct a state transition matrix $\mathbf{Q}$.

Assuming that incoming jobs have Poisson distribution with arrival rate λ, and assuming that client and server service times also have Poisson distribution with service rates respectively μ_c and μ_s, we can apply the CTMC theory to construct the set of global balance equations:

$$\mathbf{\Pi} * \mathbf{Q} = \mathbf{0}$$

where $\mathbf{\Pi}$ is a vector of steady state probabilities of system being in particular state (e.g. having a particular combination of phase and level values). Together with the normalization equation:

$$\mathbf{\Pi} * \mathbf{1} = \mathbf{1}$$

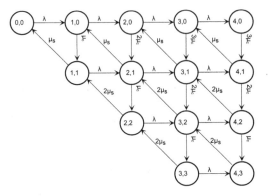

Figure 10.4 State flow diagram.

```
-λ,      0,    us,      0,     0,     0,       0,       0,       0,       0,       0,      0,      0;
λ,-uc-λ,      0,      0,     0,    us,       0,       0,       0,       0,       0,      0,      0;
 0,  uc,-us-λ,      0,      0,     0,   2*us,       0,       0,       0,       0,      0,      0;
 0,      λ,  0,-2*uc-λ,      0,     0,       0,       0,     us,       0,       0,      0,      0;
 0,      0,      λ, 2*uc,-us-uc-λ, 0,       0,       0,    2*us,       0,       0,      0,      0;
 0,      0,      0,      0,  uc,-2*us-λ,    0,       0,       0,    2*us,       0,      0,      0;
 0,      0,      0,      λ,      0,  0,-3*uc-λ,       0,       0,       0,     us,      0,      0;
 0,      0,      0,      0,      λ,  0, 3*uc,-2*uc-us-λ,0,       0,       0,   2*us,      0;
 0,      0,      0,      0,      0,  λ,      0, 2*uc,-2*us-uc-λ,0,       0,      0,   2*us;
 0,      0,      0,      0,      0,  0,      0,      uc,-2*us-λ,   0,       0,      0,      0;
 0,      0,      0,      0,      0,  λ,      0,       0,  0,-3*uc,       0,      0,      0;
 0,      0,      0,      0,      0,  0,      λ,       0,       0, 3*uc,-2*uc-us, 0,      0;
 0,      0,      0,      0,      0,  0,      0,       λ,       0,  0, 2*uc,-2*us-uc,  0;
 0,      0,      0,      0,      0,  0,      0,       0,       λ,       0,      0,  uc,-2*us
```

Figure 10.5 State transition matrix.

these equations can be solved yielding the vector Π, which fully describes stochastic properties of the analyzed tandem queue system and thus properties of the reduced UPnP network model.

As a concrete illustration of our approach, Figure 10.4 presents the state flow diagram and Figure 10.5 presents the state transition matrix that correspond to a very small tandem queue system (UPnP network model) with the following parameters:

$$L_c = 1, \quad m_c = 3, \quad L_s = 1, \quad m_s = 2$$

The global balance equations can be solved directly using Gaussian Elimination [3]. This method consists of two steps – matrix triangularization with arithmetic complexity of $n^3/3$ and the backward substitution with arithmetic complexity of $n^2/2$, where n is the size of the matrix Q which is equal to the number of states in the analyzed tandem queue system.

The number of arithmetic operations needed for a solution algorithm that has arithmetic complexity of $n^3/3$ grows rapidly with n. This makes limiting this complexity a very desirable goal. To that end, we have employed two methods – one is to reduce the number of elements that have to be eliminated in the matrix triangularization process, second is to avoid fetching and processing zero valued elements of $\mathbf{Q}$.

The first method rearranges the set of global balance equations to eliminate the leftmost non-zero element in each row of matrix $\mathbf{Q}$ thus producing new (transformed) matrix $\mathbf{Q}''$. This is accomplished by (conceptually) adding to each row the row that precedes it, starting with the second row and advancing to the last one, and then, for each phase $r \leq m_c$, restoring the last rows in that phase to its original form. In other words, given matrix $\mathbf{Q}$ with rows $\mathbf{R}_0 - \mathbf{R}_{n-1}$:

$$\mathbf{R}'_0 = \mathbf{R}_0; \qquad\qquad // \text{ e.g. unchanged}$$

for $(k = 1; k < n; k + +)$
$$\mathbf{R}'_k = \mathbf{R}_k + \mathbf{R}'_{k-1};$$
$$\mathbf{R}''_0 = \mathbf{R}'_0;$$
for $(k = 1; k < n; k + +)$
  if (phase $\leq m_c$ and k is the last row in phase)
$$\mathbf{R}''k = \mathbf{R}'_k - \mathbf{R}'_{k-1};$$
  else $\mathbf{R}''_k = \mathbf{R}'_k;$

Given that from the definition of matrix $\mathbf{Q}$, for every column j $(0 \leq j < n)$ we have [4]:

$$\sum_{k=0}^{n-1} qk, j = 0$$

and since the leftmost non-zero element in a row is the last one in its column, the above algorithm indeed eliminates this element. Figure 10.6 presents such a transformed matrix $\mathbf{Q}''$ for the original matrix $\mathbf{Q}$ from Figure 10.5.

It needs to be noted here that this is a conceptual transformation only – it does not need to be actually performed, since the structure of the transformed matrix $\mathbf{Q}''$ is rather clear, and it can be constructed directly, given l, uc, us and n!

The second method of reducing the matrix $\mathbf{Q}''$ triangularization complexity is to avoid fetching and processing its zero valued elements. Owing to the structural regularity of the phase/level process' state flow diagram, the structure of the matrix $\mathbf{Q}''$ exhibits strong patterns. The pattern of interest here is the number of elements preceding and following the diagonal one in

```
-λ,      0,    us,     0,     0,     0,     0,     0,     0,     0,     0,     0,     0,     0;
 0,-uc-λ,    us,     0,    us,     0,     0,     0,     0,     0,     0,     0,     0,     0;
 0,   uc,-λ-us,     0,     0, 2*us,     0,     0,     0,     0,     0,     0,     0,     0;
 0,    0,     0,-λ,-2*uc-λ,  us, 2*us,     0,     0,     0,     0,     0,     0,     0;
 0,    0,     0,-λ,-uc-λ, 2*us,     0,    us, 2*us,     0,     0,     0,     0,     0;
 0,    0,     0,     0, uc,-λ-2*us,     0,     0, 2*us,     0,     0,     0,     0,     0;
 0,    0,     0,     0,-λ,   -λ,-3*uc-λ, us, 2*us, 2*us,     0,    us,     0,     0;
 0,    0,     0,     0,    0,-λ,   -λ,-2*uc-λ,2*us,2*us,     0,    us, 2*us,     0;
 0,    0,     0,     0,    0,    0,-λ,-uc-λ, 2*us,     0,    us, 2*us, 2*us;
 0,    0,     0,     0,    0,    0, 0,  uc,-λ-2*us,     0,     0,     0,     0;
 0,    0,     0,     0,    0,    0,-λ,   -λ,   -λ,-3*uc, us, 2*us, 2*us, 2*us;
 0,    0,     0,     0,    0,    0, 0,   -λ,   -λ,   -λ,0,-2*uc, 2*us, 2*us;
 0,    0,     0,     0,    0,    0, 0,    0,   -λ,    0,    0,  -uc, 2*us;
 0,    0,     0,     0,    0,    0, 0,    0,    0,    0,    0,    0,    0;
```

Figure 10.6 Transformed state transition matrix.

Table 10.1 Regularities in the state transition matrix.

Phase (r)	Row within phase (n)	Number of preceding elements (p)	Number of following elements (f)
$r = 1 + m_c - 1$	$n = 0 \div r - 1$	$r - 1$	$r + 2$
	$n = r$	1	$r + 2$
$r = m_c$	$n = 0 \div r - 1$	$r - 1$	$r + 2$
	$n = r$	1	$r + 1$
$r = m_c + 1 \div m_c + L_c - 1$	$n = 0 \div m_c - 1$	m_c	$m_c + 2$
	$n = m_c$	1	$m_c + 1$
$r = m_c + L_c$	$n = 0 \div m_c - 1$	m_c	$m_c - n$
	$n = m_c$	$m_c(0)$	0

a particular row, expressed as a function of the phase this row belongs to and of the position of the row within that phase. Table 10.1 summarizes these regularities.

Utilizing these regularities, we have constructed an algorithm for the triangularization of a (transformed) matrix $\mathbf{Q}''$. A simplified C implementation of this algorithm is shown in Figs. 10.7 and 10.8. It operates on a global matrix mQ, which needs to be initialized with $\mathbf{Q}''$. To execute the procedure function gElimFRA() needs to be invoke with mc – number of service stations at the Client, and Lc – length of the Client queue, passed as parameters. The processing (triangularization) of the entire matrix is subdivided into processing of sets of rows corresponding to an individual phase, which is performed by function phsElim(). As shown in Table 10.1, four different

```
gElimFRA(mc,Lc)
{
N = (mc+1)*((mc+2)/2 + Lc);
elim1El(N-1,0,3);
fr = 1;
for(r=1;r<mc;r++) {
    phsElim(fr,r+1,r-1,1,r+2,r+2,r+1,r+1,r,N);
    fr += r + 1;
    }
phsElim(fr,mc+1,mc-1,1,mc+2,mc+1,mc+1,mc+1,mc,N);
fr += mc + 1;
if(Lc > 1)
    for(k=0;k<(Lc-1);k++) {
        phsElim(fr,mc+1,mc,mc,mc+2,mc+1,mc+2,mc+1,
            mc+1,N);
        fr += mc + 1;
        }
phsElim(fr,mc,mc,mc,-mc,0,mc+2,mc+1,mc+1,N);
}
```

Figure 10.7 UPnP network model.

types of phases can be distinguished – phases $r = 1 \div m_c - 1$, phase $r = m_c$, phases $r = m_c + 1 \div m_c + L_c - 1$ (if at all present), and phase $r = L_c$, where r is equal to the total number of jobs at the client (e.g. K_c).

Using the above algorithm followed by the standard backward substitution procedure, we can efficiently compute values of the steady state probabilities vector $\mathbf{\Pi}$. Figure 10.9 shows such values computed for a system with $L_c = 3$, $m_c = 6$, $L_s = 2$, $m_s = 4$, $\lambda = 1.0$, $\mu_c = 1.1$, $\mu_s = 1.2$ plotted as a function of the phase and the level e.g. $\mathbf{\Pi}(K_c, K_s)$.

10.4.1 Complexity Analysis

Total complexity of the triangularization algorithm presented here can be expressed as:

$$S = (L_c - 1)(m_c + 1)(m_c(m_c + 3) + (m_c + 2)) +$$
$$+ (3m_c^4 + 26m_c^3 + 69m_c^2 + 106m_c + 24)/12$$

Given that the size N of the state transition matrix $\mathbf{Q}''$ can be expressed as:

$$N = (m_c + 1)((m_c + 2)/2 + L_c)$$

it can be easily shown that $S < N^2$ for large m_c and L_c. Thus the computational complexity of the triangularization algorithm presented here is less than $O(n^2)$.

```
phsElim(lf,h,p,pl,f,fl,pf,pfl,ph,ms)
{
if(f < 0) {
    f = -f;
    dlt = 1;
    } else dlt = 0;
for(i=0;i<h;i++) {
    cr = lf + i;
    if(i == (h - 1))
        if(dlt == 1) {
            ap = p;
            af = f - dlt * i;
            }
        else {ap = pl; af = fl; }
    else { ap = p; af = f - dlt * i; }
    for(j=0;j<ap;j++) {
        cur = cr - ap + j;
        if(cur < lf)
            if(cur == (lf - 1)) cnt = pfl + 1;
            else  cnt = pf + 1;
        else cnt = f - dlt * (cur - lf) + 1;
        elim1El(cr,cur,cnt);
        }
    elim1El(ms-1,cr,af+1);
    }
}

elim1El(p,a,cnt)
{
s = mQ[p,a] / mQ[a,a];
for(nn=a;nn<(a+cnt);nn++)
    mQ[p,nn] -= mQ[a,nn]*s;
}
```

Figure 10.8 UPnP network model.

10.5 UPnP Network Simulation

In order to construct an efficient algorithm for the analytical examination of the UPnP network model we have reduced analysis of that model to the analysis of an HTTP client/server pair which we represent by a two-node tandem queue arrangement. To validate this approach we are currently implementing a UPnP network simulator which we will use to generate data that can be compared with the results produced by our optimized CTMC solution algorithm. It directly implements (simulates) the UPnP Queuing Network Model for the Control Phase Operations, as presented in Figure 10.2. Simulator employs the discrete event simulation technique; it was constructed using the OMNeT++ simulation framework.

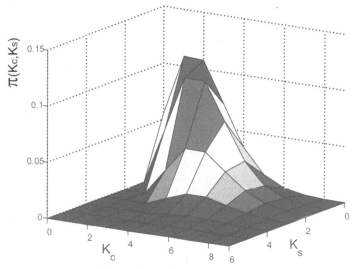

Figure 10.9 UPnP network model.

10.6 Summary and Further Work

Rapidly increasing number of networked control and monitoring applications is constructed around modern high level communication protocols that employ a combination of XML and SOAP to define the necessary data structure and implement the messaging scheme. In turn, these two standards rely on the HTTP protocol to provide the actual network communication services. Consequently, the performance of a broad class of networked applications is largely determined by the performance of the HTTP protocol operating in a control and monitoring environment.

To help investigating this performance, we have presented in this paper a model of a networked control system employing a combination of UPnP, SOAP and XML standards, which we have subsequently reduced to the model of a client/server pair that uses the HTTP protocol's request/response mechanism to communicate. We have also presented an efficient algorithm for the analytical examination of that client/server model.

To check the validity of our model simplification and the correctness of our algorithm, we are currently constructing a simulator of a UPnP/SOAP/XML based networked control application.

References

[1] *Universal Plug and Play Device Architecture*, Version 1.0, 1999–2000 Microsoft Corporation.

[2] *Linux SDK for UPnP Devices Version 1.2*, 2000–2003 Intel Corporation.

[3] T. Czachórski, *Modele kolejkowe w ocenie efektywnosci sieci i systemów komputerowych*, Pracownia Komputerowa Jacka Skalmierskiego, Gliwice 1999.

[4] G. Bolch, S. Greiner, H. de Meer and K. Trivedi, *Queueing Networks and Markov Chains: Modeling and Performance Evaluation With Computer Science Applications*, John Wiley & Sons, 1998.

[5] *OPC Foundation,OPC XML-DA Specification, Version 1.0*, July 2003.

[6] *Proposed Addendum C to Standard 135-2004, BACnet – A Data Communication Protocol for Building Automation and Control Networks*, ASHARE 2004.

[7] *Voice Extensible Markup Language (VoiceXML) 2.1*, W3C Working Draft, March 2004.

[8] K. Goldberg, S. Gentner, C. Sutter and J. Wiegley, The Mercury Project: A feasibility study for internet robots, *IEEE Robotics and Automation Magazine*, Special Issue on Internet Robotics, December 1999.

[9] T. W. Fong, H. Pangels, D. Wettergreen, E. Nygren, B. Hine, P. Hontalas and C. Fedor, Operator interfaces and network-based participation for Dante II, in *Proceedings SAE 25th International Conference on Environmental Systems*, July 1995.

PART FOUR
TCP PERFORMANCE ENGINEERING

11

Analysis of Scalable TCP

R. El Khoury[1], E. Altman[2] and R. El Azouzi[1]

[1]LIA/CERI, Université d'Avignon, Agroparc, BP 1228, 84911 Avignon, France
[2]INRIA, BP 93, 06902 Sophia Antipolis, France;
e-mail: eitan.altman@sophia.inria.fr

Abstract

In recent years, several more aggressive versions of TCP have been proposed which leave the additive increase multiplicative decrease (AIMD) paradigm. The motivation is to adapt TCP to networks with very large bandwidth delay products. In this paper, we study the scalable TCP (STCP) which is designed to be a Multiplicative Increase Multiplicative Decrease (MIMD) protocol. We analyze how it shares a bottleneck link with other CBR flows. We identify conditions under which the time varying queueing delay transforms the whole multiplicative increase dynamics of the window size into an additive increase dynamics, which implies that window evolution of STCP behaves as an AIMD protocol.

Keywords: Scalable TCP, TCP Reno, performance evaluation, congestion control.

11.1 Introduction

In high speed networks, the congestion avoidance phase of TCP takes a long time to increase the window size and fully utilize the available bandwidth. Floyd writes in [2]:

> for a Standard TCP connection with 1500-byte packets and a 100 ms round-trip time, achieving a steady-state throughput of 10 Gbps would

D. D. Kouvatsos (ed.), Traffic and Performance Engineering for Heterogeneous Networks, 235–253.

require an average congestion window of 83,333 segments, and a packet drop rate of at most one congestion event every 5,000,000,000 packets (or equivalently, at most one congestion event every $1\frac{2}{3}$ hours).The average packet drop rate of at most 2×10^{-10} needed for full link utilization in this environment corresponds to a bit error rate of at most 2×10^{-14}, and this is an unrealistic requirement for current networks.

This has motivated in recent year the design of new versions of TCP for networks with high bandwidth delay product: the HighSpeed TCP [2] as well as MIMD versions of TCP (Scalable TCP – STCP [5], High-Throughput TCP [8]), the Binary Increase Congestion Control [7] and FAST TCP [4].

In this paper, we analyze the behavior of STCP that shares a bottleneck link with CBR non-responsive traffic. We present a fluid approximation model where the discrete values of the window size are replaced by a real valued variable. We obtain the average throughput of STCP and compare it to that obtained by an AIMD congestion control. We identify two phases in the behavior of the instantaneous window size: a part of exponential increase in time that corresponds to the period in which the buffer is empty, and a linear increase during the part in which the buffer is non-empty. (The occurrence of two phases arises also in other versions of TCP, see Altman et al. [1] who study a similar setting but with the Tahoe version.) We identify conditions under which only the second phase exists, which means that the time evolution of the window size of STCP follows an AIMD dynamics. Our model is finally validated through simulations and the parameters for which the model is valid are identified. Modifications are proposed for some extreme cases.

11.2 System Model

We consider a shared bottleneck link depicted in Figure 11.1, where two streams share the bottleneck. The first is STCP, and the second one is a deterministic exogenous traffic with a constant bit rate. This latter is independent of the network load.

The parameters used for the system are:

- μ is the service rate of the queue.
- λ is the rate of the exogenous traffic. We assume $\lambda < \mu$, thus we can always have a STCP flow.
- τ_1 is the delay between the STCP source and the buffer.
- τ_2 is the delay between the buffer and the destination.
- $\tau \triangleq 2(\tau_1 + \tau_2)$ is the round trip time without the service time.

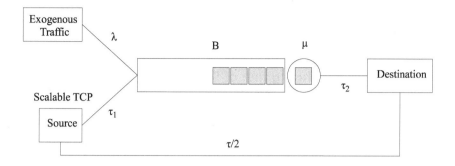

Figure 11.1 System model.

- $T \triangleq \tau + 1/\mu$ is the round trip time including the service time μ^{-1}. In our analysis, we consider a fixed T.
- B is the maximum buffer size in packets.
- $W(t)$ is the window size at time t.
- W_{loss} is the window size at the end of a congestion avoidance phase, before a loss occurs.
- W_{max} is the maximal size of $W(t)$ at the instant of the loss detection, just after a loss occurs.
- a is the increase parameter of the window.
- $\beta \triangleq 1 - b$ is the decrease parameter of the window.
- $\text{thp}_{\text{in}}(t)$ is the throughput of the controlled source between the STCP source and the buffer.
- $\text{thp}_{\text{out}}(t)$ is the throughput of the controlled source between the buffer and the destination.

When units are not given, rates will be in Mbps and time in msec. Windows and buffer size are in packets, so the rates will be converted to packets per second for some numerical results. The packet size for both the exogenous traffic and TCP is fixed and equal to 1040 Bytes. The increase and decrease factors may take, respectively, the values $a = 0.01$ and $b = 0.875$ as recommended in [5].

Motivated by the analysis and simulations of other versions of TCP, specially [1], we are focusing to study the cyclic evolution of the window size of STCP in the congestion avoidance phase, i.e., in the stationary regime. A cycle corresponds to the time between consecutive loss events. In the following section, we will derive the expression of the average throughput when STCP stays in the congestion avoidance phase. This corresponds, for example, to the transfer of large files over the network.

11.3 The Average Throughput of STCP

The long-term average throughput of STCP connection is the average number of packets successfully transmitted in a cycle, divided by the duration of a cycle. Therefore, it is given by

$$\overline{\text{thp}} = \frac{N}{C}, \tag{11.1}$$

where N is the average number of STCP packets transmitted in a cycle, and C is the average cycle duration. We need then to calculate N and C. For that, we will be using a fluid model for the window size $W(t)$ at time t, and then finding its expression as well as the throughput at the output of the buffer $\text{thp}_{\text{out}}(t)$. The average number of packets N can then be written as

$$N = \int_0^C \text{thp}_{\text{out}}(t)\, dt \tag{11.2}$$

For the calculation of $\text{thp}_{\text{out}}(t)$ we proceed as in [1] as follows.

For each acknowledgement (ACK) received in a round trip time in which congestion has not been detected, we have $W(t^+) = W(t) + a$. The rate of the window growth with arriving ACKs is $dW/d\,\text{ack} = a$, where $\text{ack}(t)$ is the number of ACKs that have returned by time t. The rate at which the ACKs are generated is $\text{thp}_{\text{out}}(t)$. Hence,

$$\frac{dW}{dt} = \frac{dW}{d\,\text{ack}} \frac{d\,\text{ack}}{dt} = \frac{dW}{d\,\text{ack}} \text{thp}_{\text{out}}$$

We see that the window grows at a rate equal to:

$$\frac{dW}{dt} = a \cdot \text{thp}_{\text{out}} \tag{11.3}$$

We distinguish two phases:

1. *When the queue is empty*:

$$\text{thp}_{\text{out}}(t) = \text{thp}_{\text{in}}(t) = \frac{W(t)}{T} \tag{11.4}$$

2. *When the queue builds*: thp_{in} is larger than thp_{out}. We assume like in [1] that the output rate of the STCP is proportional to its input rate:

$$\text{thp}_{\text{out}}(t) = \frac{\mu\, \text{thp}_{\text{in}}(t)}{(\text{thp}_{\text{in}}(t) + \lambda)} \qquad (11.5)$$

Then,

$$\text{thp}_{\text{in}}(t) = \text{thp}_{\text{out}}(t) \cdot \frac{\lambda}{\mu - \text{thp}_{\text{out}}(t)} \qquad (11.6)$$

and we know that

$$\text{thp}_{\text{in}}(t) = \frac{\mathrm{d}W}{\mathrm{d}t} + \frac{\mathrm{d\,ack}}{\mathrm{d}t} = \left(1 + \frac{\mathrm{d}W}{\mathrm{d\,ack}}\right) \text{thp}_{\text{out}}(t) \qquad (11.7)$$

We obtain from the two last equations,

$$\text{thp}_{\text{out}}(t) = \mu - \frac{\lambda}{(1 + \frac{\mathrm{d}W}{\mathrm{d\,ack}})} \qquad (11.8)$$

We let $\hat{T}$ be the time at which the queue starts to build up, and $\hat{W}$ be the corresponding window size. $\hat{W}$ corresponds to the number of packets in the network that fills up the available bandwidth of STCP. We get $\hat{W} = (\mu - \lambda)T$.

$$\text{thp}_{\text{out}}(t) = \begin{cases} \frac{W(t)}{T} & \text{if } W \le \hat{W} \\ \mu - \frac{\lambda}{1+a} & \text{if } W > \hat{W} \end{cases} \qquad (11.9)$$

Therefore,

$$\frac{\mathrm{d}W(t)}{\mathrm{d}t} = \begin{cases} a\frac{W}{T} & \text{if } W \le \hat{W} \\ a(\mu - \frac{\lambda}{1+a}) & \text{if } W > \hat{W} \end{cases} \qquad (11.10)$$

It appears from our last equation that the window grows up in two different ways in a one cycle for $W \le \hat{W}$ and for $W > \hat{W}$. These two ways depend on the value of W.

Let us now derive the expression of W_{loss}, which is the window size in packets just before a loss event (due to a buffer congestion). In other words, packets sent with a window of W_{loss} cause a congestion in the network. The window is limited by the number of packets in the pipe (the link which has the lowest Bandwidth Delay Product, BDP) and the buffer. Therefore, W_{loss} is the sum of STCP packets: in each link between the source and the destination, in the buffer and in the server. Therefore, we get

$$W_{\text{loss}} = \tau_1 \cdot \text{thp}_{\text{in}} + \left(\tau_2 + \frac{\tau}{2}\right) \cdot \text{thp}_{\text{out}} + B \cdot \frac{\text{thp}_{\text{in}}}{\text{thp}_{\text{in}} + \lambda} + 1 \qquad (11.11)$$

When the exogenous traffic rate is not null, the buffer can contain at most $B \cdot (\text{thp}_{\text{in}})/(\text{thp}_{\text{in}} + \lambda)$ STCP packets. For each ACK received, the window increments by $(1 + a)$. If we received W_{loss} ACKs as an upper bound, then W_{max} can be estimated as: $W_{\text{max}} = (1 + a)W_{\text{loss}}$.

From equations (11.8) and (11.6) in the second phase, we obtain:

$$\text{thp}_{\text{in}}(t) = \text{thp}_{\text{out}}(t).\frac{\lambda}{\mu - \text{thp}_{\text{out}}(t)} = (1 + a)\left(\frac{\mu - \lambda}{(1 + a)}\right),$$

which yields

$$W_{\text{max}} = (1 + a)\left(\frac{\mu - \lambda}{(1 + a)}\right)\left((2 + a)\tau_1 + 2\tau_2 + \frac{B}{\mu}\right) + 1 + a \qquad (11.12)$$

We can also write it by using $T = \tau + 1/\mu$:

$$W_{\text{max}} = ((1 + a)\mu - \lambda)\left(T + a\tau_1 + \frac{B}{\mu}\right) + \frac{\lambda}{\mu} \qquad (11.13)$$

11.3.1 The Window Dynamics

By integrating equation (11.10), we get:

$$W(t) = \begin{cases} k_1 \exp(\frac{at}{T}) & \text{if } W \leq \hat{W} \\ a(\mu - \lambda/(1 + a))t + k_2 & \text{if } W > \hat{W} \end{cases}$$

k_1 and k_2 are two constants. There are determined from the conditions at the limits: $W(0) = bW_{\text{max}}$ and

$$\hat{W} = k_1 \exp\left(\frac{a\hat{T}}{T}\right) = \left(\frac{\mu - \lambda}{(1 + a)}\right)\hat{T} + k_2$$

where $t = 0$ represents the time at the beginning of the cycle. We get

$$\hat{T} = \frac{T}{a} \ln\left(\frac{\hat{W}}{bW_{\text{max}}}\right)$$

So the window evolution becomes:

$$W(t) = \begin{cases} bW_{\text{max}} \exp(\frac{at}{T}) & \text{if } W \leq \hat{W} \\ a(\mu - \lambda/(1 + a))(t - \hat{T}) + \hat{W} & \text{if } W > \hat{W} \end{cases} \qquad (11.14)$$

$\hat{W}$ is given by $\hat{W} = (\mu - \lambda)T$. Hence, we obtain an expression of the window in terms of the time and depending on the parameters μ, λ, T and B. It is now clear that the window increases in two phases: for $W \leq \hat{W}$, it grows up exponentially, whereas for $W > \hat{W}$, it is linear. Furthermore, we distinguish two behaviors of the window when this latter is higher or lower than bW_{max}. Therefore, for $\hat{W} > bW_{max}$, W grows up exponentially until $\hat{W}$, and then linearly for $W > \hat{W}$. Whereas for $\hat{W} \leq bW_{max}$, W is only linear during whole the cycle of the congestion avoidance.

11.3.2 The Cycle Duration Expression

Let T_1 and T_2 be the durations of the exponential and the linear phase, respectively, in a cycle. In the following, we find the expressions of the cycle duration for the two behaviors of the window:

- For $\hat{W} > bW_{max}$: we have $C = T_1 + T_2$, where T_1 is just $\hat{T}$ and T_2 is given from the linear phase with $W(T_1 + T_2) = W_{max}$. Therefore,

$$C = \frac{T}{a} \cdot \ln\left(\frac{\hat{W}}{bW_{max}}\right) + \frac{W_{max} - \hat{W}}{a(\mu - \lambda/(1+a))} \qquad (11.15)$$

- For $\hat{W} \leq bW_{max}$: we have $C = T_2$, where T_2 is given from the linear phase with $W(T_2) = W_{max}$. Therefore,

$$C = \frac{W_{max}(1 - b)}{a(\mu - \lambda/(1+a))} \qquad (11.16)$$

11.3.3 The Average Throughput Expression

Finally, we can calculate the average throughput that is given by:

$$\overline{thp} = \frac{N_1 + N_2}{T_1 + T_2} \qquad (11.17)$$

where N_1 and N_2 are the number of packets sent in T_1 and T_2 respectively. The number of packets N_i for each phase can be written as

$$N_i = \int_{t_{i-1}}^{t_i} thp_{out}(t)\, dt$$

Here too we have to consider the two cases:

- For $\hat{W} > bW_{\max}$: in the first phase, with a duration of $\hat{T}$, we have $\text{thp}_{\text{out}} = W(t)/T$. Then,

$$N_1 = \int_0^{\hat{T}} \frac{W(t)}{T} dt = \int_0^{\hat{T}} \frac{bW_{\max} \exp(\frac{at}{T})}{T} dt$$

$$= \frac{b}{a} W_{\max} \left(\exp\left(\frac{a\hat{T}}{T}\right) - 1 \right) = \frac{\hat{W} - bW_{\max}}{a}$$

In the second phase, with a duration of $C - \hat{T}$, we have $\text{thp}_{\text{out}} = \mu - \lambda/(1+a)$. Then,

$$N_2 = \int_{\hat{T}}^C \left(\frac{\mu - \lambda}{(1+a)} \right) dt = (C - \hat{T}) \left(\frac{\mu - \lambda}{(1+a)} \right).$$

Therefore, the average throughput of the first case is:

$$\overline{\text{thp}} = \frac{\frac{\hat{W} - bW_{\max}}{a} + (C - \hat{T})(\mu - \lambda/(1+a))}{C} = \frac{W_{\max}(1-b)}{aC} \tag{11.18}$$

- For $\hat{W} \leq bW_{\max}$: there is only one linear phase. Therefore, the average throughput of the second case is:

$$\overline{\text{thp}} = \mu - \lambda/(1+a) \tag{11.19}$$

By comparing the two expressions of the throughput, we find that the second case with a linear growth gives the better throughput, since the available bandwidth is fully utilized in this case. The interesting point resides in determining which parameters of the system gives this better throughput. We discuss this point in the next section while dimensioning the buffer size.

Figures 11.2 to 11.4 present the variation of the throughput with the parameters λ, T and B with a fixed $\mu = 50$ Mbps. Figure 11.2, for $B = 10$, corresponds to the first case where the window has an exponential growth.

The STCP throughput in Figures 11.2 to 11.3 is normalized. It represents the used proportion of the available bandwidth $(\mu - \lambda)$ of STCP. By observing the figures, we see that for higher λ STCP uses a higher percentage of the available bandwidth, whereas for lower values it is the opposite. This result is independent of the round trip time and the buffer size. This is in the opposite of the TCP classical TCP behavior [1], and just for the congestion avoidance phase. It is explained by the aggressive behavior of the MIMD. Furthermore,

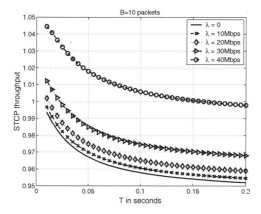

Figure 11.2 STCP throughput in presence of exogenous traffic with $B = 10$, for different λ.

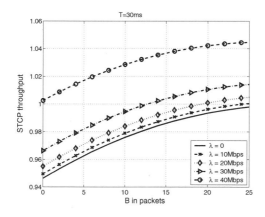

Figure 11.3 The throughput versus B, for $T = 30$ms and various values of λ.

the utilization of the bandwidth for small buffers is also good. For large λ, the throughput proportion can go beyond one. This is due to the exponential factor of growth a included in the throughput expression which lets it go beyond 1 by a factor of $1 - (\mu - \lambda/(1+a))/(\mu - \lambda)$. In this case, STCP does not allow, even slightly, the exogenous traffic to flow.

Figure 11.4 shows the throughput in terms of the buffer size. For $B < 85$, we have $\hat{W} > bW_{\max}$, and then the throughput follows equation (11.18). Whereas for large buffers beyond 85 packets, the throughput is given by equation (11.19). Therefore, we have a threshold value of the buffer size that

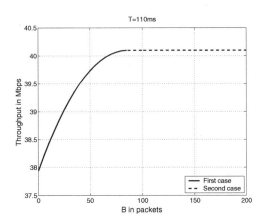

Figure 11.4 The throughput versus buffer size, for $T = 110$ ms and $\lambda = 10$ Mbps.

identifies these two cases of Figure 11.4. In the next section, we derive this threshold.

11.4 The Condition for an AIMD Type Behavior

As we have discussed previously in the window dynamics section, we have only one linear phase during whole the cycle for $\hat{W} \leq bW_{\max}$. In this such case the MIMD STCP behavior is similar to the AIMD protocol. In this section, we are interested to find the buffer threshold B_{th} beyond which we have this behavior of the STCP. This threshold is directly given from the equality $\hat{W} = bW_{\max}$. By using $W_{\max}$ from expression (11.13), and by replacing $\hat{W}$ by $(\mu - \lambda)T$, we get:

$$B_{\text{th}} = \frac{\mu[T(\mu - \lambda) - b(T + a\tau_1)((1 + a)\mu - \lambda) - b\lambda/\mu]}{b((1 + a)\mu - \lambda)}, \qquad (11.20)$$

Therefore, the condition for an AIMD behavior of the STCP is given by $B \geq B_{\text{th}}$. In fact, in this case the arrived packet at the buffer takes more and more time to be served when they found a large queue. The thp_{out} converges to a constant value which regulates the returned ACK independent of the current window size of the controlled source. Figure 11.5 shows the buffer size threshold in terms of the round trip time for different exogenous traffic. The condition $\hat{W} \leq bW_{\max}$ can be also reflected to the decreasing and increasing factors b and a respectively. For $b \geq \hat{W}/W_{\max}$ and for a given value of a and

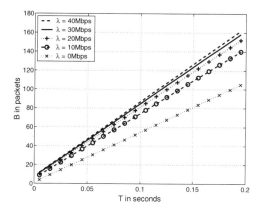

Figure 11.5 The buffer size threshold for $\mu = 50$ Mbps and λ between 0 and 40 Mbps.

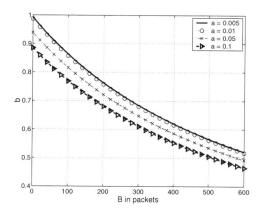

Figure 11.6 The decrease factor threshold for $\mu = 50$ Mbps, $\lambda = 10$ Mbps, $T = 0.11$ s and different increase factors a.

B, we can get the AIMD behavior of STCP. See some numerical results in Figure 11.6.

11.5 Comparison between the STCP and TCP Reno

Here, we derive the expression of the window for the TCP Reno in the congestion avoidance phase, and then we proceed like in Section 11.3 to determine the average throughput expression; we also refer to the paper [3]. In the following, we summarize our results.

11.5.1 TCP Reno Expressions

For each arriving ACK, the window grows up with:

$$\frac{dW}{d\,\text{ack}} = \frac{1}{W} \tag{11.21}$$

and therefore, the throughput leaving the buffer can be written as

$$\text{thp}_{\text{out}} = \begin{cases} \frac{W(t)}{T} & \text{if } W < \hat{W} \\ \mu - \frac{\lambda}{1+(1/W)} & \text{if } W > \hat{W} \end{cases} \tag{11.22}$$

Assuming that $W \gg 1$ (for wide area networks), the second case of the thp_{out} equation can be approximated by just $\mu - \lambda$. Let $\hat{W}$ be the necessary window size that makes the queue starts building up. It is then given by $\hat{W} = T(\mu - \lambda)$. The W_{max} expression is given by

$$W_{\text{max}} = (\mu - \lambda)\left(\frac{T+B}{\mu}\right) + \frac{\lambda}{\mu} + 2 \tag{11.23}$$

We distinguish two cases:

- Case 1: $\hat{W} > W_{\text{max}}/2$. We get

$$W(t) = \begin{cases} \frac{1}{T}t + \frac{W_{\text{max}}}{2} & \text{if } W \leq \hat{W} \\ \sqrt{2[(\mu - \lambda)(t - \hat{T}) + \frac{\hat{W}^2}{2}]^{\frac{1}{2}}} & \text{if } W > \hat{W} \end{cases} \tag{11.24}$$

- Case 2: $\hat{W} \leq W_{\text{max}}/2$. We get

$$W(t) = (2(\mu - \lambda)t)^{1/2} \tag{11.25}$$

The cycle duration is given by

$$C = \begin{cases} \frac{W_{\text{max}}^2 - \hat{W}^2}{2(\mu - \lambda)} + \hat{T} & \text{if } \hat{W} > \frac{W_{\text{max}}}{2} \\ \frac{W_{\text{max}}^2}{2(\mu - \lambda)} & \text{if } \hat{W} \leq \frac{W_{\text{max}}}{2} \end{cases} \tag{11.26}$$

where $\hat{T} := T(\hat{W} - W_{\text{max}}/2)$. Therefore, the average throughput is

$$\overline{\text{thp}} = \begin{cases} \frac{1}{C}\left(\frac{\hat{T}^2}{2T^2} + \frac{\hat{T}W_{\text{max}}}{2T} + (\mu - \lambda)(C - \hat{T})\right) & \text{if } \hat{W} > \frac{W_{\text{max}}}{2} \\ \mu - \frac{\lambda}{1+(1/W)} & \text{if } \hat{W} \leq \frac{W_{\text{max}}}{2} \end{cases} \tag{11.27}$$

A simpler approximate expression is given by

$$
\overline{\mathrm{thp}} = \begin{cases} \frac{3W_{\max}^2}{8C} & \text{if } \hat{W} > \frac{W_{\max}}{2} \\ \mu - \frac{\lambda}{1+(1/W)} & \text{if } \hat{W} < \frac{W_{\max}}{2} \end{cases} \tag{11.28}
$$

11.5.2 Throughput Comparison

As we have seen in the previous section, TCP Reno has also two behaviors depending on the two previous cases. In the first case, we have the linear (corresponding to an empty queue) and the sub-linear corresponding to a non-empty queue) growth of the window. This case also occurs when the buffer size is lower than a given threshold B_{th}, where $B_{\mathrm{th}} = \mu T - (\lambda + 2\mu)/(\mu - \lambda)$.

$$
B_{\mathrm{th}} = \mu T - \frac{(\lambda + 2\mu)}{(\mu - \lambda)} \tag{11.29}
$$

In Figures 11.7 and 11.8, we present the case where the buffer is lower than the threshold of equations (11.20) and (11.29), i.e., for the case where STCP and TCP windows grows up in two phases. In these two figures, we show the variation of the throughput ratio $(\overline{\mathrm{thp}}_{\mathrm{STCP}})/(\overline{\mathrm{thp}}_{\mathrm{TCP}})$ versus T for some values of B. We see from the figures that the STCP throughput is better than TCP, specially for small buffers and higher λ. In the second case, i.e for only one phase, the STCP is slightly higher than TCP. It depends on the value of a, refer to the previous equations.

11.5.3 Cycle Duration Comparison

Figures 11.9 and 11.10 show the difference of the cycle durations between STCP and TCP. It is clear that the cycle duration (resp. losses rate) of STCP is lower (resp. higher) than TCP Reno.

11.5.4 Link Utilization

We have seen the link utilization of the two protocols previously, but here we illustrate it with the B/BDP parameter, and present a comparison between STCP and TCP Reno. Table 11.1 shows the link utilization for different values of B/BDP with $\lambda = 0$, $\mu = 50$ Mbps and $T = 40$ ms.

From these values, we conclude that STCP uses the link more efficiently than TCP Reno. Simulations are shown in Figures 11.10 and 11.11. We plot the window variation of TCP Reno and STCP in terms of time using the

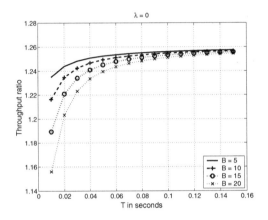

Figure 11.7 Comparison between the average throughput of STCP and TCP Reno for different buffer size and when both STCP and TCP have two different phases.

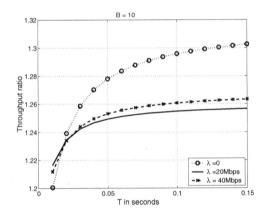

Figure 11.8 Comparison between the average throughput of STCP and TCP Reno for different λ.

same parameters. It is seen indeed that a better link utilization is obtained under STCP only when the buffer size does not limit the utilization of the pipe. This latter case is considered as an extreme case and discussed in Section 11.7. Here, B_0 is the buffer size limit under which the utilization of the pipe is limited. Figure 11.12 shows the normalized STCP throughput (or the link utilization) in the interval $[B_0/\mathrm{BDP}, B_{\mathrm{th}}/\mathrm{BDP}]$, where B_{th} is given from equation (11.20) in the case of STCP (and from equation (11.29) for TCP

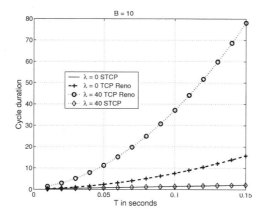

Figure 11.9 Cycle duration comparison for different λ.

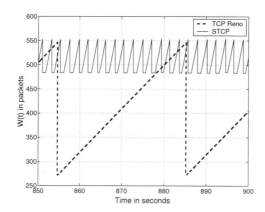

Figure 11.10 Simulation comparison of STCP and TCP Reno: window variation versus the time for $\mu = 50$ Mbps, $\lambda = 10$ Mbps, $T = 110$ ms and $B = 20$ packets. We obtain the same W_{max} corresponding to a full pipe.

Reno). For $\text{BDP} = (\mu - \lambda)T$ and by taking $\lambda = 0$, we get

$$\frac{B_{\text{th}}}{\text{BDP}} = \frac{1 - b(1 + a)(1 + a\frac{\tau_1}{T})}{b(1 + a)}, \quad \text{for STCP} \tag{11.30}$$

$$\frac{B_{\text{th}}}{\text{BDP}} = \frac{1 - b}{b} - \frac{2}{\mu T}, \quad \text{for TCP Reno} \tag{11.31}$$

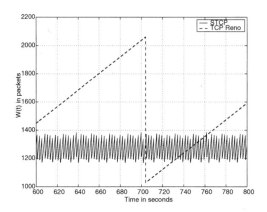

Figure 11.11 Simulation comparison of STCP and TCP Reno: window variation versus the time for $B \leq B_0$, with $B = 15$ and $B_0 = 20.4$ for $\lambda = 0$, $\mu_1 = 300$ Mbps, $\mu = 100$ Mbps, $\tau_1 = 5$ msec, $\tau_2 = 80$ msec.

Table 11.1 Link utilization with different B/BDP values.

$B/(\mu T)$	STCP	TCP Reno
0.04	0.97	0.78
0.08	0.99	0.81
0.16	1	0.85
0.32	1	0.89

From equation (11.30), we observe that the ratio B/BDP depends only on τ_1/T for a fixed parameter a and b. In the case of STCP with very small a, the ratio B_{th}/BDP keeps a constant value. For example, for $a = 0.01$, $b = 0.875$, $\tau_1 = 5$ ms and $T = 100$ ms the ratio B_{th}/BDP converge to 0.13. In TCP Reno, with that same parameters, the ratio B_{th}/BDP is near to 1. We conclude that the STCP quickly converge to 1 (or higher than 1) which means, it uses all the available bandwidth in a short period of time compared to TCP Reno.

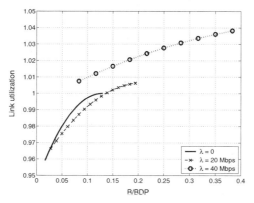

Figure 11.12 The normalized STCP throughput versus B/BDP for $T = 100$ ms, $\mu = 50$ Mbps and different λ.

11.6 Simulation Results

We simulate our system shown in Figure 11.1 with the simulator ns–2. We establish a long STCP connection with the destination by considering a transfer of bulk data with an ftp (file transfer protocol) application. We are interested in looking on the congestion avoidance phase, where the window increases by a factor of 0.01 at each reception of a successful ACK and decreases by a factor of $b = 0.875$ when it discovers a loss. The STCP and TCP used in simulations, are based on a Newreno congestion avoidance. The exogenous traffic corresponds to a constant bit rate traffic over UDP transport protocol.

In Table 11.2 we summarize all the simulation and analytic results for the case of $B > B_0$, where the available bandwidth is not limited by the buffer size. The throughput in these tables are in packets/s. The window size and buffer are in packets. λ and μ are in Mbps. Link delays are in ms. The packet size for STCP and the exogenous traffic is 1040 Bytes. Denote by "*ratio*" the ratio between the STCP and TCP Reno throughput. The analytic average throughput in Table 11.2 is given from equations (11.18), (11.19) and (11.28). Simulations in Table 11.2 confirm our analytic results providing globally less than 3% errors.

11.7 Some Extreme Cases

We observed in some simulations, especially for large buffer size and BDP, that many time-outs have occurred. We do not include this in our analysis,

Table 11.2 Simulation results for the normal case $B > B_0$.

	Parameters					thp STCP			thp TCP Reno			ratio	
Case	B	μ_1	μ	λ	τ_1	τ_2	Ana.	Simu.	%err.	Ana.	Simu.	%err.	Ana.	Simu.
1	20	100	50	10	5	50	4669	4544	2.6	3725	3646	2.1	1.25	1.24
2	15	100	50	0	5	50	5789	5636	2.6	4620	4101	11.2	1.25	1.37
3	10	100	50	0	5	50	5756	5597	2.7	4587	4498	1.9	1.25	1.24
4	10	300	50	0	5	70	5737	5266	8.9	4566	4101	10.1	1.25	1.28
5	5	100	50	0	5	10	5814	5711	1.7	4674	4508	3.5	1.24	1.26
6	25	100	50	30	5	50	2367	2325	1.8	1872	1869	0.1	1.24	1.26

which would still hold in MIMD congestion control that can recover from congestion without time-outs.

The other extreme is of very small buffers. In the case of Tahoe, small buffers resulted in losses in the slow-start. In our case, the main change with respect to our previous analysis will be that W_{max} is smaller than equation (11.12). We will estimate the correct value of W_{max} and compute the buffer size B_0 below which it holds assuming $\lambda = 0$. Let W_n be the window size at the nth RTT (round trip time), where W_0 is the initial value just after the previous loss. Then $W_{n+1} = aW_n$ till a loss occurs. In each RTT the queue builds up and then empties. While the queue is nonempty, ACKs are generated at a rate of μ so that packets in the next RTT are generated at a rate of $(1 + a)\mu$. This is larger than the link capacity μ so that the queue builds up at a rate of $a\mu$. The maximum size attained at the nth RTT is then $Q_n = aW_{n-1}$. We say that an early drop occurs if

$$B < a\mu T \tag{11.32}$$

This condition states that starting from an empty, the time $B/(a\mu)$ it takes to fill the queue (at the rate $a\mu$) is shorter than T. It will then take n^*T for the loss to occur where

$$n^* = \max_m\{Q_m < B\} = \max_m\{aW_{m-1} < B\} = \lfloor B/a \rfloor$$

Thus the loss will be detected when

$$W_{max} = \lfloor B/a \rfloor + 1 \tag{11.33}$$

When (11.32) holds then the throughput of STCP may be much lower than the one of TCP Reno. An example is given in Figure 11.11. It gives $W_{max} =$

1360 through simulations whereas (11.12) predicts a value of 2080 (42% error). Equation (11.33) in contrast gives a value of 1515 for which the error is reduced to around 10%. We classify equation (11.33) as an extreme case since for STCP it applies when B is smaller than 1% of the BDP. A more accurate expressions of (11.32) and (11.33) can be also determined.

11.8 Conclusion

We have analyzed in our work the performance of STCP that shares a bottleneck link with an exogenous non-controlled traffic. We presented a fluid analysis which led to an expression of the average throughput. We identified conditions under which the window evolution of STCP behaves as an AIMD protocol (in terms of time). A comparison between STCP and TCP Reno shows the fast link utilization of this former. We validated our model through simulations and we discussed finally some extreme cases of STCP, especially the small buffer case. An estimate condition corresponding to this case were also identified for $\lambda = 0$. This work can be extended by considering multiple STCP sources and their coexistence with other TCP versions.

References

[1] E. Altman, F. Boccara, J. Bolot, P. Nain, P. Brown, D. Collange, and C. Fenzy, Analysis of the TCP/IP flow control mechanism in high-speed wide-area networks, in *Proc. IEEE CDC*, New Orleans, USA, pp. 368–373, December 1995.

[2] S. Floyd, High-speed TCP for large congestion windows. RFC 3649, Experimental, December 2003. Available at www.icir.org/floyd/hstcp.html.

[3] T. Lakshman and U. Madhow, Window-based congestion control for networks with bandwidth-delay products and random loss: a study of TCP/IP performance, in *Proc. HPN'94 (High Performance Networking)*, Grenoble, France, pp. 133–147, June 1994.

[4] C. Jin, D. X. Wei and S. H. Low, Fast TCP: motivation, architecture, algorithms, performance, in *Proc. of IEEE Infocom*, March 2004.

[5] T. Kelly, Scalable TCP improving performance in high-speed wide area networks, in *Proc. First International Workshop on Protocols for Fast Long-Distance Networks*, Geneva, February 2003.

[6] S. McCanne and S. Floyd. ns Network Simulator. Available at http://www.isi.edu/nsnam/ns/.

[7] L. Xu, K. Harfoush and I. Rhee, Binary increase congestion control (BIC) for fast long-distance network, in *Proc. of IEEE Infocom*, March 2004.

[8] G. Vinnicombe, On the stability of networks operating TCP-like congestion control, *Proc. IFAC World Congress*, Barcelona, Spain, 2002.

12

TCP Performance in Case of Bi-Directional Packet Loss

R. E. Kooij[1,2], R. D. van der Mei[3,4] and R. Yang[3,4]

[1]*Department of Planning, Performance and Quality of Service, TNO Information and Communication Technology, Delft, The Netherlands*
[2]*Faculty of Electrical Engineering, Mathematics and Computer Science, Delft University of Technology, Delft, The Netherlands*
[3]*Department of Advanced Communication Networks, CWI, Amsterdam, The Netherlands; e-mail: mei@cwi.nl*
[4]*Department of Mathematics, Faculty of Exact Sciences, Vrije Universiteit, Amsterdam, The Netherlands*

Abstract

Performance modeling of the Transport Control Protocol (TCP) has received a lot of attention over the past few years. The most commonly quoted results are approximate formulas for TCP throughput [1] and document download times [2] which are used for dimensioning of IP networks. However, the existing modeling approaches unanimously assume that packet loss only occurs for packets from the server to the client, whereas in reality the packets in the direction from the client to the server may also be dropped. Our simulations with NS-2 show that this bi-directional packet loss indeed may have a strong impact on TCP performance. Motivated by this, we refine the models in [1, 2] by including bi-directional packet loss, also including correlations between packet loss occurrences. Simulations show that the proposed model leads to strong improvements of the accuracy of the TCP performance predictions.

Keywords: Quality of Service, TCP, download times, response times, correlated packet loss.

D. D. Kouvatsos (ed.), Traffic and Performance Engineering for Heterogeneous Networks, 255–272.

12.1 Introduction

Performance problems in IP networks are well-recognized, and many studies have been investigating performance models for TCP. Pioneering work in this field was done by Mathis et al. [3] and Ott et al. [4], who derive simple square-root formulas for the throughput of infinitely long TCP traffic flows under idealized periodic behaviour of the TCP congestion window, including the impact of the packet loss rate and the round trip time. Padhye et al. [1] propose a refinement of the models in [3, 4] by taking into account detailed packet-level dynamics of the TCP window mechanism, and show that this refined model is able to more accurately predict TCP throughput and is accurate over a wider range of loss rates. However, the Padhye model does not take into account the behaviour of the TCP slow start and fast recovery mechanisms. Based on the model in [1], Cardwell et al. [2] propose a model for the mean total data download time to transfer limited amounts of data, explicitly taking into account TCP slow start and fast recovery. This model is quite accurate in predicting the mean TDT under the assumption that losses happen only in the direction from the server to the client and losses in the successive rounds are independent; here, a round is the period of time between the departure of the first segment of the current window and the arrival of its acknowledgement. A limiting factor of the model in [2] is that it assumes that packet loss only occurs for data packets (i.e. for packets from the server to the client), whereas in reality the packets in the direction from the client to the server (e.g., ACKs) may also be dropped, which can have a significant impact on the TDTs experienced by the end users. Although in the current Internet loss of ACK's is quite rare, it can be anticipated that in the near future loss of ACKs may become more common. The first reason for this is the increase of wireless and ad hoc networks in the access, which exhibit higher loss ratios than wireline networks. A second reason is the still increasing penetration of peer-to-peer applications which, due to their high uplink bandwidth consumption, may also lead to an increase of ACK losses.

In this paper we study the performance of TCP in the presence of bi-directional packet loss. Extensive NS-2 [5] simulations demonstrate that bi-directional packet loss may indeed have a strong impact on the performance of TCP. Motivated by this, we extend the Cardwell model [2] by considering packet losses occurring in both directions. Since in practice packet losses are known to be correlated, we also include a class of correlated loss patterns into the model. Our main focus is on the interactive Web-browsing type of applications. For this type of applications, the two main factors that determine the user-perceived performance of TCP-based applications are the response time, i.e. the time from clocking on a link until the first packet arrives and something appears on the screen, and the total download time, i.e. the time between clocking on a link and the arrival of the last packet. Hence, TDT can be decomposed into two parts: (1) the response time, and (2) the data transfer time. For this reason, our model consists of two sub-models: a sub-model for the response time and a sub-model for the data transfer time. First, we develop a new model for the RT, that gives an approximation for the mean response time as a function the loss probability, the round-trip time (RTT), the timer granularity G and the initial value of the retransmission timer T_0, in the presence of bi-directional packet loss. An important feature of the model is that it includes a dynamic scheme for the Retransmission Timer, which depends on the time granularity and the RTT. Second, we develop a new model for the data transfer time is the presence of bi-directional packet loss. To this end, we use the model in [2], which in turn uses the model in [1], as the basis and extend this to include the possibility of bi-directional packet losses. Third, we evaluate the accuracy of the model extensions by NS-2 simulations for a wide range of parameter settings. The results show that the models are indeed highly accurate for a wide range of parameter settings, and outperform the performance predictions based on [2] in many situations, particularly in situations where the packet loss ratio is significant (typically 4% or more) for sustained periods of time.

TCP performance modeling is notoriously difficult, due to the bursty nature of IP networks. To this end, we need to balance accuracy and complexity: the model to be proposed in this paper is developed in such a way that on the one hand it predicts the performance of TCP quite accurately, while on the other hand the model is still simple enough to provide insight in the impact of the model parameters on the performance. To this end, the model presented in this paper is explicitly built upon the models in [1, 2] extending these models to include the specifics of bi-directional packet loss.

The contribution of this paper is three-fold. First, we demonstrate via simulations that bi-directional packet loss may indeed have a strong impact on the performance of TCP (see for example Figure 12.8). Second, the model for the RT in the case of bi-directional packet loss, including the dynamics of the Retransmission Timer, is new. Third, the extension of the Cardwell model [2] for the DTT by explicitly including the impact of packet losses occurring in both directions, and correlations between successive packet loss occurrences, has not been presented before. As such, the model proposed in this paper, which is explicitly built upon the most widely used models [1, 2] is a significant step forward in understanding the performance of TCP.

The remainder of this paper is organized as follows. In Section 12.2 we discuss the metrics that determine the end-user's perception the Quality of Service of Web browsing applications, viz., the response time (RT) and the transfer download time (TDT). These results are used for the development of simple yet accurate models for the RT and the data transfer time (DTT), which relates to the TDT via equation (12.12) discussed below; these models explicitly capture the impact of bi-directional packet loss. In Section 12.3 we evaluate the accuracy of these models by comparing the performance predictions based on the models with simulations with NS-2. In Section 12.4 we discuss a simple model to include the impact of correlated loss patterns, and evaluate the accuracy of the model by simulations. The results demonstrate that the proposed model significantly outperforms the model in [2]. Finally, in Section 12.5 we address a number of topics for further research.

12.2 Modelling

In [6] extensive research has been to done to investigate the main factors that determine the end user's perception of Quality of Service (QoS) for Web browsing applications. The results of this research shows that the following two factors dominate the user perception of web browsing quality:

1. *Response Time* (RT): time from clicking on a link until first packet arrives and something appears on the screen, and
2. *Total Download Time* (TDT): time between clicking on a link and the arrival of the last packet.

In Section 12.2.1 we develop an analytical expression for the RT, and in Section 12.2.2 we develop a model for the DTT. The models are then combined to obtain an expression for the TDT (see equation (12.12)).

12.2.1 Response Time

In this section we develop a model for the RT, i.e., the time it takes to establish a TCP connection and the additional time it takes to send the first packet containing data. From a user's point of view, the RT is simply the time from clicking on a URL link until the first packet arrives and something appears on the screen. Each TCP connection starts with a three-way handshake, in which the client and server exchange initial sequence numbers. The RT is determined by the time it takes to send four packets successfully; here, the first three packets are related to the three-way handshake while the fourth packet contains the first data.

Failures can be classified as two kinds. The first kind is due to a packet loss from the client to the server. The second kind is caused by a packet loss from the server to the client. For both kinds of failures, packets will be retransmitted from the client side. The forward data packet loss rate is denoted by p_f, and the backward packet loss rate by p_b. Moreover, suppose there are n packet losses before the first packet containing data is received. Let P_n be the probability of receiving the first data packet after exactly n packet losses, and define RT_n to be the conditional mean response time when the first data packet is received after exactly n packet losses.

Lemma 1. *Assuming that the packet loss occurrences are independent, we have: For $n = 0, 1, \ldots$,*

$$P_n = (n + 1)(p_f + (1 - p_f)p_b)^n (1 - p_f)^2 (1 - p_b)^2. \tag{12.1}$$

Proof. It is convenient to define the concept of a cycle. It is simply the RTT if both a packet sent by the client and its ACK are sent successfully. We call this a successful cycle. Based on the assumed independence of packet loss occurrences, the probability that this occurs is $p_s = (1 - p_f)(1 - p_b)$. If either the packet sent by the client is lost or its ACK then the cycle is defined as the time between sending the packet and the time it is retransmitted. This will be denoted as an unsuccessful cycle. Assuming independence, the probability that this occurs is $p_u = p_f + (1 - p_f)p_b$. If exactly n losses occur before the first data is received successfully then this implies that exactly $n + 2$ cycles have passed of which n are unsuccessful and two are successful. Obviously the last cycle needs to be successful. Therefore, for the other successful cycle there are $(n + 1)$ possible locations. Hence, $P_n = (n + 1)p_u^n p_s^2$. Substitution of the values of p_u and p_s yields the result. □

For the situation described in this section, loss is detected through a Retransmission Time Out (RTO). If a packet sent by the client is not acknowledged before the Retransmission Timer expires then the packet is retransmitted. With each retransmission the Retransmission Timer is doubled, up to a maximum value 64 times its original value. Denote by T_0 the initial value of the Retransmission Timer. The simulation package ns-2, see [6], uses $T_0 = 6$ seconds, although RFC2988 [7] recommends $T_0 = 3$ seconds. Note that TCP uses sample values of the RTT to adjust the Retransmission Timer. However, according to Karn's algorithm, this only occurs for packets that are not being retransmitted. Hence, in our situation, such an adjustment can only occur if the first two packets are sent successfully. According to [7], upon its first update the Retransmission Timer becomes

$$T_u = \max \{1, \text{RTT} + \max\{G, 2\text{RTT}\}\}, \tag{12.2}$$

where G denotes the TCP timer granularity. In many TCP implementations G is set to 500 ms. Let T_n be the average value of the Retransmission Timer, given the first data packet is received after exactly n losses. Then based on [1], it can be shown that the conditional mean response time is given by the following expression: for $n \leq 6$,

$$\text{RT}_n = (2^n - 1)T_n + 2\text{RTT}, \tag{12.3}$$

and for $n \geq 7$,

$$\text{RT}_n = [63 + 64(n - 6)] T_n + 2\text{RTT}, \tag{12.4}$$

where

$$T_n = \frac{1}{n + 1} T_u + \frac{n}{n + 1} T_0. \tag{12.5}$$

Equation (12.5) follows from the fact that, given exactly n packet losses, $n + 1$ attempts were needed to successfully transmit a packet. Consequently, with probability $n/(n + 1)$ the Retransmission Timer was set to its default T_0, whereas the timer was set to T_u with probability $1/(n + 1)$.

Lemma 2. *Assuming the packet loss occurrences are independent, the mean response time RT is given by the following expression:*

$$\text{RT} = 2\text{RTT} + \frac{A}{1 - A} \sum_{k=0}^{7} b_k A^k, \tag{12.6}$$

where $A = p_f + (1 - p_f)p_b$, $b_0 = T_0 + T_u$, $b_1 = 3T_0$, $b_2 = 6T_0 + T_u$, $b_3 = 14T_0 + 2T_u$, $b_4 = 32T_0 + 4T_u$, $b_5 = 72T_0 + 8T_u$, $b_6 = 160T_0 + 16T_u$ and $b_7 = -160T_0 - 32T_u$.

Proof. By conditioning on the value of n, it is readily seen from equations (12.3)–(12.5) that RT satisfies

$$\text{RT} = \sum_{n=0}^{\infty} P_n \, \text{RT}_n = \sum_{n=0}^{6} P_n \left[(2^n - 1)T_n + 2\text{RTT} \right]$$

$$+ \sum_{n=7}^{\infty} P_n \left[(63 + 64(n - 6)T_n + 2\text{RTT} \right]. \tag{12.7}$$

After a tedious but straightforward calculation, which was performed with the help of the Computer Algebra software Maple, this expression can be simplified to (12.6). □

12.2.2 Data Transfer Time

The Data Transfer Time (DTT) is the time between sending the first data packet and receiving the last data packet. From the previous section we know the TCP connection establishment time is the time taken to send three packets successfully. Therefore, *we approximate the connection establishment time by 3/4 times the RT*, discussed in Section 12.2.1. In [2], a model is proposed under the assumption that packet loss happens only in the direction from sender to receiver. This model directly depends on this one-way packet loss, whereas in reality not only data segments can be lost during a TCP data transmission but also the ACKs of data packets can be dropped in the direction from the receiver to the sender. Therefore, we extend the Cardwell model by including the impact of the loss of ACKs.

Packet loss may occur in one of the following two cases:

Case 1: Packet is lost during transmission from the sender to the receiver
As a first step, we discuss in which situation a data packet is considered lost by TCP. We focus on a single data packet. We assign to this data packet an index k indicating the position of the data packet within the current window. Obviously, if k is lost during transmission from the sender to the receiver, then this packet is considered lost by the sender TCP. The probability of the occurrence of Case 1 is equal to p_f.

Case 2: Packet is sent successfully, but the ACK is lost
Case 2 occurs when packet k is sent successfully, but the ACK for packet k is lost. The probability of the occurrence of Case 2 is $(1 - p_f)p_b$. For Case 2, we

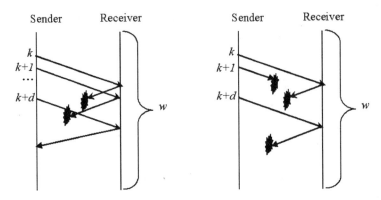

Figure 12.1 Packet k considered sent (left) and lost (right).

cannot determine immediately whether data packet k is considered lost by the sender TCP or not. In fact, if packet $k + 1$ and its ACK are sent successfully then the sender does not notice that the ACK of packet k was lost, hence packet k is not considered lost. If either packet $k + 1$ or its ACK is lost then we have to take into account packet $k + 2$ and its ACK. We have to repeat this analysis until we have considered all packets in the current window. In conclusion, in Case 2, if there exists $m \in \{1, \ldots, w - k\}$, the ACK of packet $k+m$ is received by the sender, then the sender is ensured that packet k is sent successfully. Here, w denoted the window size. For this situation, Figure 12.1 (left) illustrates an example. If for all $m \in \{1, \ldots, w - k\}$ the ACK of packet $k + m$ is not received by the sender, packet k is considered lost by the sender TCP, see Figure 12.1.

Denote p_k as the probability that packet k is considered lost by the sender. Then we have,

$$p_k = \left[p_f + (1 - p_f)p_b\right]^{w-k} . \qquad (12.8)$$

Therefore, in Case 2 the expected probability that a packet is considered lost by the sender can be expressed as follows:

$$\sum_{k=1}^{w} \frac{p_k}{w} = \frac{1}{w} \sum_{k=1}^{w} \left[p_f + (1 - p_f)p_b\right]^{w-k} . \qquad (12.9)$$

Combining the results for Case 1 and Case 2, the complete expression for p, denoting the probability that a packet is considered lost by the sender, can be deduced to:

$$p = p_f + \frac{(1 - p_f)p_b}{w} \sum_{k=1}^{w} \left[p_f + (1 - p_f)p_b \right]^{w-k}$$

$$= p_f + \frac{p_b}{w(1 - p_b)} \left[1 - (p_f + p_b - p_f p_b)^w \right]. \quad (12.10)$$

Expression (12.10) includes the unknown variable w, the current window size. In order to complete the DTT model for bi-directional packet loss we substitute w by the minimum of $W_{\max}$ and the average window size given in [1]. Thus, we approximate w as:

$$w = \min \left\{ W_{\max}, \frac{2 + b}{3b} + \sqrt{\frac{8(1 - p)}{3bp} + \left(\frac{2 + b}{3b} \right)^2} \right\}. \quad (12.11)$$

Our model for the TDT is now complete:

$$\text{TDT} = (3/4)\text{RT} + \text{DTT}(p), \quad (12.12)$$

where RT denotes the mean response time derived in Section 12.2 and DTT(p) is the Cardwell formula for the DTT where we use the packet loss probability p given in (12.10) where w satisfies (12.11). The Cardwell formula for DTT(p), extensively discussed in [2], is briefly described in the Appendix.

12.3 Validation

To assess the accuracy of the models developed in Section 12.2, we have performed extensive simulations. The results are outlined below. The topology of our simulation is depicted in Figure 12.2. In Section 12.3.1 we assess the accuracy of the model for the RT developed in Section 12.2.1, and in Section 12.3.2 we validate the model for the TDT developed in Section 12.2.2.

12.3.1 Response Times

For the validation of the model for the mean RT, the packet loss probability at the aggregation link n1-n2 has been varied between 0–10%. It is assumed that the packet loss is random (i.e. without correlation between consecutive lost

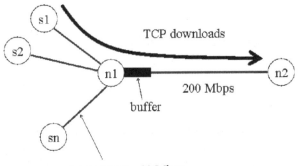

Figure 12.2 Simulation topology.

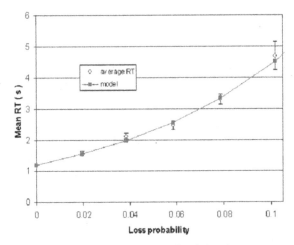

Figure 12.3 Comparison between model and simulation for mean response time.

packets) In our simulation packets in both up- and downlink directions suffer from this packet loss. For ease of the discussion, we assume that the packet loss rate in both directions is the same, i.e. $p_f = p_b$. The (minimum) RTT is set to be 600 ms. The NS-2 script has recorded actual values of the packet loss, RTT and the mean RT. Figure 12.3 shows the mean RT as a function of the loss probability, for the simulations and for the model.

We conclude from Figure 12.3, and many additional simulation results not shown here, that our model for the mean RT is highly accurate over the whole range of loss rates considered. In fact, analytical result for the mean RT always falls within the 95%-confidence interval of the simulated values.

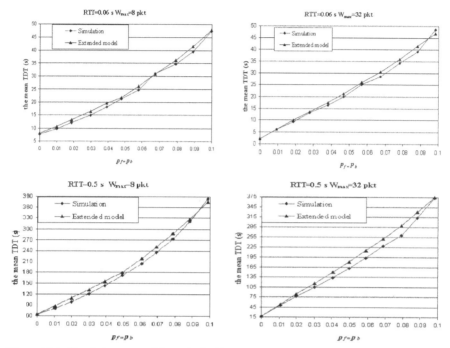

Figure 12.4 Simulation vs. model results for bi-directional packet loss: access link = 30 Mbps.

12.3.2 Total Download Times

In this section we report NS-2 simulation experiments that have been run as a validation for our model for the mean TDT. The simulation topology is given by Figure 12.2. We ran several simulations where we varied the access link rate, the RTT, the maximum window size W_{max} and the file size. For all simulations we have set the number of data packets acknowledged by one ACK equal to 1 while the initial slow-start window size is 1 packet. In Figures 12.4 and 12.5 the Maximum Segment Size (MSS) is 1640 Bytes and the link capacity equals 200 Mbps. The file size was taken to be 1000 packets. The forward loss probability is set equal to the backward loss probability (i.e., $p_f = p_b$), and the delayed acknowledgment option is disabled.

Figures 12.4 and 12.5 and additional simulations show that our model works very well for predicting the mean TDT for large files. The relative errors under these new situations are very low; in most cases they are within $\pm 6\%$. For small files our model is very accurate under the condition that the forward and backward loss probabilities are at most 5%.

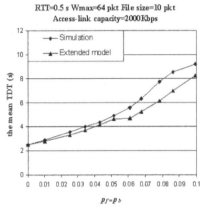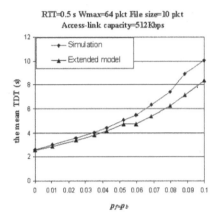

Figure 12.5 Simulation vs. model results for bi-directional packet loss: access link = 2 Mbps (left), access link = 512 kbps (right).

The experiments mentioned above are all under the assumption that the forward loss probability is equal to the backward loss probability. Obviously, in a realistic network environment this is not necessarily the case. Therefore we simulated another scenario in which the loss probabilities in both directions are not necessarily equal. For this experiment we set MSS = 1000 Bytes with a link capacity of 100 Mbps. The files size is set to 500 packets, and the delayed acknowledgment option is disabled. Furthermore, we assume that RTT = 0.3 seconds, access-link capacity = 2 Mbps, $W_{max} = 32$ packets, while the backward loss probability is fixed at $p_b = 2\%$. The forward loss probability p_f is varied between 0–10%.

From Figure 12.6, and similar simulation results not presented here, we conclude that our model is also accurate for predicting the mean TDT under the situation that the forward packet loss probability is unequal to the backward packet loss probability.

12.4 Correlated Packet Loss

According to Boutremans and Le Boudec [8], the correlation structure of the packet loss process can be modeled with an underlying Markov chain. In particular, the two-state Gilbert model was found to be an accurate model in many studies, see for instance [9, 10]. Therefore we use the Gilbert model to simulate packet loss patterns over links. For simplification, we only consider packet losses in the forward direction, thus it is assumed that ACKs are

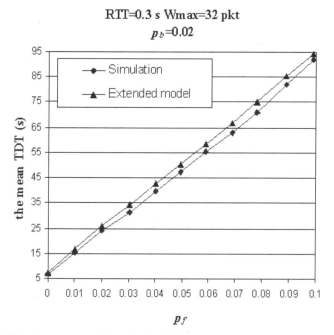

Figure 12.6 Simulation vs. model results for bi-directional packet loss with fixed ACKs loss rate.

not lost i.e. $p_b = 0$, and moreover, the delayed acknowledgment option is disabled.

In the Gilbert model, one state represents a lost packet which is called state B and the other state represents the situation when a packet is successfully delivered to the destination which is referred to as state G. Let m denote the probability of going from state G to state B, and n is that of staying in state B. On the link between the sender and the receiver, the average packet drop rate then satisfies

$$\pi_B = \frac{m}{1 + m - n}. \tag{12.13}$$

Substituting this packet loss probability defined in our model for TDT (replace p_f by π_b), we get the model for predicting the mean TDT in the situation when packet losses are correlated according to the two-state Gilbert loss model. To validate the accuracy of the model for the case of correlated packet loss, we ran a variety of additional NS-2 simulations for the simulation topology depicted in Figure 12.2. For this experiment, MSS = 1000 Bytes while the link capacity was set to 200 Mbps. Furthermore, we assume that

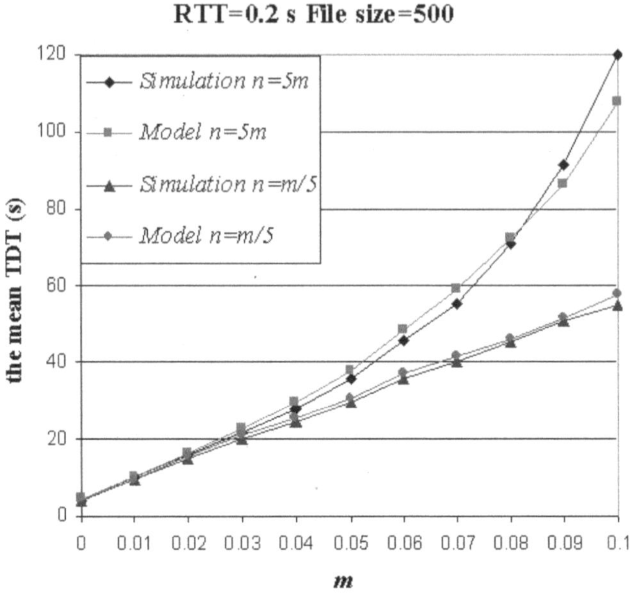

RTT=0.2 s File size=500

Figure 12.7 Simulation vs. model results for correlated packet loss.

RTT = 0.2 seconds, access-link capacity = 10 Mbps, W_{max} = 32 packets. In the forward direction the probability m of going from state G to state B: $\{0, 0.01, \ldots, 0.1\}$, while the probability n of going from state B to state G is in $\{m/5, 5m\}$. Figure 12.7 shows the mean TDT as a function of m.

Figure 12.7 demonstrates that when packet loss on links is correlated and the correlation of packet losses is known, we can apply our model to predict the mean TDT by substituting the packet loss probability p_f by π_B.

To justify the relevance of the inclusion of bi-directional packet loss in the model, we again have performed various simulations. Figure 8 below shows the mean TDT as a function of the packet loss ratio for the following two scenarios. In Scenario 1 (left-hand side), the RTT = 60 ms and the maximum TCP window size 8 packets, while in Scenario 2 (right-hand side) we have the RTT = 500 ms and the maximum window size 32 packets. In both cases, the file size was set to 1000 packets.

The results shown in Figure 12.8 demonstrate that our extended model outperforms the Cardwell model [2], especially when the packet loss becomes significant. Moreover, it shows that the inclusion of bi-directional packet loss is indeed justified, and leads to more accurate performance predictions. As

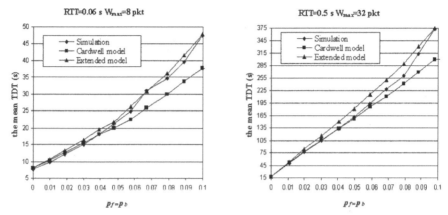

Figure 12.8 Simulation vs. model results for correlated packet loss.

such the model extension discussed in the paper leads to more accurate TCP performance predictions.

12.5 Conclusions and Topics for Further Research

In this paper we have extended the commonly used TCP performance model for the download times by Cardwell [2] by including the impact of bi-directional packet loss. Extensive simulations with NS-2 show that this bi-directional packet loss may have a strong impact on TCP performance, and that the proposed model refinement accurately captures the impact of bi-directional packet loss, and correlations between the loss occurrences. As such the model refinement is highly valuable for IP link dimensioning purposes.

The results lead to a number of challenges for further research. First, the model considered in this paper is focused on TCP performance over a single network domain. However, next-generation communication services (e.g., online consumer services, E-commerce applications) will typically have highly distributed architecture, crossing multiple administrative domains. Extension of the model towards the inclusion of multiple domains is a challenging topic for further research. Second, in the present paper we considered a simple correlation structures between packet loss occurrences, whereas in reality the loss patterns may be far more complicated. Extension of the model including more realistic packet loss patterns is a challenging area for further research. Finally, in many communication networks TCP-based applications and UDP-based applications will be integrated. Inclusion of the impact of UDP streams on the performance of TCP is also a challenging area for further research.

Appendix: The Cardwell Formula

Below we briefly describe the Cardwell formula for the Data Transfer Time (DTT), considered as a function of the packet loss rate. The reader is referred to [2] for a more in-depth discussion. Adopting the terminology introduced in [2], the Cardwell formula for the DTT can be decomposed as follows:

$$DTT(p) = E[T_{ss}] + E[T_{loss}] + E[T_{ca}] + E[T_{delack}], \quad (12.14)$$

where T_{ss} is the latency for the initial slow start, T_{loss} is the delay due to any Retransmission Time Out (RTO) or fast recovery that happens at the end of the initial slow-start, T_{ca} is the time to send the remaining data (i.e., the amount of data after slow start or any following loss recovery), and T_{delack} is the delay between the reception of a single segment and the delayed ACK for that segment. Expressions for each of the four terms in (12.14) are detailed out below.

First, $E[T_{ss}]$ is approximated by the following expression:

$$E[T_{ss}] = \begin{cases} RTT\left[\log_\gamma\left(\frac{W_{max}}{w_1}\right) + 1 + \\ \quad + \frac{1}{W_{max}}\left(E[d_{ss}] - \frac{\gamma W_{max} - w_1}{\gamma - 1}\right)\right] & \text{when } E[W_{ss}] > W_{max} \\ \\ RTT\log_\gamma\left(\frac{E[d_{ss}](\gamma - 1)}{w_1}\right) & \text{otherwise,} \end{cases}$$

$$(12.15)$$

where $E[d_{ss}]$, the number of data segments we expect the sender to send before losing a segment, is given by the following expression:

$$E[d_{ss}] = \frac{(1 - (1 - p)^d)(1 - p)}{p} + 1, \quad (12.16)$$

with p the data segment loss rate and d the number of data segments to be transmitted. Moreover, W_{max} is the maximum window size, γ is the rate of exponential growth of the congestion window during slow start, and w_1 is the number of segments in the initial congestion window.

Second, $E[T_{loss}]$ can be expressed as follows:

$$E[T_{loss}] = l_{ss}\left(Q(p, E[W_{ss}])E[Z^{TO}] + (1 - Q(p, E[W_{ss}]))RTT\right), \quad (12.17)$$

where

$$l_{ss} = 1 - (1 - p)^d \tag{12.18}$$

is the probability that initial slow start phase ends with the detection of a packet loss, which can occur either via retransmission timeouts (RTOs) or triple duplicate ACKs. The probability that a sender in congestion avoidance will detect a packet loss with an RTO can be expressed in terms of the the packet loss rate p and the window size w as follows (cf. [1]):

$$Q(p, w) = \min\left(1, \frac{1 + (1 - p)^3(1 - (1 - p)^{w-3}))}{(1 - (1 - p)^w)/1 - (1 - p)^3)}\right). \tag{12.19}$$

Moreover, the expected cost of an RTO, $E[Z^{TO}]$ is given by (cf. [1]):

$$E[Z^{TO}] = \frac{G(p)T_0}{1 - p}, \tag{12.20}$$

where T_0 is the average duration of the first timeout in a sequence of one or more successive timeouts, and $G(p)$ is given by

$$G(p) = 1 + p + 2p^2 + 4p^3 + 8p^4 + 16p^5 + 32p^6. \tag{12.21}$$

Third, $E[T_{ca}]$ is given by

$$E[T_{ca}] = \frac{E[d_{ca}]}{R(p, \text{RTT}, T_0, W_{max})}, \tag{12.22}$$

where

$$E[d_{ca}] = d - E[d_{ss}] \tag{12.23}$$

is the mean number of data segments left after slow start and any following loss recovery, and

$$R = \begin{cases} \dfrac{\frac{1-p}{p} + \frac{w}{2} + Q(p, w)}{\text{RTT}\left(\frac{b}{2}w + 1\right) + \frac{Q(p,w)G(p)T_0}{1-p}} & \text{if } w < W_{max} \\[4ex] \dfrac{\frac{1-p}{p} + \frac{W_{max}}{2} + Q(p, W_{max})}{\text{RTT}\left(\frac{b}{8}W_{max} + \frac{1-p}{pW_{max}} + 2\right) + \frac{Q(p,W_{max})G(p)T_0}{1-p}} & \text{otherwise} \end{cases} \tag{12.24}$$

and where w is given by equation (12.11). Finally, $E[T_{delay}]$, the expected delay between the reception of a single segment and the delayed ACK for that segment is Operating System specific, and typically take a value in the range of 100–150 milliseconds.

References

[1] J. Padhye, V. Firoiu, D. Towsley and J. Kurose, Modeling TCP throughput: A simple model and its empirical validation, *IEEE/ACM Transactions on Networking*, vol. 8, pp. 133–145, 2000.

[2] N. Cardwell, S. Savage and T. Anderson, Modeling TCP latency, in *Proceedings of INFOCOM 2000*, Tel Aviv, Israel, 2000.

[3] M. Mathis, J. Semke and J. Mahdavi, The macroscopic behavior of the TCP congestion avoidance algorithm, *Computer Communication Review*, vol. 27, pp. 67–82, 1997.

[4] T. Ott, J. Kemperman and M. Mathis, The stationary behavior of ideal TCP congestion avoidance. Preprint.

[5] Network Simulator 2 (software), http://www.isi.edu/nsnam/ns.

[6] J. Beerends, S. van der Gaast and O. K. Ahmed, Web browse quality modeling. White contribution COM12-C3 to ITU-T Study Group 12, November 2004.

[7] V. Paxson and M. Allman, Computing TCP's retransmission timer, Internet Archives RFC 2988, November 2000.

[8] C. Boutremans and J. Y. Le Boudec, Adaptive joint playout buffer and FEC adjustement for internet telephony, in *Proceedings of IEEE Infocom*, pp. 652–662, 2003.

[9] E. Altman, K. E. Avrachenkov and C. Barakat, TCP in presence of bursty losses, in *Proceedings ACM Sigmetrics*, Santa Clara, CA, pp. 124–133, 2000.

[10] E. Altman, K. E. Avrachenkov, C. Barakat and P. Dube, TCP over a multi-state Markovian path, in *Performance and QoS in Next Generation Networking*, K. Goto, T. Hasegawa, H. Takagi and Y. Takahashi (Eds.), Springer, pp. 103-122, 2000.

13

Access Network Buffer Dimensioning for Bulk TCP Traffic

Zlatka Avramova[1], Danny De Vleeschauwer[1,2],
Sabine Wittevrongel[1] and Herwig Bruneel[1]

[1]*SMACS Research Group, TELIN Department, Ghent University,*
Sint-Pietersnieuwstraat 41, B-9000 Gent, Belgium;
e-mail: {kayzlat, sw, hb}@telin.ugent.be
[2]*Network Strategy Group, Alcatel-Lucent Bell Copernicuslaan 50,*
B-2018 Antwerpen, Belgium; e-mail: danny.de_vleeschauwer@alcatel-lucent.be

Abstract

It is largely recognised in the research community that access and edge network buffers need special attention with respect to dimensioning, since those are often attached to bottleneck links in contrast with the core Internet network. The quality of experience of the end user – as a result of the performance of its applications – will depend on the adequate construction of these buffers. The capacity of the access networks keeps increasing. However, there is recently a well-defined share in the network traffic mix of more and more (audio and video) streamed applications. Also on top of traditional web applications, the bandwidth-greedy peer-to-peer elastic (TCP-controlled) traffic is growing. In this paper we consider a best-effort queue with TCP traffic, through which no stringent quality guarantee can be provided to streamed multimedia (non-responsive and time-critical) applications; hence a different buffer is recommended for the real-time applications but this is subject to a different study (e.g., [27]). Given that the last-mile connection is not high-speed, the maximised utilisation of its (download) capacity is an important issue, as well as that each flow gets its fair share. Long-lived, also called

D. D. Kouvatsos (ed.), Traffic and Performance Engineering for Heterogeneous Networks, 273–296.

"persistent" or bulk traffic flows, create the heaviest buffer demand (and not the short-lived (web) flows). We develop a heuristic formula to calculate the required buffer size, such that both requirements can be fulfilled by choosing the buffer large enough so that there is no packet loss incurred by it. Unfortunately, this buffer size may turn out to be too large to be implemented in state-of-the-art access equipment. So, we also show how efficiency and fairness requirements are met in a smaller buffer in case the buffer acceptance discipline is Drop-tail or RED. We show that Drop-tail cannot achieve fairness and that RED, if appropriately dimensioned, can (and we also present a methodology for its configuration).

Keywords: Persistent TCP flows, DSL access networks, buffer dimensioning.

Abbrevations

ADT	ADapTive sizing of buffers
AIMD	Additive Increase, Multiplicative Decrease
BDP	Bandwidth-Delay Product
BE	Best Effort
BRAS	BRoadband Access Server
BSCL	Buffer Sizing for Congested Links
CA	Congestion Avoidance
DSL	Digital Subscriber Line
DSLAM	DSL Access Multiplexer
NDVR	Networked Digital Video Recorder
QoE	Quality of Experience
QoS	Quality of Service
RED	Random Early Drop
RTT	Round Trip Time
TCP	Transport Control Protocol
VoD	Video on Demand
VoIP	Voice over IP

13.1 Introduction

For a long time the TCP (Transport Control Protocol) buffers in network nodes, such as routers, have been dimensioned according to the "rule of thumb" found by Villamizar and Song [5] back in the mid-nineties, which postulates that buffers should be set to the bandwidth-delay product (BDP) of the (bottleneck) link (in fact this rule has been suggested before them by others). In the late nineties, Morris [13, 14] showed that this rule is not universally applicable. More recently two new buffer dimensioning models have been put forward, referred to as the "Stanford" model [6] and the "Georgia Tech" model [2]. They are somewhat contradictory, leading to different conclusions and solutions, but this can be partially justified by the fact that the approaches and the scopes of their applicability are different. These models deal with buffers through which pass flows of the order of hundreds or thousands.

In reality, it is usually the access or edge network that proves to be the TCP bottleneck and on the access link less flows run concurrently. Little is known to determine the buffer in the case of a few competing flows. In this paper we present our formula for access network buffer dimensioning. It not only allows to calculate the buffer size required to ensure that there is no packet loss, that the link is efficiently used and that each flow gets its fair share, but also calculates the expected throughputs of the separate flows, based on the propagation delay and the advertised window (also referred to as "TCP socket buffer"). The formula was validated by simulations run on the ns-2 simulator [21], a common tool for study in the network research community.

This paper is organised as follows. The next section gives an overview of related work. In Section 13.3 we describe the set-up, i.e., the network topology we consider and the settings for our studies. Section 13.4 presents the heuristic formula we developed for buffer dimensioning without packet loss. In Section 13.5 we consider the case when the buffer is set to a value smaller than the value calculated in Section 13.4, for a Drop-tail and RED (Random Early Drop) queue-management algorithm respectively. In the last section the conclusions of our study are presented.

13.2 Related Work

In this section we give an overview of the existing models and their approaches to the problem and we place our work in this framework. All these models pertain to the case where long-lived TCP flows, reaching the CA (Congestion Avoidance) phase, account for the larger part of the buffer need.

According to the "rule of thumb" [5] discovered back in the mid-nineties, a buffer should be dimensioned to the product of its capacity and the *average* RTT (Round Trip Time) of the flows passing through it (the BDP rule). The basic idea behind it is that the TCP flows behaviour is governed mainly by the AIMD (Additive Increase, Multiplicative Decrease) algorithm in CA, i.e., the flows increase their throughput linearly until they sense congestion, upon which they halve their throughput. In order to still attain full link utilisation after congestion, the sum of their halved congested windows should come up to the BDP of the link. Therefore, their congestion windows will double in most cases until a new congestion event occurs. The excess of data "in transit" above the capacity of the link must be stored in a buffer and this reasoning leads to the rule. As its name suggests, it is just a rule of thumb, not a refined and fundamental method to calculate precise buffer requirement in any case. Applying it to core routers is difficult as the hundreds or thousands of flows passing through such a router can have very different RTTs (from some ms to a second or even more). Therefore, the conception of the *average* RTT can be very inaccurate.

In the late nineties Morris [13, 14] showed some shortcomings of this rule of thumb. The rule was concerned mainly with link utilisation, but Morris discovered that when the number of flows increases, an unacceptable amount of packet loss is observed (more than 3–5%) as some flows monopolise the buffer space (he particularly noticed that effect when the number of flows exceeds the number of places in the buffer). Therefore, as he came to the conclusion that a flow needs to have a congestion window of at least 6 packets to work properly and to recover quickly, he suggested that the buffer be dimensioned proportionally to the number of flows by a multiplicative factor of 6. He was the first one to put forward the idea of buffer sizing proportionally to the number of flows, which was further developed in the "Georgia Tech" model [2], and which lies in the basis of our model too.

More recently two new models were put forward [20]. First, the main assumption of "Standford" model [6] is that for a large number of flows de-synchronisation (between flows) in packet loss leads to the fact that the aggregate window follows a Gaussian distribution of which the standard deviation decreases as the number of flows grows. It results in a square root of the number of flows in the denominator of the BDP. Secondly, the "Georgia Tech" model [2] assumes that the loss or utilisation constraint is the leading one for dimensioning the buffer depending on the number of flows, and thus two corresponding formulas are presented (a simple queueing delay constraint is also considered).

Both models rely on some statistical properties of a large number of flows, and hence, apply mainly to high-speed link routers. Some of their underlying assumptions do not apply to access networks. Our model addresses exactly this problem and is in fact the generalisation of the loss-constraint formula of the Georgia Tech model.

As indicated above, in [2] Dhamdhere et al. present several formulas to dimension a minimum required (bottleneck) buffer with respect to a link utilisation, or buffer-incurred loss rate or queueing delay constraint. Those formulas are summarised under a Buffer Sizing for Congested Links (BSCL) model, which according to the authors is more applicable to access/edge networks than to core networks. We note however that the BSCL model is still for a large number of flows, which is not the case for an access residential link, for instance. In the model, distinction is made between flows that are bottlenecked in that specific network node or elsewhere in the network. Both the Georgia Tech and the Standford models rely on a metric accounting for the number of active (and bottlenecked) flows. This generally would require to keep a per-flow status list, or to use other more elegant tracking and estimating techniques, e.g., the mechanism proposed in [16] (used also for flow admission control to ensure a given quality of service, QoS level). The results of [2] apply as long as 80–90% of the running (TCP) connections are long-lived (persistent) flows. We will elaborate further on the three considered constraints.

Under the utilisation constraint, for a single flow bottlenecked in the given buffer the authors arrive simply at the BDP formula. This holds in cases when the flow reacts to a loss by halving the congestion window (and not by further reducing it at a congestion event), which is likely to be the case if the SACK option of TCP is used. For a heterogeneous flows scenario, two cases are considered: a global synchronisation of the losses between the concurrent connections and a partial de-synchronisation when also the burst-length is accounted for. For the global synchronisation case, the authors arrive at a variant of the BDP formula where the harmonic mean of the flows' RTTs is taken. The harmonic mean is always smaller than the arithmetic one, and this comes up to the fact that flows with smaller RTT bring along higher buffer requirement (an observation we also arrive at later in this paper). For the case of partial de-synchronisation, a factor q is introduced (for which a probabilistic model is needed), and this factor is 1 if all flows are synchronised (all flows experience loss at the same time). When the flows are de-synchronised (i.e., $q < 1$), a fraction q of them will reduce the congestion window by half at a congestion event, while the rest will augment it further by one packet. Thus, the backlog needed to keep the link saturated (effectively utilised) at congestion is smaller, which leads to a smaller buffer requirement in the de-synchronisation scenario. This phenomenon is also accounted for in the

Standford model for high-speed (core links) Internet buffers, even if it may need further refinement.

The loss-constraint formula can be viewed as a variant of our heuristic formula for a lossless buffer presented below. Instead of taking an averaged congestion window of K packets and multiply it by the number of flows, in the no loss scenario the advertised windows of all the flows are summed up. The acceptable loss rate, beyond which TCP goodput and throughput are reduced (because a multitude of congestion events and of timeouts), is 1–2%, also according to other authors (e.g., [26]).

A delay constraint is subject to the following trade-off between delay and loss: on the one hand, if the buffer is small, the RTTs of the flows are not increased significantly and the flows are more responsive, also having a chance to complete quickly (having small RTT) provided that they see no high loss rate; on the other hand, the loss rate is reduced with larger buffers but the introduced queueing delay and jitter have a negative effect on transport efficiency. Moreover, if some time-critical applications are transported through the same buffer, this clearly poses well-defined delay constraints.

Finally, Dhamdhere et al. [2] suggest to take the more stringent constraint of the three described above for buffer dimensioning. When the number of flows is smaller than a given number N_b, the aim is to fill in the link and the utilisation constraint formula is selected; else, if the number of flows exceeds N_b, the loss should not become more than acceptable and the loss-constraint formula is taken into account. To keep the loss constraint, the queueing delay will be growing proportionally with the number of bottlenecked active flows. The buffer size, B must be sufficiently large so that every flow can have a window of K packets stored either in it or elsewhere in the network (accounted for by the BDP factor in the formula).

In [1], Dhamdhere and Dovrolis confront their proposed buffer dimensioning methodology to the proposed Standford model. They argue that whether a small or large buffer will be selected depends on the performance objective. For high-speed links, which can hardly become saturated, small buffers could be a good solution to accommodate bursts, but this must be applied cautiously. While "mice", short-lived flows will take advantage of the small delay incurred and complete quickly, the persistent flows, the bulk TCP traffic can degrade significantly in performance, as demonstrated in the paper.

The letter [26] by the Standford researchers tries to reconcile the contradictory buffer models proposed so far and is an indirect answer to (the critics of) [1]. Additionally, it considers the "tiny buffer rule", proposed in [7, 10], where buffer dimensioning is based on a logarithmic function of the maximum congestion window of the flows passing through a buffer. Ganjali and McKoewn [26] admit that the proposed buffer sizes for a given example can differ by several orders of magnitude! However, the key observation is that the different rules apply to different parts of the network and under different constraints. They provide an experimental/test proof that their rule allows to keep good performance: full link utilisation and no packet drops. However, they stay cautious since this rule may keep a congested link fully utilised and with small incurred delays, but utilisation is not necessarily a user-friendly metric ensuring quick flow-completion times if too many drops occur. Moreover, de-synchronisation is not well-defined and the number of flows is also a difficult assessment task. Considering that in core routers the number of flows is of the order of tens of thousands, reducing the buffer by two orders of magnitude with respect to the BDP could be proposed. However, the authors recommend reducing it only by one order of magnitude since their model needs further research work on the effect in operating routers.

The test proof presented in [26] is not accepted by Vu-Brugier et al. [8]. The experiment cited in [26] demonstrates that there have been no losses for the buffers above 2.5 ms during the time of an experiment of approximately a week, while at the same time the Standford model relies on de-synchronisation between concurrent connections in experienced losses.

Ganjali and McKoewn [26] agree that their "small buffer rule" will probably not hold for edge and access networks and as proved in [1] larger buffers will be needed in the peripheral parts of the network.

Finally, the "tiny buffer rule" by Raina and Wischik [7] and Enachescu et al. [10] is suggested to be applicable for optical buffers (of the order of 20–50 packets) since sacrificing a small amount of link capacity there (10–15%) is not an issue in such very high-bandwidth links. Also, it is believed that the ingress traffic to those, coming from the access links, is naturally smoothed and does not expose much burstiness.

In [8], Vu-Brugier et al. suggest to apply an adaptive sizing of buffers, ADT (see [17]), that maximises the aggregate flow's throughput up to a given utilisation factor of the output link. Conducting tests with it on a 10 Mb/s link (along with other tests) with long-lived connections, they prove that with growing number of flows the buffer demand decreases. However, we believe that for a small number of flows the graph is not enough relevant. First, there is an initial increase in the buffer requirement with growing number of flows (for a Drop-tail buffer) but this graph depicts buffer demand only from the moment a (partial) de-synchronisation occurs. This validates the conclusions so far: during periods of light utilisation larger buffers are needed and vice versa. The authors also explore the contribution in buffer demand from short-lived (web) TCP flows. They give an example of approximately 500 web sessions running with five persistent flows which leads to more than doubling the required buffer for only five persistent TCP connections.

Although the link is still effectively used during heavy load periods of the test run, a high loss rate is observed, which according to the authors is due to the reduced buffer still maximising the utilisation of the link. They conclude that although it may be tempting to choose a small buffer and to sacrifice some throughput at light congestion periods, under heavy congestion conditions those could significantly degrade the performance of some applications and hence reduce the user quality of experience (satisfaction).

A distinction is made between high-speed core Internet networks on the one side, where losses rarely occur and buffering is hardly needed, and access network buffers on the other, which are supposed to be bottlenecks and therefore need special attention. An experiment is conducted where the egress link's capacity is reduced from 100 to 1 Mb/s to emulate an access/edge network link and their conclusion that the Standford model is not suitable is expected. When capping the capacity of the egress link is removed, thus emulating core links, no queueing is needed anymore and the TCP flows behaviour is insensitive to the buffer size employed. This once again stresses the need to dimension properly the buffers in the access networks.

A more different approach on buffer sizing is taken by Prasad et al. [15]. Instead of considering persistent TCP flows as a starting point for buffer provisioning rules, the authors consider TCP flows with heavy-tailed size distribution. Also, the performance metrics for the dimensioned buffer are not link/buffer specific, such as link utilisation and drop rate, but the maximisation of the flow's throughput. A new line is drawn between small (short-lived, "mice") and persistent flows. Depending on the network conditions and settings (including buffer size and occupancy), some "persistent" flows may complete even without going out of the Slow Start TCP phase and reaching the CA phase. Thus, whether a TCP flow can be classified as persistent depends not only on its size, but must be considered coupled with buffer size (at least). As a consequence, some of the "persistent" TCP flows do not account much for the buffer demand, in the same way as short-lived (web) flows do (a widely recognised fact). In order to properly dimension a buffer not only its placement within the network is important (discussed above), but also the type of elastic flows feeding it. The authors develop two models: the S and L models, respectively for connections of not heavy-tailed or of heavy-tailed

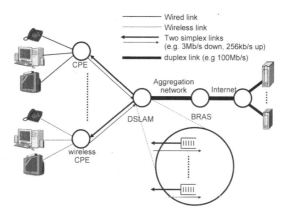

Figure 13.1 Access network and buffers under study.

size distribution. Also, the models proposed account for the output/input link capacity ratio. In case this is smaller than 1, significant buffering is needed since the loss rate can be approximated as a power law function of the buffer size (especially for the L-model). This is the case of edge/access links. On the other hand, when this ratio is larger than or close to 1, hardly any buffer is needed since the loss probability drops almost exponentially with the buffer size. Therefore, the case with the ratio smaller than 1 is interesting for both the S and L models, and an optimal buffer maximising throughput can be found. For the L-model, the authors find the optimal buffer to be around 0.83 of the BDP, assuming the decay power law constant 0.5. However, in the formula, there is a parameter accounting for the RTT of the flow (which needs better definition).

13.3 Set-up

We consider a DSL (Digital Subscriber Line) network for the sake of example of an access last-mile network in our study. However, the dimensioning formula we derive is general and is not limited in application to a specific peripheral network.

Figure 13.1 depicts a typical DSL access scenario. The DSLAM (DSL Access Multiplexer) terminates some hundreds of copper pairs. The BRAS (BRoadband Access Server) serves tens of DSLAMs and acts as an edge router to the Internet.

While initially web browsing spurred the popularity of the Internet, nowadays more and more applications are transported over the Internet: VoIP (Voice over IP), streamed video, VoD (Video on Demand), peer-to-peer file transfers, etc. Streaming services, generating non-responsive (non-elastic) traffic, will have their own dedicated queues and dimensioning these queues is a separate problem.

In this paper we concentrate on applications that generate long-lived TCP-controlled flows that are transported over the BE (Best Effort) service class, e.g., peer-to-peer file transfers, mails with large attachments, time-shifted TV where the content is downloaded during off-peak-hours in a NDVR (Networked Digital Video Recorder), and so on. These flows put the heaviest demand on the BE buffers in the access network.

We use the NewReno and SACK TCP flavours, which are commonly used nowadays, and a packet size of 1500 bytes [3] (also used in [2, 6, 8, 15]). The sources of the information are on some servers or peer nodes and the DSL user downloads this information via a TCP connection.

We assume that neither the DSL uplink nor the aggregation link between the DSLAM and the BRAS are the bottleneck. The former is justified by the fact that in our considered scenarios the uplink only needs to carry the TCP acknowledgements and in most cases this generates a traffic flow that can easily be supported by the uplink. For the latter we assume that the service access provider should have dimensioned its aggregation network between the DSLAM and the BRAS to avoid bottlenecks there, otherwise the user will not be able to use its access link efficiently. We model the rest of the network (outside the access network) mainly as links and servers introducing a scenario-specific delay, but these links have large enough bandwidth and are never bottlenecks. In the rest of the paper we refer to "propagation delay" as the sum of all delays accumulated outside the bottleneck buffer under study (i.e., in the links and the buffers, downstream and upstream).

For the purposes of our study we assume that the customer premises network is a wired network, otherwise if wireless, the focus of our study should have been shifted.

As performance parameter we consider the average throughput y_n (expressed in packet/s or Mb/s) of the n-th persistent flow. There is a well-known relation between the average throughput, the average TCP congestion window size w_n (expressed in packets or bits) and the average RTT R_n (expressed in s) of the n-th flow:

$$y_n = \frac{w_n}{R_n}, \quad 1 \leq n \leq N, \tag{13.1}$$

where N is the number of competing long-lived flows. We will try to choose the downstream buffer size B such that the link (of downstream link rate c) is efficiently used:

$$\sum_{n=1}^{N} y_n = \rho c, \tag{13.2}$$

with ρ as close to 1 as possible, under the restriction that TCP should give every flow a fair share. Our definition of a fair share of two flows is that the ratios of the average throughputs should be inversely proportional to the ratios of the RTTs:

$$\frac{y_n}{y_m} = \frac{R_m}{R_n}. \tag{13.3}$$

13.4 Buffer Dimensioning for No Packet Loss

In this section we answer the question how large the buffer must be so that there is no packet loss incurred from it. In such a case, the persistent TCP sources open their window to the maximum value: $w_n = a_n$, with a_n the receiver-announced window (TCP socket buffer, advertised window). The average RTT of a flow is then given by

$$R_n = d_n + \frac{\bar{Q}}{c}, \tag{13.4}$$

where $\bar{Q}$ is the average buffer content and d_n is twice the propagation delay (including the access network). We consider the propagation delay d_n to be constant as we assumed that there are no bottlenecks outside the access network and hence more queues to account for in equation (13.4).

Based on our requirement that the link should be fully exploited (see equation (13.2)), we have that

$$\sum_{n=1}^{N} \frac{a_n}{d_n + \frac{\bar{Q}}{c}} = c. \tag{13.5}$$

As the left-hand side is a decreasing function of $\bar{Q}$ and the sum of all a_n/d_n is assumed to be larger than c, this is an implicit relation that determines the (average) buffer occupancy $\bar{Q}$ uniquely. This implicit equation can easily be solved numerically to give a value $\bar{Q}_0$. If we choose the buffer size just larger than this value, the buffer occupancy in steady state only fluctuates gently (as the senders do not change their window), and hence, this is the minimum buffer size B we need to provision. The throughput for a given flow can be calculated by

$$y_n = \frac{a_n}{d_n + \frac{\bar{Q}_0}{c}}, \tag{13.6}$$

and the RTT can be calculated by equation (13.4). According to our definition of equation (13.3) each flow gets its fair share provided all announced windows a_n are the same.

Notice that by using equation (13.5) and the fact that the sum of all throughputs amounts to the capacity c, we have that

$$\sum_{n=1}^{N} a_n - \sum_{n=1}^{N} d_n y_n = \bar{Q}_0. \tag{13.7}$$

This relation has the following interpretation: the buffer is the sum of all advertised windows minus the information that is in transit ("on the flight"), defined by the capacity of the access network and the delays.

Table 13.1 and Figure 13.2 show some example scenarios. Three cases are studied in Table 13.1 with the calculation results for RTTs, throughputs and buffer size in the left column of the table and with results from simulation run throughputs and queue evolution found in the right column. The smaller the value of d_n, the (proportionally) larger the flow's throughput is, but this leads also to higher buffer requirement (see also Figure 13.2). It can be noticed in the simulation results that initially the buffer and RTTs grow, but that the throughputs converge to the predicted value in steady state. Remark that if the advertised windows a_n are larger, the transient takes longer (see e.g. case 3).

We can consider the required buffer space $\bar{Q}_0(a_n, d_n)$ as a function of the window sizes a_n and the propagation delays d_n. The first derivative of this function to d_n is negative and shows the inverse relation between the required queue length and the propagation delay. Therefore, the smaller the propagation delay, the greater is the buffer need (a similar conclusion as in [2]). On the other hand, the first derivative of this function to a_n is positive, which means that large advertised windows are not desirable from a buffer dimensioning point of view.

The sensitivity of the buffer to the propagation delay d_n and the advertised windows a_n is illustrated in Figure 13.3. We fix one of the two parameters and vary the other till eight-fold of its smallest value X (which are $a_n = 45$ packets, $d_n = 25$ ms). It can be seen that the buffer demand is much more sensitive

Table 13.1 Required buffer space as calculated with equation (13.7) (shaded cells) and the buffer evolution (right) for the case of three FTP flows filling up a link of 3 Mb/s for several combinations of a_n and d_n.

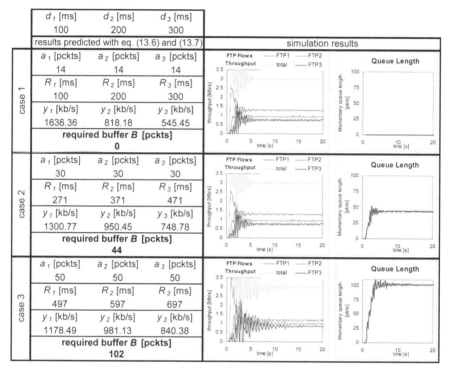

	d_1 [ms]	d_2 [ms]	d_3 [ms]
	100	200	300
	results predicted with eq. (13.6) and (13.7)		

case 1	a_1 [pckts]	a_2 [pckts]	a_3 [pckts]
	14	14	14
	R_1 [ms]	R_2 [ms]	R_3 [ms]
	100	200	300
	y_1 [kb/s]	y_2 [kb/s]	y_3 [kb/s]
	1636.36	818.18	545.45
	required buffer B [pckts]		
	0		
case 2	a_1 [pckts]	a_2 [pckts]	a_3 [pckts]
	30	30	30
	R_1 [ms]	R_2 [ms]	R_3 [ms]
	271	371	471
	y_1 [kb/s]	y_2 [kb/s]	y_3 [kb/s]
	1300.77	950.45	748.78
	required buffer B [pckts]		
	44		
case 3	a_1 [pckts]	a_2 [pckts]	a_3 [pckts]
	50	50	50
	R_1 [ms]	R_2 [ms]	R_3 [ms]
	497	597	697
	y_1 [kb/s]	y_2 [kb/s]	y_3 [kb/s]
	1178.49	981.13	840.38
	required buffer B [pckts]		
	102		

to the advertised windows values than to the propagation delays. Figure 13.2 shows typical buffer sizes needed in realistic situations. From this figure it can be seen again that the buffer demand increases slightly when the propagation delay d_n decreases, but increases dramatically when the announced window a_n increases. The lines in Figure 13.2 pertain to the homogeneous cases, while the dots in between show results for two heterogeneous cases: propagation delays of the flows respectively of 50, 100, 200 ms (upper circles) and 100, 200, 250 ms (lower circles).

If the *window scaling* option is not implemented, the maximum window is 65536 bytes (64 kB) or 45 packets (for TCP packets of 1500 bytes carrying a payload of 1460 byte [3]). However, for high-speed links and large RTTs, equation (13.6) (with the buffer size set to 0) shows that with this value, it is likely that the link cannot be filled by one flow. Therefore, on high-speed

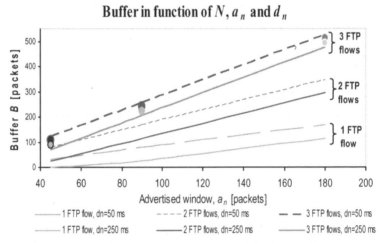

Figure 13.2 Buffer demand for several homogeneous (straight lines) and heterogeneous (dots) cases (upper circles: $d_n = 50$; 100; 200 ms; lower circles: $d_n = 100$; 200; 250 ms).

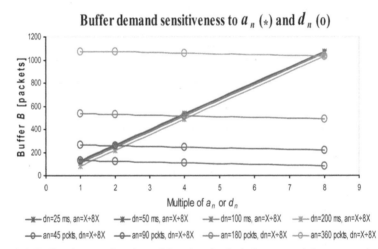

Figure 13.3 Buffer size as a function of the advertised window a_n and the propagation delay d_n (for the homogeneous case of three flows).

links, the window scaling option is often enabled. This makes that the required buffer size as calculated by equation (13.7) is often larger than typical buffer sizes in modern state-of-the-art access line equipment. Therefore, we investigate in the next sections if we can achieve the aims of efficiency (see equation (13.2)) and fairness (see equation (13.3)) with a smaller buffer.

13.5 Smaller Buffer Size Than Required for a Lossless Operation

In the previous section we defined a formula allowing to set the buffer size such that no losses would occur (from buffer overflow) at given network conditions. Unfortunately, as we saw from the examples and the graphs, sometimes the required buffer is too large for implementation or the dynamic network conditions may turn it to be too small. Therefore, we pose the question what happens if a buffer (at a bottleneck) is limited so that it incurs losses on the flows passing through it.

If the buffer is smaller than the value calculated by equation (13.7), the TCP sources will experience packet loss. According to [9] the precise average throughput, taking into account timeouts, is given by

$$y_n = \frac{\alpha}{R_n \sqrt{\frac{2bp_n}{3}} + T_o \min(1, 3\sqrt{\frac{3bp_n}{8}}) p_n (1 + 32 p_n^2)}, \qquad (13.8)$$

where p_n is the packet loss seen by the n-th flow, T_o is the average timeout period duration, b is the number of acknowledged packets by one sent-back acknowledgement (which is normally 1 unless the *delayed acknowledgements* option is activated, in which case its value is usually 2), and the factor α depends on the loss process (see [11]), but is in the neighbourhood of 1. Note that if there are no timeouts, equation (13.8) comes down to equation (13.1) (therefore equation (13.1) is a particular case of formula (13.8). So, fairness can only be reached, if each of the flows sees the same amount of packet loss (and the same timeout interval). In the sections to follow we demonstrate that this is not necessarily the case.

13.5.1 Drop-tail

In our simulation studies we observed that the packet loss p_n (and timeout intervals T_o) for the different flows are different when using Drop-tail buffers, and hence, very often unfairness is an issue.

First, we demonstrate this for a homogeneous case of two flows. Such a scenario is illustrated in Figure 13.4 for two flows of $a_1 = a_2 = 90$ packets and $d_1 = d_2 = 25$ ms. Figure 13.4a shows that for a buffer of 100 packets, smaller than the value calculated by equation (13.7) of 174 packets, there is unfairness. One of the flows gets "stuck" in timeout and never recovers. From [9] we know that the probability of timeouts is approximately equal to $\min(1, 3/w_n)$. So for flows with large opened windows the chance that

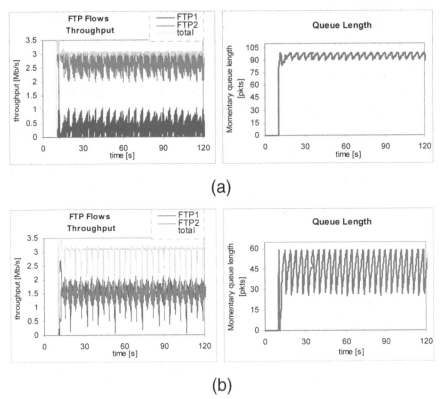

Figure 13.4 Throughput for the homogeneous case of two flows (a) with a buffer just smaller than the one calculated by equation (13.7), (b) with the buffer so small that it cannot accommodate a maximum window of any of the flows.

timeouts occur is small, but for flows that do not "inject" much data in the network in the Slow-Start phase, timeout events are highly probable.

Figure 13.4b shows that fairness is restored, if the buffer is smaller than a certain boundary value. The boundary under which fairness is restored was found to be approximately equal to the minimal difference of the advertised window and the largest BDP, i.e., $\min_n\{a_n-cd_n\}$. The reasoning behind this is that even if all flows, but one, "stumble" in timeouts, the remaining flow still suffers from losses because the buffer is not able to accommodate the data in transit generated by the flow, and hence the others do get a chance to recover. For the considered case reducing the buffer length from 100 to 60 packets leads to restoring the throughput fairness (demonstrated in Figure 13.4b).

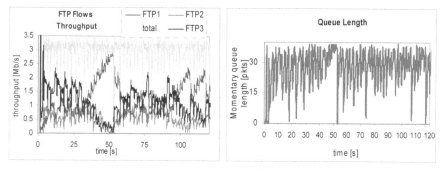

Figure 13.5 Buffer size that can accommodate at least the maximum window size of one flow, but still less than the value calculated by equation (13.7).

Secondly, we consider a heterogeneous case: $a_1 = a_2 = a_3 = 90$ packets, $d_1 = 50$ ms, $d_2 = 100$ ms, and $d_3 = 200$ ms. If the buffer length is larger than the value to accommodate the window of one flow, but still smaller than the one calculated by equation (13.7), then many stable situations can occur, including one or more flows becoming "stuck" in timeout forever, depending on the initial conditions. The system can even pass from one state to another as illustrated in Figure 13.5. Such a scenario is subject to many research efforts and the broad discussion around does not fit in our paper length and focus limitations.

13.5.2 Random Early Drop (RED)

In the previous section we proved that if the Drop-tail buffer acceptance mechanism is implemented at a bottleneck, there is no guarantee about fairness between the competing flows. Another disadvantage of Drop-tail is the synchronisation between flows, which can occur especially in access links where the number of simultaneous connections is small and buffers are not too large. RED has been exactly conceived to meet these weak points of Drop-tail. However, because there are many parameters to tune, it is a priori difficult to predict whether the impact of RED will be favourable or detrimental [12] because of the difficulty in setting its parameters properly. A buffer smaller than the value calculated with equation (13.7) can offer fairness in the case of a RED buffer acceptance mechanism, provided that the system is stable, and therefore, the losses for all flows are the same, i.e., $p_n = p$.

A huge number of studies have been executed on the RED stability (e.g. [4, 19, 22–25]). It is well understood that RED is very sensitive to proper parameter setting. RED, if properly implemented, can lead to good performance and address the problems outlined at Drop-tail mode, i.e., fairness of the bandwidth share for the competing flows since synchronisation leads to large oscillations of the buffer content. On the other hand, if not operating in the optimal regime, RED can cause unnecessary large packet loss or lead to link under-utilisation. The main parameters defining the working point of a RED-managed buffer are: the minimum Q_{min} and maximum Q_{max} buffer thresholds, the maximum drop probability p_{max} (the drop probability at Q_{max}) and the queue weight q_w (defining how closely or loosely the average queue estimate follows the evolution of the buffer contents, the momentary queue that builds up). The first study on RED dates even back to 1993 [19]. The recommended values in [18] are: a minimum threshold of 5, maximum threshold of 15, maximum drop probability of 0.1 and queue weight of 0.002. First it was recommended that the maximum threshold must be at least twice the value of the minimum threshold; then it was corrected to at least three times. Most largely used is the assumption that RED works in a stable way if the maximum threshold is four times the minimum threshold [22, 23] and we assume this also as given. Another assumption we make (also proved in other RED-related papers) is that the stable operating point is at equal space from the two RED thresholds (i.e., around 2.5 times the minimum threshold Q_{min}). This also implies that the average loss ratio of the buffer must be half the maximum drop probability p_{max} (which will be checked below). As for the other two parameters to be set for a RED buffer, i.e., the queue weight probability and maximum drop probability, it was proved in [23] that their settings must be scenario dependent and have no recommended values like the above quoted ones. We calculate the queue weight by an algorithm proposed in [24]. The choice of the maximum drop probability p_{max} is one of the most crucial choices in this RED buffer configuration process. It depends on whether the buffer will be working according to a drop-conservative policy or a delay-conservative policy where more losses are incurred on the supported flows through the buffer. The choice of p_{max} may thus lead to undesirable effects, such as timeouts (in general a TCP flow can bear up to around 2% of losses without too many timeouts) or if the average queue is around the lower RED boundary Q_{min}, under-utilisation of the link may occur as well as large buffer contents oscillations.

In what follows we describe a methodology for properly choosing the RED parameters, similar to the one presented in [22, 23]. We prove that the formula presented here also holds for environments with losses and thus comes up to the one in [22, 23], but only in cases where all flows suffer (approximately) the same loss p. In this case we combine equation (13.2) and equation (13.8) (instead of equation (13.6)) to arrive at:

$$\sum_{n=1}^{N} \frac{\alpha}{R_n \sqrt{\frac{2bp}{3}} + T_o \min\left\{1, 3\sqrt{\frac{3bp}{8}}\right\} p(1 + 32p^2)} = c. \qquad (13.9)$$

Taking into account that the operating point is in the middle of the interval maximum threshold-minimum threshold, and that the average drop probability p is half the maximum drop probability p_{max}, the following equations can be written:

$$R_n = d_n + (Q_{min} + Q_{max})/2c, \qquad (13.10)$$

$$T_o = R_n + 2(Q_{min} + Q_{max})/c, \qquad (13.11)$$

$$\sum_{n=1}^{N} \frac{\alpha}{R_n \sqrt{\frac{bp_{max}}{3}} + T_o \min\left\{1, 3\sqrt{\frac{3bp_{max}}{16}}\right\} \frac{p_{max}}{2}(1 + 8p_{max}^2)} = c. \qquad (13.12)$$

This loss function (equation (13.12)) is depicted in Figure 13.6 for a 3 Mb/s access link, for 3 homogeneous flows (four scenarios of different d_n values), for a heterogeneous scenario ($d_1 = 50$ ms, $d_2 = 100$ ms, $d_3 = 200$ ms) and one scenario with 6 homogeneous flows of $d_n = 200$ ms. The dark thick dot-ticked line represents the RED function: i.e., the discarding probability as a function of the (average) buffer content. Notice, for example, that for the homogeneous case of 150 ms propagation delay, the operating point lies around 50 (at the intersection of the RED discarding function and the loss function of equation (13.12)) and the buffer will be performing optimally for such a load since this is the optimal point on the RED curve.

In Figure 13.7, the theoretical curves (calculated with equation (13.12)) are compared to simulation results (for three flows). The simulation results are depicted with dash-dotted lines, while calculation results – with full lines. The simulation results are obtained at RED implementations with proper setting of the parameters according to the described algorithm. It can be seen that experimental and theoretical results correspond nicely. If the average number of flows through the buffer is known, as well as their RTT distribution, a curve for a heterogeneous case losses function can be calculated (e.g., like the light curve in the middle) and the buffer can be dimensioned according to it. This process must meet two contradictory requirements: reasonable losses for all the flows (less than 1%, maximum 2%) and at the same time the buffer at the access link must not introduce unacceptable delays. Thus, if the average queue $\bar{Q}_0$ is chosen around 25 (then $Q_{min}=10$ and $Q_{max}=40$ and maximum buffer is $B = 60$), the buffer contributes around $\bar{Q}_0/c = 100$ ms to the average delay of the packets and the average loss rate is around 0.005. In the case of $\bar{Q}_0 = 37.5$, the loss probability is smaller, i.e., around 0.003, but then the delay incurred in the buffer is 150 ms. For TCP flows, this loss gain is not significant as they can still perform well in even higher losses conditions (they have a feedback mechanism), but this difference of 50 ms can be crucial for time-critical applications if those queue their packets in the same best-effort buffer queue. For example, we can choose a compromise value in between, let it be $\bar{Q}_0 = 30$ ($Q_{min} = 12$, $Q_{max} = 48$, maximum buffer space is 72, $p_{max} = 0.0083$, queue weight is 0.0055). In general, however, it is a better strategy to transport time-critical applications through another (real-time) queue and use a scheduler. In this case, delays through the best-effort buffer are not of major concern to the network operator as they are not critical for the TCP performance. The buffer can be chosen larger up to a value acceptable from an equipment engineering point of view. For large buffers, loss for any propagation delay converges to one value (as can be seen in Figure 13.6) and thus the system is even more stable and fair, being less sensitive to the dynamic network conditions. Finally, remark that even if the system is dimensioned for a certain scenario, it will be unstable when more flows than the system was designed for, compete through the buffer, as e.g. illustrated by the dotted curve in Figure 13.6.

To summarise, the buffer dimensioning algorithm in the RED case is as follows:

1. Given the capacity of the link, the average number of simultaneous persistent TCP flows and their RTT distribution, the loss probability function is defined according to equation (13.12).
2. An operating RED buffer point is selected (considering the buffer engineering constraints, the requirements for loss and eventually delay) and as a consequence Q_{min} and Q_{max} are fixed.
3. Queue weight is defined in accordance with the model developed in [24].

13.6 Conclusions

In this paper we considered performance issues of an access link scenario. We assumed that the downstream direction of the last-mile link constitutes the bottleneck by its comparatively narrow bandwidth and limited buffer ca-

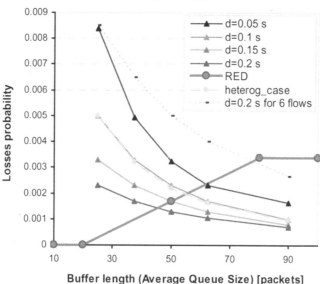

Figure 13.6 Loss function curves and RED algorithm curve, defining the operating point at the intersection.

pacity at the downlink buffer (e.g., the individual link buffer at the DSLAM). The aim is to achieve efficient transport (i.e., the downstream link capacity fully exploited), while TCP should offer a fair share to the flows (i.e., a share inversely proportional to the RTT). We developed a formula, which was endorsed by simulations, for dimensioning the bottleneck buffer in case there are no packet losses in the buffer, given the number of flows, their RTT distributions and TCP advertised windows. We showed that the outlined aims can be achieved, if the buffer is chosen so large that there is no packet loss (i.e., greater than the calculated value), such that the TCP flows open up their window to their maximum value. By this heuristic formula we can calculate not only the required buffer size but also the associated throughputs of all the flows and their corresponding RTTs. The main idea behind this formula was that a buffer needs to be able to accommodate the sum of the TCP maximum windows minus the amount of information in transit accommodated by the capacity of the access network. Unfortunately, for this goal to be achieved, the required buffer space is often too large to be implemented in state-of-the-art access equipment.

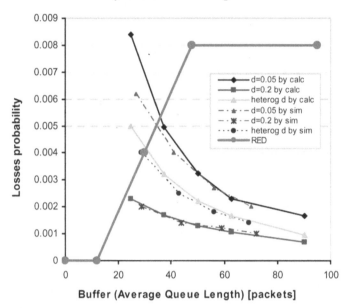

Figure 13.7 Calculated (full lines) vs. simulated (dash-dotted lines) loss function curves and choice of operating point.

Therefore, we further studied the issue whether or not the efficiency and fairness aim can be attained with smaller buffer size. The Drop-tail buffer acceptance mechanism cannot guarantee fairness and sometimes efficiency (e.g., the flows experience different loss rates, also some flows can get "stuck" in the TCP timeout phase and practically never get out of it, synchronisation effects are common for the considered links). The RED buffer acceptance mechanism on the other hand can offer fairness and efficiency provided it is configured properly (which is not an easy task as the settings depend on the number and the parameters of the TCP flows). Therefore our formula holds for RED as well and thus we came up to Ziegler's results [22, 23] on RED and also outlined a methodology for properly configuring a RED buffer.

References

[1] A. Dhamdhere and C. Dovrolis, Open issues in router buffer sizing, *ACM SIGCOMM Computer Communication Review*, vol. 36, no. 1, pp. 87–92, January 2006.

[2] A. Dhamdhere, H. Jiang and C. Dovrolis, Buffer sizing for congested internet links, in *Proceedings of IEEE INFOCOM 2005*, IEEE Communications Society, New York, pp. 1072–1083, 2005.

[3] A. Medina, M. Allman and S. Floyd, Measuring the evolution of transport protocols in internet, *ACM SIGCOMM Computer Communication Review*, vol. 35, no. 2, pp. 37–52, April 2005.

[4] C. Hollot et al., A control theoretic analysis of RED, in *Proceedings of IEEE INFOCOM 2001*, IEEE Computer and Communications Societies, vol. 3, pp. 1510–1519, 2001.

[5] C. Villamizar and C. Song, High performance TCP in ANSNET, *Computer Communications Review*, vol. 24, no. 5, pp. 45–60, October 1994.

[6] G. Appenzeller, I. Keslassy and N. McKeown, Sizing router buffer, in *Proceedings of ACM SIGCOMM 2004*, Portland, August 30–September 3, 2004.

[7] G. Raina and D. Wischik, Buffer sizes for large multiplexers: TCP queueing theory and instability analysis, in *Proceedings of EuroNGI*, Rome (Italy), pp. 173–180, April 2005.

[8] G. Vu-Brugier, R. Stanojevic, D. Leith and R. Shorten, A critique of recently proposed buffer-sizing strategies, *ACM/SIGCOMM Computer Communication Review*, vol. 37, no. 1, pp. 43–48, January 2007.

[9] J. Padhye, V. Firoiu, D. Towsley and J. Kurose, Modeling TCP throughput: A simple model and its empirical validation, In *Proceedings of the ACM SIGCOMM 1998 Conference on Applications, Technologies, Architectures, and Protocols for Computer Communication*, Vancouver (Canada), August 31–September 4, 1998.

[10] M. Enachescu, Y. Ganjali, A. Goel, N. McKeown and T. Roughgarden, Routers with very small buffers, in *Proceedings of the IEEE INFOCOM 2006*, Barcelona (Spain), pp. 1–11, April 2006. Also available at http://yuba.stanford.edu/tr.html as Technical Report TR05-HPNG-060606.

[11] M. Mathis, J. Semke, J. Mahdavi and T. Ott, The macroscopic behavior of TCP congestion avoidance algorithm, *Computer Communications Review*, vol. 27, no. 3, July 1997.

[12] M. May, J. Bolot, C. Diot and B. Lyles, Reasons not to deploy RED, in *Proceedings of the 7th International Workshop on Quality of Service*, London (UK), June 1999.

[13] R. Morris, Scalable TCP congestion control, in *Proceedings of IEEE INFOCOM 2000*, vol. 3, Tel Aviv (Israel), pp. 1176–1183, March 2000.

[14] R. Morris, TCP behavior with many flows, in *Proceedings of the International Conference on Network Protocols (ICNP '97)*, Atlanta (USA), pp. 205–213, 1997.

[15] R. Prasad, C. Docrolis and M. Thottan, Router buffer sizing revisited: The role of the output/input ratio, in *Proceedings of CoNEXT*, 2007.

[16] R. Prasad, M. Thottan and T. Lakshman, A stateless and light-weight bandwidth management mechanism for elastic flows, in *Proceedings of the IEEE INFOCOM 2008*, pp. 1481–1489, April 2008.

[17] R. Stanojevic, R. Shorten and C. Kellett, Adaptive tuning of drop-tail buffers for reducing queueing delays, *IEEE Communications Letters*, vol. 10, no. 7, pp. 570–572, July 2006.

[18] S. Floyd, Discussions on setting RED parameters, available on-line at www.aciri.org/floyd/red.html, November 1997.

[19] S. Floyd and V. Jacobson, Random early detection gateways for congestion avoidance, *IEEE/ACM Transactions on Networking*, vol. 1, no. 4, pp. 397–413, August 1993.

[20] S. Gorinsky, A. Kantawala and J. Turner, Link buffer sizing: A new look at the old problem, in *Proceedings of IEEE Symposium on Computers and Communications (ISCC'05)*, pp. 507–514, June 2005.

[21] The Network Simulator ns-2, http://www.isi.edu/nsnam/ns/.

[22] T. Ziegler, C. Brandauer and S. Fdida, A quantitative model of RED with TCP traffic, in *Proceedings of 9th IEEE/ACM/IFIP International Workshop on Quality of Service (IWQoS)*, Karlsruhe (Germany), June 6–8, 2001.

[23] T. Ziegler and S. Fdida, C. Brandauer, Stability criteria of RED with TCP traffic, in *Proceedings of the 9th IFIP Conference on Performance Modelling and Evaluation of ATM and IP Networks (IFIP9)*, Budapest (Hungary), June 27–29, 2001.

[24] V. Firoiu and M. Borden, A study of active queue management for congestion control, in *Proceedings of IEEE INFOCOM 2000*, Tel Aviv (Israel), vol. 3, pp. 1435–1444, March 2000.

[25] V. Misra, W. Gong and D. Towsley, Fluid-based analysis of a network of AQM routers supporting TCP flows with an application to RED, in *Proceedings of SIGCOMM 2000*, pp. 151–160, 2000.

[26] Y. Ganjali and N. McKeown, Update on buffer sizing in internet routers, *ACM/SIGCOMM Computer Communication Review*, vol. 36, no. 5, pp. 67–70, October 2006.

[27] Z. Avramova, D. De Vleeschauwer, K. Laevens, S. Wittevrongel and H. Bruneel, Modelling H.264/AVC VBR video traffic: Comparison of a Markov and a self-similar source model, *Telecommunication Systems*, vol. 39, no. 2, pp. 91–102, October 2008.

14

On TCP NCR Improvement Using ECN-Capable Routers

Agnieszka Brachman and Lukasz Chrost

Institute of Informatics, Silesian University of Technology, ul. Akademicka 16, 44-100 Gliwice, Poland; e-mail: {agnieszka.brachman, lukasz.chrost}@polsl.pl

Abstract

It has been shown that TCP NCR protocol is much more robust to non-congestive events than other traditional TCP implementations, resulting in higher throughput in wireless environments. By delaying congestion response and congestion window reduction it allows achieving better performance in face of packet reordering and high bit error rate typical for wireless networks. However, in wireless-cum-wired networks with bottleneck link present, TCP NCR suffers from throughput degradation due to congestion occurrences. In this paper we propose a technique based on the explicit congestion notification (ECN) allowing to distinguish losses caused by congestion to losses caused by wireless transmission errors. Furthermore we introduce new functionality to AN-AQM queue management algorithm enabling packet marking utilization. Through numerous simulations we show that the proposed combination of techniques improves overall throughput and network stability in particular cases.

Keywords: TCP NCR, AN-AQM, CMAN-AQM, ECN, congestion control.

List of Abbreviations

AQM	Active Queue Management
TCP	Transport Control Protocol

D. D. Kouvatsos (ed.), Traffic and Performance Engineering for Heterogeneous Networks, 297–313.

ACK   Acknowledgment
CW   Congestion Window
ECN   Explicit Congestion Notification
AN-AQM   Adaptive Neuron Active Queue Management
CMAN-AQM   Congestion Marking Adaptive Neuron Active Queue
               Management
TCP NCR   TCP Non-Congestion Robustness
RTT   Round Trip Time

14.1 Introduction

In current TCP/IP-based networks the TCP holds major role in all reliable services such as WWW, remote shell, electronic mail, file transfer, etc. Except for being a reliable transport protocol, TCP is also responsible for congestion reduction in the network along with Active Queue Management (AQM) schemes applied at routers. TCP adjusts its sending rate basing on the information about congestion events. The AQM mechanisms deal with the congestion occurrences by notifying the data sources about the congestion by dropping or marking superfluous packets. In this paper we investigate how TCP NCR can benefit from the explicit congestion notification (ECN) performed by the ECN-capable routers.

In the recent years wireless links have become more and more popular. Their commonness induced the rapid development of wireless data transmission technology and caused partial ousting of wires. Wireless medium is less efficient, less predictable and much harder to control than the wired one. It is more prone to interferences, what results in higher error rate and bursty transmission. The throughput cannot be easily calculated due to fading and diversity of the wireless channel quality. The shortcomings limit the capability of network to provide reliable services. Losses caused by the transmission errors can be often recovered by the link layer mechanisms, however in such case the packets will be delivered in out of order manner. Most TCP implementations recognize packet reordering as congestion occurrence and reduce the congestion window size which results in throughput degradation.

TCP receiver begins in slow start phase and increases congestion window CW exponentially until it reaches value of slow start threshold (*ssthresh*). The threshold reflects current network overload and indicates appropriate window size. After exceeding the value of *ssthresh* the sender enters congestion avoidance phase and the congestion window is increased approximately by one segment every RTT (*Round Trip Time*). The TCP receiver sends acknow-ledgements indicating the next sequence number expected form the sender. The native implementations of TCP as Tahoe, Reno, New Reno interpret reception of three duplicate acknowledgements as congestion and respond to it by retransmitting lost packets and decreasing sending rate. The solution is proved to work well and achieve high channel utilization when the only cause of packet losses originates from a congestion event. However, when this assumption is false, the TCP suffers from serious throughput degradation.

There are several TCP implementations designed to distinguish segment losses caused by the congestion events from the losses caused by the trans-mission errors, namely TCP Westwood [13], Veno [14] or NCR [9]. TCP Westwood (TCPW) estimates the bandwidth in order to set the congestion window size and the value of *ssthresh* which is used after the congestion occurrence, that is after the reception of three duplicate acknowledgements or timeout of the retransmission timer. The algorithm takes into account the bandwidth used at the time the congestion occurred. The experimental performance evaluation of the algorithm can be found in [13]. The TCPW estimation proved to work poorly in the presence of bidirectional traffic. The TCPW uses ACK packets to derive the throughput related measurements and therefore it is fragile to the temporary disturbances.

TCP Veno [14] intends to distinguish the packet losses caused by the network congestion from the losses caused by the transmission errors. If the losses occur in noncongestive state the sending rate is only slightly reduced, as opposed to the situation when the network congestion is detected. This approach allows performing diminution of the congestion window accord-ingly to the cause of the packet losses. To distinguish the congestion-related drops from the congestion-unrelated losses, TCP Veno estimates the values of expected and actual bandwidth. It uses accordingly minimized and smoothed round trip time for the estimation and therefore it can be deceived when the packets are reordered hence delayed.

Basing on the idea that if we delay the triggering of the congestion re-sponse algorithm for a small period of time, the link level retransmission will recover lost packets, the TCP NCR (Non-Congestion Robustness) algorithm has been proposed [9]. The idea works well if the packet losses are caused by the channel errors. At the same time if the bit error rate is low, TCP NCR is similar to New Reno [11] or TCP SACK [12] while considering the performance and link utilization. TCP NCR is a solution based on TCP-DCR protocol [8]. It is described in the detail in Section 14.3.

The network congestion control mechanisms consist not only of TCP algorithm at the end points but also of the queue management mechanisms applied at the network routers. Several passive queue management schemes e.g. Drop Tail have numerous flaws including link underutilization, queue length fluctuation, global synchronization of competing TCP flows and unfairness among them. The AQM mechanisms reduce the single link congestion and global synchronization by dropping selected packets before actual queue buffer overflow to prevent heavy congestion occurrence. They also provide better fairness and more stable buffer occupancy due to adjustment of the packet drop probability. The routers supporting ECN notify the data source about the congestion by marking forwarded packets instead of dropping them. ECN was introduced in [1] and its usability has been widely discussed in [4–7].

We investigate two AQM schemes, namely Adaptive RED [2] and AN-AQM [3]. In RED the superfluous packets are randomly dropped from the buffer queue when the buffer occupancy exceeds certain threshold. When the ECN is enabled the packets are marked rather then dropped. It allows sender to decide whether the network becomes overloaded and reduce the sending rate accordingly. We propose a modification to AN-AQM enabling packet marking. The modified AN-AQM scheme is introduced in Section 14.3.

The rest of the paper is organized as follows. Section 14.2 describes an overview of related work. Section 14.3 depicts enhancements proposed for TCP NCR and ECN-capable routers using RED or AN-AQM algorithms. In Section 14.4 the performance evaluation under different scenarios is outlined and finally conclusions are presented in Section 14.5.

14.2 Related Work

Various Active Queue Management algorithms are used in order to prevent the excessive buffer usage impacting both in better link utilization and fairness among competing flows. ECN (Explicit Congestion Notification) has been proposed as a possible improvement of congestion responsiveness. The routers supporting ECN notify the sender about the congestion by marking forwarded packets instead of dropping them as the average queue length exceeds the defined threshold. ECN-capable transmission protocols identify the congestion-experienced packets basing on the packet header. The network routers set a Congestion Experienced (CE) bit of the header with varying probability basing on current network state. When the receiver gets the marked packet it sets an ECN-Echo bit in outgoing packets. When the sender gets ECN-Echo marked packet it reduces its congestion window each.

The reduction does not occur more often than once per each RTT. Several solutions based on the packet marking idea have been proposed to improve the TCP performance in wired-cum-wireless networks. The most popular are presented below.

In [4] the authors developed an analytical model to estimate the optimal value of the maximum threshold of a RED buffer to achieve zero congestion loss level at the gateway. Basing on the achieved solution they proposed a TCP variant to enhance the TCP throughput in the presence of non-congestion losses in a lossy networks. The solution has been proved to work well, however it requires that the buffer size is always large enough to store the packet overload therefore it generates long delays when the network congestion level is high.

In [5] Shagdar et al. depicted that the main adventage of using ECN is that it allows avoiding unnecessary retransmissions and packets' drops. They showed that ECN significantly improves the TCP performance over wired network however its inability to distinguish transmission errors from the network congestion makes it unsuitable for wired-cum wireless networks. The authors have proposed EWLN mechanism to improve ECN performance in lossy links and have confirmed its effectiveness via numerous simulations. The EWLN is deployed at the link layer.

Shao et al. [6] evaluate the effect of the ECN on TCP Westwood. The authors proved that TCP Westwood interoperability with RED/ECN is as beneficial as TCP New Reno's. In this paper we investigate TCP Westwood performance with and without the ECN in the environment where packet reordering rate is high.

The described proposals of TCP performance improvements in wireless networks through the use of ECN-based mechanisms to distinguish the losses caused by the transmission errors and congestion, usually required changes in the AQM algorithms at the network routers or at the access points level.

14.3 Algorithm Description

As the research was dedicated for the exploration of ECN functionality provided by various TCP implementations as well as different AQM schemes, AN-AQM modification enabling ECN-wise packet marking has been introduced.

14.3.1 Congestion Marking Adaptive Neuron AQM (CMAN-AQM)

AN-AQM is a recently proposed robust AQM scheme believed to outperform other well-known AQM schemes [3] but does not utilize ECN mechanisms. AN-AQM is based on estimation of dropping probability of incoming packets by an adaptive neuron. The neuron uses both integrated Hebbian Learning and Reinforced Learning based on two main error factors – the queue length error and the normalized rate error described by the following equations:

$$e(k) = q(k) - Q_t \tag{14.1}$$

$$y(k) = \frac{r(k)}{C} - 1 \tag{14.2}$$

where $e(k)$ is the queue length error, $q(k)$ is the queue length and Q_t is target queue length, $y(k)$ is the normalized rate error, $r(k)$ is the input rate at the bottleneck link and C is the capacity of the bottleneck link. There are six inputs of the AN-AQM scheme embracing both error factors and their historical values.

AN-AQM scheme acts by dropping incoming packets with the probability given by an adaptive neuron. In this article we propose a new AQM scheme based on the AN-AQM principles and providing ECN functionality.

The CMAN-AQM algorithm is based on two adaptive neurons. The neurons act as 2 separate AN-AQM schemes with different target queue lengths $(Q_{T1}, Q_{T2} : Q_{T1} < Q_{T2})$ where Q_{T1} is the real target queue length and Q_{T2} stands for target maximum queue length. According to equations provided in [3] the scheme outputs two values:

$$p_1(k) = p(k, Q_{T1}) \tag{14.3}$$

$$p_2(k) = p(k, Q_{T2}) \tag{14.4}$$

where $p_1(k)$ is the probability of dropping of the ECN-incapable or marking the ECN-capable packets and $p_2(k)$ is the probability of dropping the incoming ECN-capable packets.

Marking the packet as congestion experienced does not prevent it from entering the queue thus ECN-capable packets would receive much higher benefit in the environment of mild congestion and a queue overflow could occur in the case of heavy congestion. Therefore the second drop-only mechanism has been provided.

14.3.2 TCP NCR with ECN

TCP NCR introduces a set of simple modifications to improve robustness to transmission errors that are much more frequent in wireless environment than in a wired one. In a lossy environment the reception of three duplicated acknowledgements may be insufficient to determine a packet loss occurrence. The fundamental idea is to delay the retransmission, allowing the link layer to recover the lost packet. The delay in the congestion response can influence the network performance, therefore the threshold must be properly selected to provide a minimal impact when the losses are caused by the network congestion control mechanisms at the router layer. According to Bhandarkar et al. [9] the retransmission should be delayed until approximately a congestion window of data have left the network. Since the congestion window represents amount of data that can be transmitted during one round trip time (RTT), the fast retransmission is delayed for roughly one RTT. The choice of the delay determines the performance of the TCP NCR. A large delay would result in poor performance in the face of the network congestion, on the other hand too short period would not provide enough time for the link layer to retransmit the lost segments. The analytical results confirm that the choice of the delay equal to the RTT is the most optimal. If a packet is not recovered then the loss is interpreted as caused by the network congestion and the fast retransmission and recovery algorithms are used.

TCP NCR modifies the SACK-based loss recovery algorithm introduced in [12]. The changes are required only at the sender side assuming that the receiver applies SACK option in the TCP header, otherwise TCP NCR should not be used. TCP NCR extends the TCPs Limited Transmit scheme [10] during which new data is sent until a decision is made weather the loss is caused by the reordering of the packets or a network congestion. The Extended Limited Transmit scheme is proposed in two variants: Careful Limited Transmit and Aggressive Limited Transmit. During Careful Limited Transmit the sending rate is halved immediately and restored to the previous state if a packet is recovered before entering the loss recovery phase. One packet is sent for every two segments that are known to have left the network. The retransmissions are suspended for approximately one RTT. When using the Aggressive Limited Transmit variant the sending rate is preserved. The congestion response is delayed to decide whether the loss was caused by a network congestion control mechanism at the router layer or a simple packet reordering. The aggressive approach allows achieving better throughput in

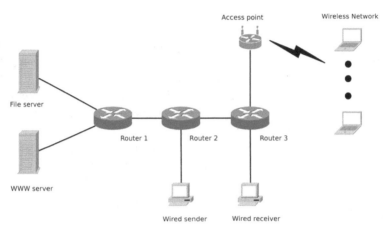

Figure 14.1 Network topology.

the face of packet reordering since it delays the reduction of the sending rate. In further investigation we consider only aggressive version of the TCP NCR.

We have added a modification to the TCP NCR that allows to differentiate whether the loss was caused by a packet reordering or a congestion event. We presume that if the CE bit is not set, the packet loss is caused by a non-congestion event and therefore we do not reduce the congestion window even if we retransmit the packet and enter the fast recovery phase. In case of ECN-Echo marked packet we reduce the congestion window immediately and do not delay the congestion response. The proposed modifications in TCP NCR require the changes are made only at end points and do not interfere with the router algorithms or the network layer functionality.

14.4 Performance Evaluation

The section discusses the results of simulations in the wireless local-area networks. To study TCP and ECN-capable routers performance the ns-2 simulator [15] was used. We have investigated two TCP variants designed for wired networks: TCP New Reno and TCP SACK as well as TCP variants intended for the wireless networks: TCP Westwood and aggressive TCP NCR. We have compared the proposed CMAN-AQM algorithm to the representative AQM scheme, namely, ARED.

The network topology used in simulations is shown in Figure 14.1. Wireless receivers are connected to a 11 Mbps, 802.11b access point. The file and WWW servers and the wired sender and receiver are connected with

100 Mbps links, the link delay is 2 ms. The links between routers contribute a 20 ms delay and their bandwidth is set to 10 Mbps. The packet error rate in the wireless network varies from 0 to 10%. The AQM scheme is applied at the router 3 while the maximum queue length is set to 200 packets. All of the adaptive neuron configuration parameter values have been set accordingly to the [3] except for: $Q_{T1} = (1/3)Q_l$, $Q_{T2} = 1/2Q_l$ where Q_l is the maximum queue length. CMAN-AQM scheme used in the "non-ECN" environonment acts like the traditional AN-AQM scheme. Therefore CMAN-AQM scheme is used in place of AN-AQM in the "non-ECN" simulations. ARED is configured according to recommendations presented in [2]. Simulations are run for 100 seconds, statistics are collected from 10th second. We evaluate proposed algorithms using several scenarios.

14.4.1 Network Overload

In this scenario only FTP transfer is performed. Five wireless stations are connected to file server, moreover 15 simultaneous connections exist between wired sender and receiver. Packet error rate is negligible. All connections use 1500 bytes packet size, they all start around 10th second, statistics are collected for 90 seconds.

The aggregate throughput for all wireless clients is similar regardless of AQM mechanism used and oscillates around 2 Mbps. Since the bottleneck link is fully utilized throughout the simulation, reduced throughput on the wireless link results in a higher throughput for the wired connections. ARED allows the wireless stations to consume more wired, bottleneck bandwith therefore the fairness between wired and wireless client is higher. Nevertheless the wireless nodes receive less bottleneck bandwith than the wired ones, it is mainly due to the longer RTT (*Round Trip Time*) and significantly lower maximum, achievable throughput. Enabling the ECN significantly improves the throughput stability and the fairness among the competing flows in the wireless network what we can observed in figure 14.2. AN-AQM with packet marking provides maximum achievable fairness, much better then fairness offered by ARED with ECN. TCP Westwood benefits most from ECN especially when considering fairness among flows. Figure 14.3 presents difference between throughput and goodput for wireless stations. To calculate throughput we counted all sent packets, goodput calculation takes into account received packets, discarding duplicated. When packets are marked the differences between throughput and goodput are much lower then in scenario where packets are dropped. ARED allows occupying wider bottleneck

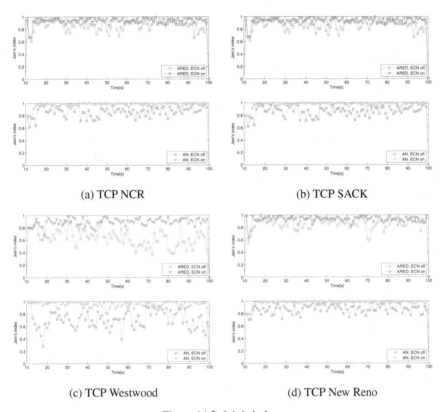

(a) TCP NCR   (b) TCP SACK

(c) TCP Westwood   (d) TCP New Reno

Figure 14.2 Jain's index.

bandwith for all TCP protocols except for TCP Westwood. For this protocol AN-AQM without packet marking seems the best choice. Packet drop rate on bottleneck link is depicted on Figure 14.4.[1] Once again TCP algorithm does not influence it significantly except for TCP Westwood. AN-AQM is distinguished by higher drop rate which results in lower throughput. With ECN enabled drop rate is close to 0% at the cost of higher packet delay and packet delay variance. AN-AQM offers shorter mean queue length (Figure 14.5) and much lower queue length variance (Figure 14.6). Queue length variance influences stability of packet transmission delay directly thus minimization is very desirable. The queue length variance is comparable for TCP NCR, TCP SACK and TCP New Reno regardless of AQM scheme used. Still AN-AQM

---

[1] In the figure, the successive bars indicate the results for the following TCP protocols: TCP NCR, TCP SACK, TCP Westwood and TCP New Reno.

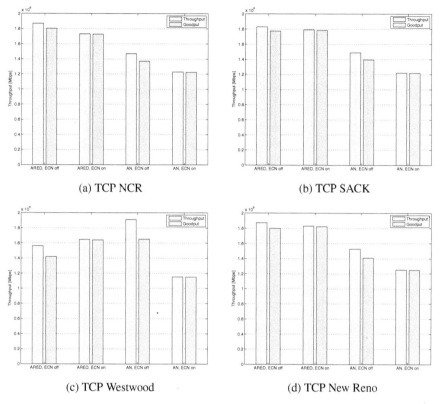

(a) TCP NCR   (b) TCP SACK

(c) TCP Westwood   (d) TCP New Reno

Figure 14.3 Throughput and goodput comparison.

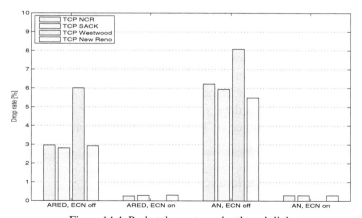

Figure 14.4 Packet drop rate on bottleneck link.

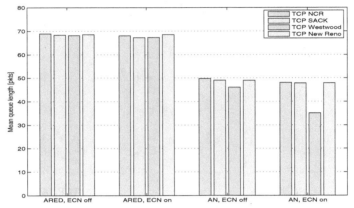

Figure 14.5 Mean queue length on bottleneck link.

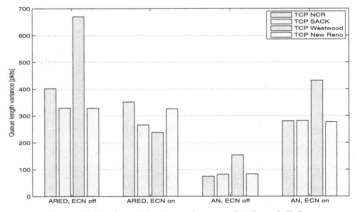

Figure 14.6 Queue length variance on bottleneck link.

offers up to three times better queue length stabilization. The impact of TCP implementation on subjacent AQM scheme performance highly depends on the selection of the AQM scheme itself. CMAN-AQM copes with varying environment similarly regardless of superimposing TCP algorithm and ECN utilization while the influence of TCP implementation on RED performance is very high.

14.4.2 Transmission Interference

In this experiment the conditions are very similar to the scenario presented in previous section. We added error model to the wireless network. The

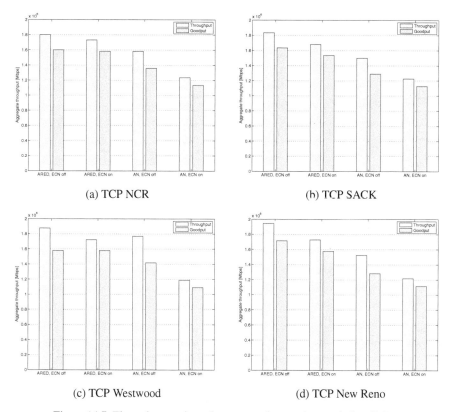

(a) TCP NCR

(b) TCP SACK

(c) TCP Westwood

(d) TCP New Reno

Figure 14.7 Throughput and goodput comparison on lossy wireless link.

packet reordering rate is set to 10%. 802.11's link layer implements recovery mechanism therefore corrupt packets will be retransmitted at MAC level between access point and receiver. The retransmission limit is set to three attempts. Retransmissions cause packet reordering. We investigate influence of packet reordering on throughput and goodput comparison as well as on retransmission rate.

Figure 14.7 presents throughput and goodput comparison. The bottleneck link is congested because of the download to wired receivers, however all wireless connections manage to perform their transmission. Due to the congestion, disparity in overall throughput on wireless network among different TCP implementation is minor. The differences between goodput and throughput are lower when ECN is enabled, especially for NCR and Westwood. CMAN-AQM with packet marking offers lower throughput however causes

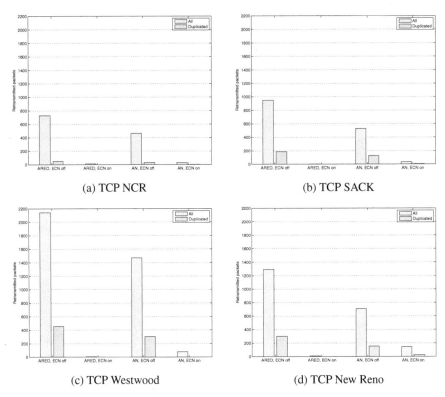

(a) TCP NCR   (b) TCP SACK

(c) TCP Westwood   (d) TCP New Reno

Figure 14.8 Retransmission rate.

lower differences in throughput and goodput. If we compare unnecessary retransmission rate (Figure 14.8), we can observe that TCP NCR rarely retransmits packet unjustifiably and this is the main advantage of this protocol above other implementations.

14.4.3 ECN Unaware Traffic

In previous scenarios all flows were either ECN-capable or ECN-unaware. In the last scenario we introduce UDP traffic. Five wireless clients perform FTP transfer from file server. Five connections exist between WWW server and wired receiver, five additional connections performing CBR transfer between wired sender and receiver are established. Packet size for UDP connections is set to 1000 B.

Figure 14.9 presents aggregate throughput for TCP receivers on wireless network vs increasing UDP transmit rate. Introducing ECN unaware transfer

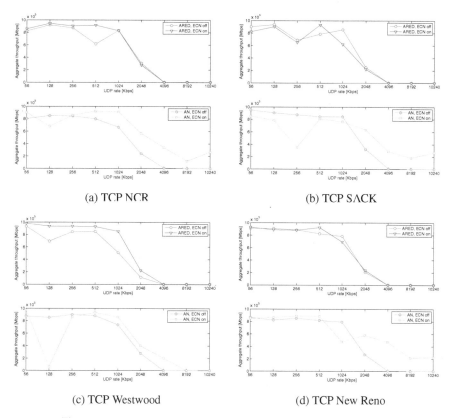

Figure 14.9 Aggregate throughput on wireless network vs UDP rate.

highly degrades bandwith used by wireless connections. Transmit rate higher than 2 Mbps makes it impossible for wireless stations to acquire any channel rate. Enabling packet marking does not prevent UDP sources from utilizing almost the whole bottleneck link. While in congested scenario with only TCP traffic ARED allowed achieving higher throughput for wireless stations insomuch when UDP traffic is added AN-AQM outperforms ARED especially when UDP transmit rate is high. It allows wireless stations to hold minimal transmission rate while ARED blocks both wired and wireless TCP connections. Queue length is similar for all TCP and AQM algorithms. It is due to the fact that ECN unaware, congestion non responsive UDP traffic occupies majority of the bottleneck link. From the same reason queue length variance is rather low. UDP is very aggressive traffic, when all sources transmit with 2 Mbps rate, UDP consumes 80% of the whole bandwidth.

14.5 Conclusion

We have introduced a novel ECN-capable AQM scheme called CMAN-AQM based on the AN-AQM algorithm. We have modified TCP NRC to identify ECN-marked packets and use the information for discrimination of packet loss causes. Through numerous simulations we have demonstrated that TCP NRC acts more efficiently than other TCP implementations in the environment where the packet reordering rate is high. We have demonstrated that CMAN-AQM is capable of stabilizing queue length around the given target while utilizing ECN scheme. The simulations also demonstrate low impact of used TCP implementation on both AN-AQM and CMAN-AQM performance and high influence of overlying TCP implementation on ARED performance concerning queue length stabilization. We have also proved that in the presence of UDP traffic, packet marking performed by ARED does not prevent ECN unaware sources from consuming the whole bandwidth, at the same time AN-AQM enables TCP senders to preserve minimum transmission rate.

Acknowledgment

This work was partially supported by the Ministry of Science and Higher Education under Grant N517 025 31/2997.

References

[1] K. Ramakrishnan, S. Floyd and D. Black, *The Addition of Explicit Congestion Notification (ECN) to IP*, RFC 3168, September 2001.

[2] S. Floyd, R. Gummadi and S. Shenker, Adaptive RED: An algorithm for increasing the robustness of RED's active queue management, Preprint, available at http://www.icir.org/floyd/papers.html, August 2001

[3] J. Sun and M. Zukerman, An adaptive neuron AQM for a stable Internet, in *NETWORKING 2007. Ad Hoc and Sensor Networks, Wireless Networks, Next Generation Internet*, I. F. Akyildiz, R. Sivakumar, E. Ekici, J. Cavalcante de Oliveira and J. McNair (Eds.), Lecture Notes in Computer Science, vol. 4479, Springer, p. 844, 2007.

[4] H. Bai and M. Atiquzzaman, Enhancing TCP throughput over lossy links using ECN-capable RED gateways, in *Proceedings Vehicular Technology Conference*, VTC 2003-Fall, 6–9 October 2003, IEEE 58th vol. 4, pp. 2721–2725, 2003.

[5] O. Shagdar, M. N. Shirazi and Z. Bing, Improving ECN-based TCP performance over wireless networks using a homogeneous implementation of EWLN, *Proceedings 10th International Conference on Telecommunications, ICT 2003*, 23 February–1 March, vol. 1, pp. 812–817, 2003.

[6] S. Shao, M. Y. Sanadidi and M. Gerla, A simulation study of the interoperability of TCPW with RED and ECN, *Proceedings SPECTS 2003*, Montreal, Canada, July 2003.

[7] K. Pentikousis, H. Badr and B. Kharmah, On the performance gains of TCP with ECN, in *Proceedings of the Second European Conference on Universal Multiservice Networks (ECUMN)*, Colmar, France, April, pp. 82–91, 2002.

[8] S. Bhandarkar, N. Sadry, A. L. N. Reddy and N. Vaidya, TCP-DCR: A novel protocol for tolerating wireless channel errors, Technical Report TAMU-ECE-2003-01, 2003.

[9] S. Bhandarkar, A. L. N. Reddy, M. Allman and E. Blanton, Improving the robustness of TCP to non-congestion events, Network Working Group Request for Comments: RFC 4653, 2006.

[10] M. Allman, H. Balakrishnan and S. Floyd, Enhancing TCP's loss recovery using limited transmit, RFC 3042, January 2001.

[11] S. Floyd, T. Henderson and A. Gurtov, The NewReno modification to TCP's fast recovery algorithm, RFC 3782, 2004.

[12] E. Blanton, M. Allman, K. Fall and L. Wang, A conservative selective acknowledgement (SACK) – Based loss recovery algorithm for TCP, RFC 3517, April 2003.

[13] S. Mascolo, C. Casetti, M. Gerla, M. Y. Sanadidi and R. Wang, TCP Westwood: Bandwidth estimation for enhanced transport over wireless links, in *Proceedings of ACM Mobicom 2001*, Rome, Italy, July 16–21, pp. 287–297, 2001.

[14] C. P. Fu and S. C. Liew, TCP Veno: TCP enhancement for transmission over wireless access networks, *IEEE Journal on Selected Areas in Communications*, vol. 21, no. 2, 216–228, February 2003.

[15] The ns-2 network simulator: http://ww.isi.edu/nsnam/ns.

15

An Analysis of TCP-Tolerant Real-Time Multimedia Distribution in Heterogeneous Networks

Agnieszka Chodorek[1] and Robert R. Chodorck[2]

[1]*Department of Telecommunications and Photonics, Kielce University of Technology, al. Tysiaclecia Państwa Polskiego 7, 25-314 Kielce, Poland; e-mail: a.chodorek@tu.kielce.pl*
[2]*Department of Telecommunications, The AGH University of Science and Technology, al. Mickiewicza 30, 30-059 Kraków, Poland; e-mail: chodorek@kt.agh.edu.pl*

Abstract

TCP-tolerance is an ability of real-time multimedia flow to optimal share network resources with the TCP. "Optimal" means, that both the multimedia transmission is carried out preserving real-time conditions (quality of service is assured) and TCP flows are able to use remaining bandwidth. In the TCP-tolerant circumstances, the utilization of bottleneck link is close to 100%. Typically, real-time multimedia flows are not TCP-tolerant, due to large bursts, which causes short-time increments of buffer occupancy, which are misinterpreted as local congestions. Because the TCP characterizes by large congestions-sensitivity, these bursts cause decrement of TCP throughput, larger than it results from the difference between throughput of the bottleneck link and the target bit rate of real-time multimedia traffic. In this article an analysis of TCP-tolerant real-time multimedia distribution in heterogeneous networks is presented. The TCP-tolerance was achieved using the burst control mechanism. Simulation analysis was carried out using the Berkeley's ns-2 simulator.

D. D. Kouvatsos (ed.), Traffic and Performance Engineering for Heterogeneous Networks, 315–336.

Keywords: Real-time multimedia transmission, TCP, congestion avoidance, TCP-tolerance.

15.1 Introduction

The modern Internet is a heterogeneous network, which utilizes different network technologies. These technologies are optimized to specific environments. They differ in range of transmission, transmission medium (wired, wireless), achievable throughput, costs, etc. Thanks to usage of different network technologies, access to the Internet is possible almost all over the world – from large metropolises to sands of Sahara. However, usage of heterogeneous networks creates many problems, which are unresolved up until now.

Transmission in heterogeneous networks denotes that a packet is sending via networks, which differ in quality of service. It involves congestions at point of contact of different technologies, what leads to packet losses. Network heterogeneity involves also difficulties in preserving real-time characteristics of transmission (delay, jitter).

Links in heterogeneous networks are shared by transmissions carried out by different applications (bulk data transfer, WWW, multimedia applications, etc.). Modern applications are different in the type of generated traffic (elastic, inelastic) and mechanisms of protocols and architectures used during transmission. As an effect, we observe strong interactions between individual transmissions, which can result in degradation of one or more competing flows. Solutions, which are intended for counteracting these degradation are TCP-friendly protocols [1, 5, 6, 10] and TCP-tolerant mechanisms [2]. The last one will be analyzed in this chapter.

One of possible TCP-tolerant mechanisms is burst control, proposed in [2]. This low-complexity mechanism is based on burst fragmentation and assures avoidance of TCP congestion collapse when competing with real-time transmission, but does not allow for transmission rate reduction when congestions appear.

The aim of the chapter is to analyze TCP-tolerant real-time multimedia distribution in heterogeneous networks. The simulational analysis was carried out in Berkeley's ns-2 environment. The particularly heavy emphasis was placed on quality of service of competing streams, measured by bandwidth, loss and delay parameters.

The rest of the chapter is organized as follows. Section 15.2 presents problems of coexistence of elastic and inelastic traffic in heterogeneous networks.

Section 15.3 provides an overview of burst control mechanism. Section 15.4 describes simulational experiment, while Section 15.5 presents an analysis of burst controlled real-time multimedia transmission. Section 15.6 presents an analysis of buffer occupancy in the case of burst controlled real-time multimedia transmission. Section 15.7 summarizes our experiences.

15.2 Coexistence of Elastic and Inelastic Traffic in Heterogeneous Networks

In IP networks, there are two kinds of traffic, which are widely different in required Quality of Service (QoS) parameters. The criterion of identity is sensitivity to transmission delay and bandwidth. The first, elastic traffic is, generally, insensitive to transmission delay and bandwidth. It is generated by such applications, as ftp, WWW, electronic mail, peer-to-peer file transfer. These applications typically use the TCP transport protocol.

The second, inelastic traffic is, generally, sensitive to both, transmission delay and bandwidth. This kind of traffic, also known as streaming traffic, is generated by applications such as Voice over IP (VoIP), Video on Demand (VoD), Internet Protocol Television (IPTV), Internet radio, Internet television, videoconferencing tools. These applications typically use the RTP or the UDP transport protocols.

15.2.1 The TCP Transport Protocol

Transmission Control Protocol (TCP) is the primary transport protocol of the Internet. The *Encyclopedia of Internet Technologies and Applications* defines the TCP as "the connection-oriented, multi-purpose transport protocol, designed for reliable data transfer" [5]. The TCP has several reliability-oriented mechanisms. The most important ones are:

- error control, intended for reliability assurance of TCP connections,
- flow control, intended for prevention of receiver buffers from overflowing,
- congestion control, intended for counteracting and avoiding congestions.

From a sharing traffic point of view, the most important TCP's mechanism is the congestion control. This mechanism allows the TCP connection to react to variable network circumstances, and, as a result, to share the bottleneck link nearly equally. If we look into the problem of the QoS of TCP connections, we will find out that the equality of throughputs of TCP flows in shred link determines fair bandwidth allocation between competing TCP flows.

15.2.2 The UDP Transport Protocol

The User Datagram Protocol (UDP) was designed for early IP networks, which were unicast in nature. It was designed as a very simple transport protocol, completing the TCP/IP protocol suite. However, in contrast to TCP whose mechanisms were optimized from the point of view of unicast transmission, the simplicity of the UDP protocol allows it to be used for multicasting as well.

Ascetic UDP's mechanisms include CRC-based error detection and multiplexing of transport connections (using socket interface). 16-bit checksum secures the UDP datagram (both UDP header and user's data) and the flow state information (IP addresses of end systems, port numbers used by end systems and the protocol identifier). No flow control, congestion control or error control are provided.

Although UDP is not a "modern protocol" – it was defined a quarter of century ago – the protocol is still in common use. Just like TCP, at the present time UDP is an element of all popular operating systems. However, it is usually used for non-professional or semi-professional applications or (most frequently) serves as an underlying protocol for session-layer or application-layer control protocols or other transport protocols.

15.2.3 The RTP Transport Protocol

The Real-time Transport Protocol (RTP) [8] was designed for real-time transmission of multimedia information. Its practical application covers, among other things, interactive audio/video services and real-time distribution of audio and (or) video data to a large number of recipients. In contrast to UDP, RTP was designed as a multicast (multipoint-to-multipoint) protocol, and unicast (point-to-point) transmission is treated as particular case of general multicasting.

Nowadays, version 2 of the RTP protocol is in common use. This version was introduced by RFC 1889 in 1996 and extended by RFC 3550 in 2003 [8]. Extensions made the protocol better for multimedia transmission to large multicast groups.

Typically, RTP protocol occupies the upper sublayer of the fourth (transport) layer of the OSI model, while the lower sublayer is occupied by UDP.

Running RTP on top of UDP allows utilization of the multiplexing and checksum services of the lower protocol.

Real-time multimedia transmission needs fixed (constant or variable) bandwidth and cannot be window controlled. To meet these requirements RTP does not provide any typical window-based mechanism such as flow control or congestion control (instead, RTP-level translator-based congestion control is supported). The real-time transmission does not require reliable transmission (however, small BERs are preferred), so the RTP protocol does not ensure reliability. However, it detects out-of-order delivery and packet loss (both based on the sequence numbers included in RTP header), as well as packet damage (using checksum service of underlying UDP protocol). Because the receiver is able to repair out-of-order errors, unordered packets are delivered to application in sequence.

RTP protocol carries real-time data and, usually, also conveys additional information allowing the receiver to identify the type of delivered content (e.g. audio, video) and to identify the method of encoding (PCM, ADPCM, MPEG-4, etc.). The timestamp header field, which reflects the sampling instant of the first octet in the carried data, allows the protocol to support synchronisation of audio and video data and reconstruction of the timing. The RTP header also includes a field which identifies synchronization source (SSRC) of the data carried as the payload. If the synchronisation source is a mixer, the packet header will contain the list of identifiers of the contributing sources (CSRC).

RTP itself does not provide any control functions. For the purpose of control, the RTP Control Protocol (RTCP) was developed. RTCP is an integral part of the RTP specification (both RFC 1889 and RFC 3550). The RTCP functionality includes monitoring the quality of service, conveying information about the participants in a RTP session, supporting scalability of large multicast sessions and, optionally, conveying additional control information for the purpose of session management.

Each RTP transmission is associated with an RTCP transmission. In teleconferencing systems, where several data streams are sent independently, each contributing stream is augmented with RTCP flow, allowing individual control of each stream.

15.2.4 Elastic and Inelastic Traffic in Shared Link

In heterogeneous networks, the inelastic, multimedia traffic and the elastic, TCP traffic, are transmitted via the same link. The inelastic, multimedia traffic is self-limited (or rate-limited), i.e. it has strictly defined, constant or variable, bit rate. As a result, the real-time multimedia stream has hard bandwidth requirements, but, from the other side, it never occupies more bandwidth than it results from the bit rate of the traffic source. The elastic, TCP traffic is unlimited (in practice, limited by flow and congestion control) and its bit rate strongly depends on network circumstances (bandwidth of bottleneck link, number of competing flows, etc.). As an effect, the TCP flow does not have hard bandwidth requirements, and, in favorable circumstances, it can occupy whole bandwidth of the bottleneck link.

Although real-time transmission usually does not occupy the whole link capacity, in some circumstances, competing TCP flow can collapse and achieve significantly lower throughput than the network can deliver. This unexpected collapse of TCP connection is identified as TCP-intolerance of real-time transport protocol [2, 9]. Moreover, it is possible to observe a satellite phenomenon – real-time intolerance, where real-time stream also achieves significantly lower throughput than it results from a simple difference of link throughput and TCP throughput [3].

Similar term, TCP-unfriendliness, denotes such behavior of inelastic traffic, where one or more inelastic (typically: multimedia) streams are not able to equally share the bottleneck link with one or more TCP flows. In contrast, TCP-friendly streams are able to fairly (equally) share link with the TCP. The well-known definition of TCP-friendliness goes as follows: "We say a flow is TCP-friendly if its arrival rate does not exceed the arrival of a conformant TCP connection in the same circumstances" [6]. Because TCP-unfriendliness is caused by lack of TCP-like congestion control mechanism [6], it can be eliminated using protocols, which apply this method of congestion control. Such protocols are called TCP-friendly protocols [1, 5, 6]. Three main properties of TCP-friendly protocols are [2, 9]:

- avoidance of TCP congestion collapse,
- fairness toward competing flows,
- congestion control integrated with the transport protocol.

TCP-friendly protocols change bit rate of the flow and do not change (neither physically nor virtually) the bit rate of the source of real-time traffic [2]. Previous research shows that if fair bandwidth allocation does not correspond with real-time requirements, the last two advantages of TCP-friendly protocol may cause degradation of real-time characteristics of the streaming traffic [4]. In result, it is impossible to assure QoS parameters of the real-time stream (throughput, latency, jitter, error rate) at acceptable level when TCP-friendly

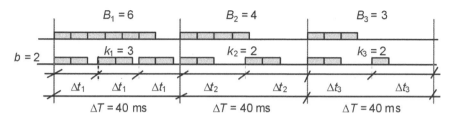

Figure 15.1 Burst control.

protocols are used. However, QoS parameters of the TCP flow (throughput, error rate) and link utilisation stays at high level.

In contrast to the TCP-friendliness, which is intended to assure fairness, TCP-tolerance is intended to preserve (as good as possible) primitive character of the traffic in the best-effort service. TCP-tolerance can by defined as such behaviour of transport protocol, which allows TCP to utilize bandwidth unoccupied by real-time multimedia transmission [2]. In other words, it is an ability of real-time transport protocol to reasonably share network resources with TCP flows. TCP-tolerance can be understood as the first attribute of TCP-friendly protocols (avoidance of TCP congestion collapse).

Unlike the TCP-friendliness, TCP-tolerance does not implement congestion control. However, it can be used with specialized congestion control mechanisms, optimized for multimedia transmission (as adaptive coding, translators, receiver-based layered multicast or receiver-based multicast stream replication).

15.3 Burst Control for TCP-Tolerant Applications

Burst control spreads out burst over time, while the source's bit rate stays unchanged. It results in debursting, similar as obtained as an effect of traffic smoothing (or traffic shaping). However, in contradiction to traffic smoothing, burst control does not limit the rate of transmission of data.

If encoder generates video frames at 25 Hz (25 frames per second), the period ΔT between two successive frames will be constant ($\Delta T = 40$ ms). Due to usage of real-time transport protocol, at the output of the sender, we observe burst of packets. Let us assume that the i-th burst consists of B_i packets, where B_i is the smallest integer $\geq f_i/p$ (where f_i is the i-th frame size and p is a payload length).

Burst control mechanism allows the sender to transmit multimedia stream at original bit rate (constant or variable), with fully controlled burstiness.

The mechanism subdivides the period ΔT on k non-overlapping blocks (time slots) of size Δt. In each time slot, the sender sends burst of b packet (Figure 15.1), except the last, where burst can be smaller or equal to b. Because burst is spreading out over ΔT, the source's bit rate (typically evaluated over ΔT period) stays unchanged.

The number of time slots k_i and the duration of time slot Δt_i depend on video source and the method of encoding. In particular:

$$\Delta t_i = \Delta T \cdot \frac{1}{C\left(\frac{f_i}{p \cdot b}\right)} \qquad (15.1)$$

where $C(x)$ is a round toward infinity (the smallest integer greater than or equal to x) and p is a data packet's payload length.

Parameters p and b depends on applications. The p is set according to separate specifications of used real-time protocols (e.g. according to RTP profiles). The value of b should be estimated as trade-off between application requirements and TCP-tolerance. Tests carried by the authors show that b set to 1 usually assures the best TCP-tolerance [2].

Burst fragmentation does not assure avoidance of synchronization of bursts generated from different sources. To avoid synchronization, two complementary mechanisms have been introduced:

- duty factor,
- randomization of duration of time slots.

Duty factor *duty* reduces the period ΔT to $\Delta T \cdot duty$ (Figure 15.2a). As a result, the duration of time slot Δt_i can be expressed by

$$\Delta t_i = \Delta T \cdot \frac{1}{C\left(\frac{f_i}{p \cdot b}\right)} \cdot duty \qquad (15.2)$$

This method is especially useful in the case of synchronized CBR sources, which are located in the same node (e.g. layered multicast, stream replication multicast). To avoid synchronization of bursts (and preserve synchronization of streams), streams (layers) should be transmitted using different values of *duty*. The duty factor can be set from 0 to 1, where 1 denotes usage of nominal time ΔT and 0 denotes $\Delta t_i = 0$ (thus, $b_i = B_i$). In practice, values of *duty* close to 1 are preferred [2].

In some cases, duty factor insufficiently reduces dangerous of synchronization. Thus, the other method – randomization of duration of time slots

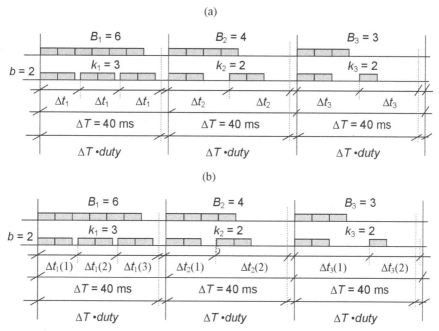

Figure 15.2 Burst control with duty factor: (a) without randomization, (b) with randomization of duration of time slots.

(Figure 15.2b) – should be applied together with duty factor method. Time slots can be randomized, for example, by addition of small random value (positive or negative) or by addition of small percentage (e.g. 1%) of original value Δt_i computed from equation (15.2). Here, randomized time slots $\Delta t_i'$ have been computed according the equation:

$$\Delta t_i' = \Delta t_i \cdot (1 + U\,(-0.01,\ 0.01)) \tag{15.3}$$

where $U(-0.01,\ 0.01)$ is a uniformly distributed random number between -0.01 and 0.01.

The duty factor *duty* allows for easy randomization of duration of time slots. It gives the margin necessary to reduce the effects of cumulation of positive random components from equation (15.3), which can appear because of relatively small population of random variables (see Δt_3 in Figure 15.2b).

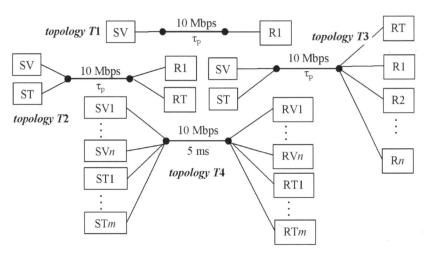

Figure 15.3 Topologies used in experiments.

15.4 Experiments

Burst control mechanism was implemented in Berkeley's *ns*-2 network simulator [7]. Experiments have been carried out using typical single-bottleneck topology, which consists of two routers and a set of senders and receivers. In all experiments throughput of bottleneck link was large enough to ensure live video transmission.

The transmission scheme has been simulated in four different topologies (Figure 15.3), to explore the performance issues as well as scalability. Senders, as well as receivers, were connected to the nearest router at 100 Mbps and 1 μs delay. Senders' buffers are theoretically unlimited (5000 packets). Routers were connected with a bottleneck link at 10 Mbps and τ_p propagation delay (default value of τ_p is 5 ms). In routers, Drop Tail queue was used and queue size was set to 5000 packets (infinite buffer size) or to 13 packets (finite buffers; queue assure buffering of 10 ms TCP traffic).

Topology $T1$ consists of a single video source (SV) and single receiver (R1). Topology $T2$ consists of a video source (SV), a TCP source (ST), video receiver (RV), and TCP receiver (RT). Topologies were used for analysis of mechanism's performance with ($T2$) and without ($T1$) background traffic. Topology $T3$ extends topology $T2$ with multiple video receivers. It was used for exploration of the scalability of the multicast transmission with respect to session size.

Topology *T*4 extends topology *T*2 with multiple sessions. The number of senders/receivers was varied from 1 to 10 to investigate the intra-session behaviour of burst controlled multimedia.

Both, elastic (TCP) and inelastic (video) traffic was modeled using build-in models. As the source of elastic traffic (ST), FTP over TCP (SACK version) was used. TCP packets have a maximum segment size (MSS) of 1000 bytes. As a model of video (inelastic traffic), constant bit rate (CBR) streams were used. CBR stream was transmitted using RTP [8]. Maximum payload size of RTP packets was set to 188 bytes (small packets) or 1000 bytes (large packets).

Each experiment consists of six phases:

- 3 Mbps CBR over RTP, without background TCP traffic (topology *T*1), infinite buffers,
- 3 Mbps CBR over RTP, with background TCP traffic (topology *T*2), infinite buffers,
- 3 Mbps CBR over RTP, with background TCP traffic (topology *T*2), finite buffers,
- 6 Mbps CBR over RTP, without background TCP traffic (topology *T*1), infinite buffers,
- 6 Mbps CBR over RTP, with background TCP traffic (topology *T*2), infinite buffers,
- 6 Mbps CBR over RTP, with background TCP traffic (topology *T*2), finite buffers.

The simulation was run for 5 (simulated) minutes. As measures of quality of transmission, bandwidth, loss and delay parameters were used. In particular, QoS of the real-time stream was measured by throughput, goodput, end-to-end delay, delay variation and packet error rate. QoS parameters of the TCP flow includes throughput, goodput and error rate.

15.5 An Analysis of Burst Controlled Real-Time Multimedia Transmission

In the section, intra- and inter-session behavior of burst controlled multimedia will be discussed. In intra-session experiments we explore, how the RTP payload length and propagation delay impact the QoS measures (obtained for both, elastic and inelastic traffic). In inter-session experiments QoS measures were observed as a function of a number of competing flows and as a function of bit rate of single multimedia stream.

15.5.1 QoS Measures as a Function of RTP Payload Length

In experiment, we vary RTP payload length from 100 B to 64000 B, while TCP's maximum segment size was set to 1000 B. The propagation delay at the bottleneck link was set to default value 5 ms. Because the buffer size is expressed in packets, RTP payload length has strong impact in TCP-unfriendly behavior of RTP protocol [3].

In the experiment we observe the general ability of burst controlled RTP to transmit video traffic in real-time.

For all analyzed conditions, TCP-tolerant system behaves stable for any value of the RTP payload length. Achieved goodput have arisen directly from the source's bit rate and does not depend on RTP payload length (Figure 15.4a). Larger RTP throughput in the case of smaller values of payload length (Figure 15.4b) is caused by larger influence of overheads. Overheads have also influence on end-to-end delay.

In contrast, TCP-intolerant system (without burst control mechanism – Figures 15.4c and 15.4d) behaves unstable for small and medium values of RTP payload length. This instability is characterized by low goodput/throughput, which results from typical for TCP-intolerance (and TCP-unfriendliness) large packet losses. For example, in the burst controlled system, packet losses (RTP+TCP) never exceeds 3%, while in the TCP-intolerant system we observe losses larger than 80%. It is worth to remark that applying of burst control also decreases the delay variation. The graph of delay variation vs. RTP payload length shows that delay variation of the TCP-tolerant system is usually at least 10 times smaller than delay variation of the system without burst control (even if observed in the TCP-tolerant regions of characteristics of the system without burst control). It is caused by less asynchronous character of burst-controlled traffic.

However, the suggestion that burst control is unnecessary – it will be sufficient to increase RTP payload to achieve TCP-tolerance – is only theoretical. In Figures 15.4c and 15.4d, 3 Mbps CBR stream achieves stability for RTP payloads equal to or larger than 1500 B, while 6 Mbps CBR stream achieves stability for RTP payloads equal to or larger than 5000 B. In practice, RTP payloads are limited by network MTUs and practical limitation (taken from multiple field trials) of RTP payload (as well as TCP's MSS) is about 1400 B in the case of long-distance transmission and no more than 1500 B for site-local transmission.

Characteristics of TCP throughput vs. RTP payload length (Figures 15.4e and 15.4f) shows that burst control gives good TCP-tolerance for any value of the RTP payload length. In the case of finite buffers (13 packets) TCP's throughput/goodput is always larger than in the case of the system without burst control. Even in the worst case (6 Mbps CBR, RTP payloads equal to 100–150 B) it is possible to achieve TCP transmission (0.54–1.2 Mbps), while in the system without burst control TCP transmission totally collapses (6–9 kbps).

In the second experiment we analyze multicast session scalability. Video session size was varied according to topology *T3*. In the three experiments ($n = 5$, $n = 10$ and $n = 30$) we obtained the same graphs as in the case of single receiver (topology *T2*).

15.5.2 QoS Measures as a Function of Propagation Delay

We repeated experiments of the previous subsections but vary the propagation delay of bottleneck link from 1 ms to 1 s, while RTP payload length was set to 1000 B (and is equal to TCP's MSS).

Although propagation delay have, generally, small influence on RTP transmission, longer delays at the bottleneck link increases delays in the TCP's congestion control loop. TCP becomes less aggressive, what can be important in the case of investigations of RTP/TCP coexistence in the same link.

Results show that TCP-tolerant system behaves stable for any value of the propagation delay (Figure 15.5a). As previously, achieved goodput/throughput of RTP arises directly from the source's bit rate and does not depend on the propagation delay. Small packet losses (less than 5%) are observed only for small and medium delays and for delays greater than or equal to 20 ms packet losses are close to 0 (0.0004% and less). In contrast, if we test TCP-intolerant system, RTP packet losses are stable and keep at constant level (about 50% for 3 Mbps CBR traffic and about 75% for 6 Mbps CBR traffic – typical behavior of TCP-intolerant/TCP-unfriendly system).

Propagation delay has strong dependency on time-based QoS measures. End-to-end delay is directly proportional to propagation delay (Figure 15.5b) and there is no visible difference between characteristics of TCP-tolerant and TCP-intolerant system. However, for propagation delays less than 20 ms, the delay variance ranges from about 2×10^{-7} to about 7×10^{-6} and is about 10 times smaller than in the case of TCP-intolerant transmission. For propagation delays greater than or equal to 20 ms, delay variance of TCP-tolerant system is always smaller than 2×10^{-7} (10–100 times smaller than observed for TCP-intolerant multimedia).

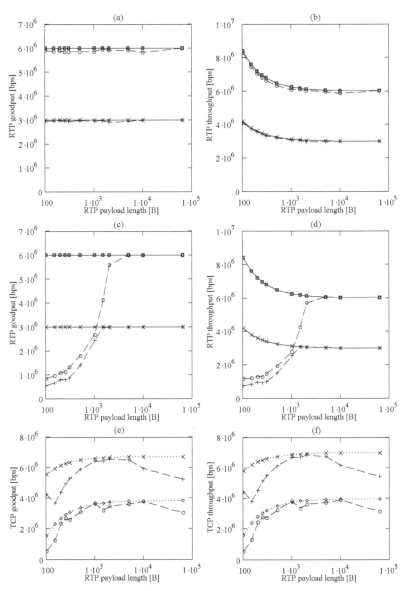

Figure 15.4 Goodput and throughput as a function of RTP payload length: (a), (b), (c), (d) RTP protocol; (e), (f) TCP protocol; (a), (b), (e), (f) with burst control; (c), (d) without burst control. Legend: 3 Mbps CBR over RTP transmission: without background TCP traffic, infinite buffers (solid line, no additional symbol), with background TCP traffic, infinite buffers (dotted line, ×), with background TCP traffic, finite buffers (dashed line, +); 6 Mbps CBR over RTP transmission: without background TCP traffic, infinite buffers (solid line, *), with background TCP traffic, infinite buffers (dotted line, ◊), with background TCP traffic, finite buffers (dashed line, o).

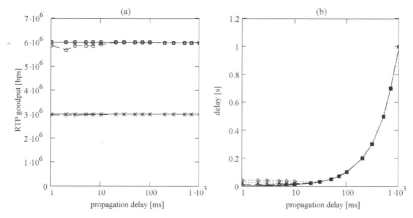

Figure 15.5 QoS measures of RTP transmission: (a) RTP goodput versus propagation delay, (b) mean end-to-end delay versus propagation delay. Legend: as in Figure 15.4.

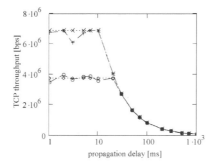

Figure 15.6 TCP throughput vs. propagation delay. Legend: as in Figure 15.4.

TCP's QoS characteristics have shape typical for TCP transmission (Figure 15.6). In the case of TCP-tolerant system transmission was stable for all tested propagation delays, while in the case of TCP-tolerant system stability was achieved only for very large delays (greater than or equal to 100 ms). However, TCP's performance in this region of characteristics is, typically, very low.

As previously, in the second experiment we analyze multicast session scalability. In the network of topology $T3$ we randomly chose propagation delays of links between receivers and their nearest router. In the three experiments ($n = 5$, $n = 10$ and $n = 30$) we observe that characteristics of QoS measures vs. propagation delay are strictly the same as in the case of single receiver (topology $T2$).

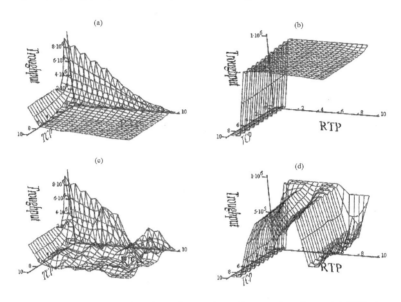

Figure 15.7 QoS measures as a function of a number of competing flows: (a) TCP through-put in system with burst control, (b) RTP throughput in system with burst control, (c) TCP throughput in system without burst control, (d) RTP throughput in system without burst control.

15.5.3 Inter-Session Experiments

In the first experiment, i TCP flows ($i = 1, 2, \ldots, 10$) compete for 10 Mbps bandwidth with j CBR video streams ($j = 1, 2, \ldots, 10$), 1 Mbps each. It is worth to remark that if $j = 10$, the network is not well dimensioned for live video transmission.

Results shows that for all tested i and j, burst control is able to preserve real-time character of the multimedia stream (Figure 15.7b). Moreover, burst control always manifest strong TCP-tolerant behaviour – TCP is able to per-form reasonably fair and stable transmission (Figure 15.7a). In the case of multimedia transmission without burst control (Figures 15.7c and 15.7d), we observe a degradation of both, TCP and RTP, flows.

In the second experiment, one ftp flow competes for 10 Mbps band-width with one CBR video stream. The target bit rate of video source varies from 1 to 10 Mbps. Results depicted in Figure 15.8, show that TCP-tolerant transmission scheme is able to preserve real-time character of multimedia transmission and, simultaneously, is able to avoid collapse of TCP transmis-sion (Figure 15.8a). Such a situation is practically impossible if the real-time

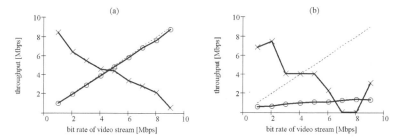

Figure 15.8 Throughput of RTP (○) and TCP (×) as a function of bit rate of video stream: (a) system with burst control, (b) system without burst control. Dotted line – throughput equal to bit rate of video stream.

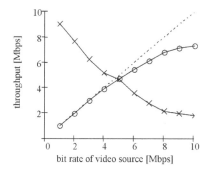

Figure 15.9 Throughput of TFRC (○) and TCP (×) as a function of bit rate of video stream [4]. Dotted line – throughput equal to bit rate of video stream.

system does not implement TCP-tolerance (Figure 15.8b). These properties of burst-controlled multimedia are observed even if bit rate of video source is close to the capacity of the bottleneck link.

It is worth to notice that obtained results are better than results of typical TCP-friendly transmission (Figure 15.9). TCP-friendliness achieved by TFRC protocol [10] will not influence on TFRC real-time characteristics only if there is possibility to fair bandwidth allocation (bit rate of video source is equal or less than half the bottleneck link capacity).

15.6 Problems of Buffer Occupancy in Multimedia Content Distribution

In this section we discuss a problem of buffer occupancy during the burst controlled multimedia transmission. In experiments we explore, how the RTP payload length, and propagation delay and bitrate of video stream impact the mean and variance of buffer occupancy.

15.6.1 Mean and Variance of Buffer Occupancy as a Function of RTP Payload Length

As in Section 15.5.1, we vary RTP payload length from 100 B to 64000 B, while TCP's maximum segment size was set to 1000 B. The propagation delay at the bottleneck link was set to default value 5 ms.

In the case of CBR video over RTP without background TCP traffic and infinite buffers on bottleneck link, mean buffer occupancy is smaller for the system, which implements burst control (Figures 15.10a and 15.10b). It is especially visible for small values of RTP payload length, where buffer occupancy is hundreds of times smaller for burst controlled RTP than for transmission without burst control. In the case of large values of RTP payload length (equal to or larger than 10000 B), difference between system with and without burst control is rather small. However, RTP payloads are limited by network MTUs and practical limitation (taken from multiple field trials) of RTP payload (as well as TCP's MSS) is about 1400 B in the case of long-distance transmission and no more than 1500 B for site-local transmission.

In the case of CBR video over RTP with background TCP traffic and both, infinite (Figures 15.10c and 15.10d) and finite buffers on bottleneck link, mean buffer occupancy of the system with burst control is close to the mean buffer occupancy of the system without burst control.

For all six experiments, variance of buffer occupancy always was smaller for burst controlled multimedia transmission, if the RTP payload length does not exceed 10000 B (Figures 15.11a and 15.11b). This is caused by less asynchronous character of burst-controlled traffic.

15.6.2 Mean and Variance of Buffer Occupancy as a Function of Propagation Delay

As in Section 15.5.2, we vary the propagation delay of bottleneck link from 1 ms to 1 s, while RTP payload length was set to 1000 B (and is equal to TCP's MSS).

Let us remind that although propagation delay have rather small influence on RTP transmission, longer delays at the bottleneck link increases delays in the TCP's congestion control loop. Thus, in the case of longer delays TCP

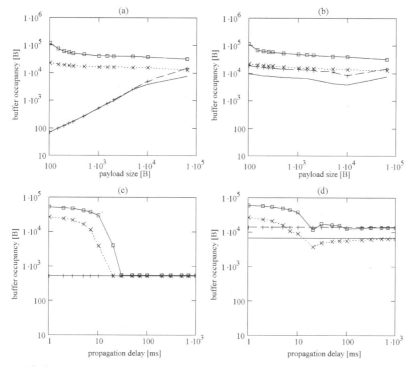

Figure 15.10 Mean buffer occupancy versus: (a), (b) RTP payload length; (c), (d) propagation delay; (a), (c) with burst control; (b), (d) without burst control.
Legend: 3 Mbps CBR over RTP transmission: without background TCP traffic (solid line, no additional symbol), with background TCP traffic (dotted line, ×); 6 Mbps CBR over RTP transmission: without background TCP traffic (dashed line, +), with background TCP traffic (solid line, ∗).

becomes less aggressive, what is observed in Figures 15.10 and 15.11 as a stabilization of characteristics at the level determined by transmission without background traffic and with infinite buffers on bottleneck link. In the case of all experiments, burst controlled transmission leads to:

- comparable mean buffer occupancy for smaller values of propagation delay (Figures 15.10c and 15.10d),
- smaller mean buffer occupancy for larger values of propagation delay (Figures 15.10c and 15.10d),
- significantly smaller variation (Figures 15.11c and 15.11d),

when compared with the system, which does not implement burst control.

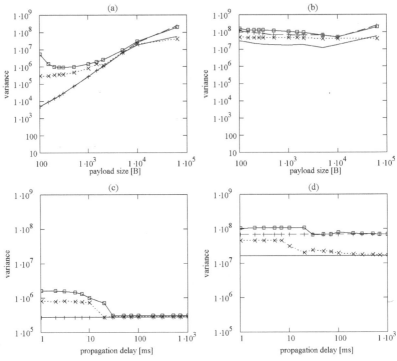

Figure 15.11 Variance of buffer occupancy versus: (a), (b) RTP payload length; (c), (d) propagation delay; (a), (c) with burst control; (b), (d) without burst control. Legend: as in Figure 15.10.

15.6.3   Mean and Variance of Buffer Occupancy as a Function of Bit Rate of Video Stream

In the last experiment, one ftp flow competes for 10 Mbps bandwidth with one CBR video stream. The target bit rate of video source varies from 1 to 10 Mbps.

Results depicted in Figure 15.12 show that TCP-tolerant transmission scheme gives comparable mean buffer occupancy as the system without burst control. However, if the network is well dimensioned for live video transmission (in Figure 15.12 – the bit rate of video stream should be less or equal to 9 Mbps), the burst controlled multimedia transmission is characterized by significantly lower variation of buffer occupancy.

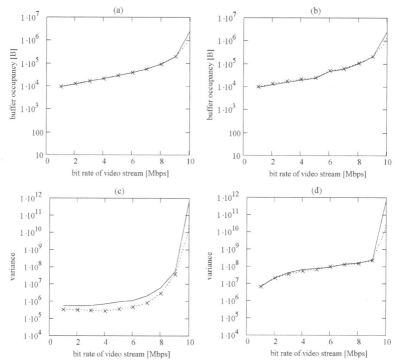

Figure 15.12 Statistical characteristics of buffer occupancy as a function of bit rate of video stream: (a), (b) mean buffer occupancy, (c), (d) variance; (a), (c) with burst control, (b), (d) without burst control. Legend: RTP payload length equal to 1000 B (solid line, no additional symbol), RTP payload length equal to 188 B (dotted line, ×).

15.7 Conclusions

TCP-tolerance can by defined as such behavior of transport protocol, which allows TCP to utilize bandwidth unoccupied by real-time multimedia transmission. TCP-tolerant transmission scheme is able to preserve real-time character of multimedia transmission and, simultaneously, is able to avoid collapse of TCP transmission. One of possible TCP-tolerant mechanisms is burst control, which spreads out burst over time, while the bit rate of the source stays unchanged.

An analysis of bust-controlled TCP-tolerant transmission, presented here, shows that applying burst control to real-time transport protocols results in growth in the quality of service of competing streams (if compared with the best-effort), measured by bandwidth, loss and delay parameters.

Moreover, the TCP-tolerance smoothes buffer occupancy, which is observed as a significantly lower variation of buffer occupancy when compared with TCP-intolerant system. Last but not least, one advantage of burst control is a very good scalability of multicast connections.

Acknowledgement

This work is partially supported by the Polish Government under grant No. N517 012 32/2108 (years 2007–2009).

References

[1] Braden B., D. Clark, J. Crowcroft, B. Davie, S. Deering, D. Estrin, S. Floyd, V. Jacobson, G. Minshall, C. Partridge, L. Peterson, K. Ramakrishnan, S. Shenker, J. Wroclawski and L. Zhang. L., Recommendations on queue management and congestion avoidance in the internet, RFC 2309, April 1998.

[2] A. Chodorek and R. R. Chodorek, Burst control, in *Proc. SympoTIC'04*, Bratislava, Slovakia, 24–26 October 2004, IEEE Press Cat. No. 04EX877, 2004.

[3] A. Chodorek, Coexistence of elastic and inelastic traffic for different buffer size in congested node, in *Contemporary Problems in Computer Networks. Applications and Security*, A. Crzywak and A. Kwiecien (Eds.), WNT, Warsaw, pp. 137–144, 2004 [in Polish].

[4] A. Chodorek, Streaming video with TFRC – Simulation approach, in *Proc. SympoTIC'04*, Bratislava, Slovakia, 24–26 October 2004, IEEE Press Cat. No. 04EX877, 2004.

[5] A. Chodorek, TCP and TCP-friendly protocols, in *Encyclopedia of Internet Technologies and Applications*, M. Freire and M. Pereira (Eds.), Information Science Reference, pp. 612–618, 2008.

[6] S. Floyd and K. Fall, Promoting the use of end-to-end congestion control in the internet, *IEEE/ACM Transactions on Networking*, vol. 4, pp. 458–472, August 1999.

[7] http://nsnam.isi.edu/nsnam/index.php/Main_Page

[8] H. Schulzrinne, S. Casner, R. Frederick and V. Jacobson, RTP: A transport protocol for real-time applications, RFC 3550, July 2003.

[9] A. Chodorek and R. R. Chodorek, An analysis of TCP-tolerant real-time multimedia distribution, in *Proc. HETNETs 2008*, Karlskrona, Sweden, February 2008.

[10] S. Floyd, M. Handley and J. Padhye, TCP friendly rate control (TFRC): Protocol specification, Internet-draft draft-ietf-dccprfc3448bis-06.txt, March 2008.

PART FIVE

CONGESTION CONTROL

16

Router-Assisted Congestion Control with Probabilistic Marking

António Almeida and Carlos Belo[†]

Instituto Superior Técnico/IT, Torre Norte, Av. Rovisco Pais, 1049-001 Lisboa, Portugal; e-mail: antonio.almeida@lx.it.pt

Abstract

Currently, the Internet lacks the native support for explicit-rate algorithms and congestion control has been limited to algorithms relying on implicit binary signaling. In spite of the improved performance of many proposals for explicit-rate algorithms, their implementation has found significant constraints in current networks, in part because these algorithms rely on a multi-byte congestion notification per packet. In this paper, we present a new approach based on binary probabilistic marking where congestion control can be efficiently exercised by router algorithms originally devised for explicit-rate marking. The proposed procedure includes a source algorithm and a router marking mechanism. We show, by simulation, that the behavior of the flow rates for sources subject to the proposed binary feedback is close to the behavior of sources subject to explicit-rate feedback.

Keywords: congestion control, binary signaling, probabilistic marking, binary source behavior, explicit-rate algorithms.

16.1 Introduction

In this paper we present a novel approach to probabilistic marking in the context of binary congestion control. We propose a new function for adapting the

---

[†] Deceased May 2007.

D. D. Kouvatsos (ed.), Traffic and Performance Engineering for Heterogeneous Networks, 339–369.

source rates under explicit binary feedback and a marking procedure for the routers. In earlier contributions, we have been looking to the aspects of stateless active queue management and also to the associated congestion signaling mechanisms among routers and sources. In [1] we present a router algorithm that performs congestion control without keeping per-flow information. In [2] we present a variant of this algorithm deriving its stability regions, along with extensive simulations showing its performance. More recently [3] we tested the algorithm of [2] in the control of modified TCP sources (with a conversion, at the sources, of flow rates into window sizes). Router algorithms of this kind, i.e. stateless active queue management schemes, are attractive in terms of implementation, in particular in networks where a large number of flows is expected to compete for bandwidth. The proposed algorithms were primarily devised for *explicit-rate* congestion control, where routers send multi-byte feedback messages to sources. However, we have also presented alternative mechanisms relying on *binary feedback* that can be used by these algorithms in an efficient way, providing close performance to that obtained with explicit-rate sources [4, 5]. In particular, the results obtained with these binary schemes show that the oscillations of source rates are very small, when compared to other binary schemes, and sources are able to sense the network variations in an efficient way. The rationale behind these mechanisms lies on the tracking of the values of fair share calculated by the routers by looking at the stream of binary indications arriving at the source. This way, sources adapt their source parameters dynamically in order to adjust the rates.

The probabilistic marking scheme presented here improves the performance of our previous mechanisms, in particular in terms of reaction times to the events of severe congestion. While the approach in [4, 5] is deterministic (both in the marking procedure at the routers and the source adaptation functions), the mechanism proposed here is based on a new strategy involving probabilistic marking and leads to a less complex source algorithm.

16.1.1 Related Work

The Internet, today, lacks the native support for explicit-rate signaling at the IP layer. Still, a significant number of contributions in recent years explore the router-assisted explicit-rate approach as an efficient method for congestion control, e.g. [6–10]. Even though a large number of proposals for explicit-rate feedback algorithms have been made, most of those proposals have found barriers in terms of implementation in the Internet, partly because these algorithms rely on the assumption of having multi-byte feedback mechanisms in the network. The ATM ABR protocol [11] is the only framework that included, from its conception, built-in mechanisms for explicit-rate congestion signaling. Several earlier contributions had significant importance on the inclusion of such mechanisms in the ABR protocol, such as e.g. [12–14], and many proposals thereafter have shown the clear advantages and effectiveness of having explicit-rate feedback for congestion control. However, congestion control in the Internet has been dominated by the mechanisms implemented in TCP [15], which rely on *implicit binary* feedback carried out through packet losses. A significant effort has been made though, for the inclusion of *explicit binary* signaling mechanisms in IP and TCP, through the ECN (*explicit congestion notification*) approach, which is an intermediate solution easier to deploy than multi-byte explicit-rate signaling. ECN was first proposed in [16] and has been accepted as a suitable alternative to the implicit signaling mechanisms currently available in the Internet [17, 18].

The subject of congestion signaling is strongly connected to the subject of fairness and stability of congestion control algorithms. We can trace back some performance studies of binary algorithms, along with early proposals for new algorithms. In [19] the authors present a model for binary congestion control and analyze existing schemes, looking in particular at the fairness and flow rate oscillations of these schemes. This work raises the fundamental concerns about binary feedback mechanisms, namely the short and long term fairness, the amplitude of oscillations of source rates, the transient behavior and the discrimination of flows under different propagation delays (a latent problem in TCP, identified in [20]). In the case of the Internet, since the first RED (Random Early Detection) proposal of [21], many others have been made to improve the TCP source behavior, in particular the aggregate behavior of many flows competing at the same router. The DUAL scheme [22] is one such early contribution, pointing out the concerns about the oscillatory patterns in the sending window size of TCP. This proposal includes both timeout and delay thresholds as congestion signals. TCP Vegas [23] detects the incipient stages of congestion before losses actually occur, improving the utilization of the network.

Other proposals have been made, this time seeking implementations based on ECN feedback, or variants of it. In an early contribution [24], the widespread versions of TCP are compared to a modified mechanism where sources receive explicit binary indications of congestion. Although the results were encouraging, the effectiveness of this approach depends heavily on the efficiency of the router algorithms, which raised the possibility of providing TCP/IP with explicit-rate signaling generated by properly designed router algorithms. In [25] the authors propose a modification to TCP, to work above ECN, in order to overcome the bias against connections with long round trip delays. In [26] methods for controlling directly the rate of TCP were proposed, using a rate-based algorithm that estimates the fair share of flows and translates this rate to a window size. The calculated window size is then explicitly fed back to the sources. Their results show improvements on the fairness, throughput and end-to-end delay properties for TCP applications. The authors in [27] also suggest the inclusion of a rate-based algorithm in TCP, by sending the value of the estimated rate in the IP or TCP headers.

Binary notification mechanisms have been closely associated to the *additive increase, multiplicative decrease* (AIMD) source rate adjustment scheme. This source algorithm is very simple in terms of implementation, particularly when routers do not intervene in an active way (e.g. the TCP end-to-end approach). In [28] a simple analysis is presented showing its fairness, using a simplified model where the flows have the same increase and decrease factors and share the same bottleneck. However, in general, these assumptions do not hold and there are scenarios where this source algorithm leads to unfair rate allocations. In an early contribution [29], the AIMD algorithm was proposed to inter-operate with an implicit signaling mechanism based on the variations of the round-trip delays. The DECbit protocol [30] was implemented with the same source algorithm, this time using explicit binary indications of congestion. The DECbit variant with selective notifications [31] was shown to exhibit a better behavior concerning the beat-down effect of binary schemes. In the context of proposals for ATM ABR, the AIMD mechanism was also considered as the suitable source behavior function for binary congestion control, although subsequent studies have shown that the tuning of the source parameters had a major impact in the effectiveness of the control schemes. The contributions by Vojnović and Le Boudec [32] and Ritter [33] address the oscillatory behavior, buffer requirements, and the fairness of ABR algorithms subject to binary feedback. A performance analysis for these algorithms was also carried out in [34, 35], showing that the buffer requirements and allocation fairness are closely dependent on the oscillatory behavior of the flow rates.

The concerns about the oscillatory behavior of TCP have been reported in many proposals for new strategies in the increase and decrease actions of TCP sources. In [36] the authors present a general additive increase, multiplicative decrease scheme, with a formulation where the control is performed at the sources, as opposed to having a control centered at the routers. Although the scheme in [36] was proposed to be implemented at sight of a TCP-friendly behavior, it provides ground for consideration of dynamic adjustment parameters at the sources. In terms of objectives for the source behavior, our proposal in [5] is similar, although with control actions centered at the routers. The mechanism proposed in this paper is also an alternative to the AIMD algorithm, in the sense that sources adapt dynamically their parameters for rate adjustment, according to the received feedback.

The idea of *probabilistic marking* was introduced in early proposals involving active queue management schemes for the Internet, namely in the context of RED algorithms and variants of it. In most of these proposals, probabilistic marking is not intended for the explicit communication of a particular parameter from the router to the sources. Instead, the objective is to signal a state of congestion to sources by marking packets in proportion to their flow rates. The marking probability on these schemes depends usually on the average queue length of the router or other measured parameters. Also, in most of the proposed schemes, sources implement the AIMD source algorithm.

The proposal of [37] is, to our knowledge, the first scheme intended to communicate multi-valued messages to sources using some sort of probabilistic marking. The authors present a model involving just one bottleneck in the network and two synchronized sources. Although it is not clear how the scheme behaves in the presence of other intervening routers and the granularity of communicated values is too coarse, their proposal is close to ours in terms of objectives. Probabilistic marking has also been suggested in the context of algorithms formulated as optimization problems involving source utility functions. In the scheme of [38] a fraction of packets is marked by the routers according to a function of the total arrival rate and the virtual capacity of the router. The total marks are then distributed among the users in proportion to their flow rates. The random marking scheme of [39, 40] is based on the distributed algorithm described in [41] where sources receive the sum of all prices calculated by the routers in their flow's path. In order to communicate this aggregate feedback, the authors in [39,40] propose marking packets, at each router, with a probability that depends on the congestion price. This way, it is expected that the total fraction of marked packets can be used by the sources to estimate their path prices. The authors in [42] propose also to send aggregate feedback to sources, originating from all routers, using ECN marks that would be produced by an OR operation in their flows' path.

The previous proposals involving aggregate feedback from the routers share a common problem. They are based on a fluid flow model, where it is assumed that there is an infinite number of packets flowing from one source to the routers. With this assumption, the probability of a packet marked by one router being considered as a target packet to be marked by another router is negligible. In this case, it is reasonable to assume that the total marks provide a good estimate of the path's marking probability. However, the fluid flow approximation is not suitable for most real scenarios where the frequency of packets available for marking may be small. The measured probability at the source may provide only a coarse estimate of the aggregate probability of the path, usually much less than the real value. This problem was recognized in [42], where the authors have proved that their algorithm might not converge in the case of single bit feedback. Instead, they have proposed to use 1 byte for feedback, involving a SUM operation at the routers. These results raise concerns about the implementation of distributed algorithms that are formulated under a global optimization approach, since these algorithms require the communication of aggregate shadow prices calculated at the routers (i.e. the optimization parameters exchanged among the entities involved in the distributed optimization). It is not clear how to implement these algorithms efficiently using a binary feedback infrastructure.

16.2 Probabilistic Marking Procedure

Let us consider a class of router algorithms that calculate, periodically, the value of *fair share*, denoted by p_k, which is a parameter to be transmitted to sources. Assume that the values of p_k are calculated at the end of router evaluation intervals, I_k. Sources are controlled by one router (the bottleneck in their path) and the overall operation can be viewed as a control loop, with the forward path of data packets containing a CI (*congestion indication*) bit, and the backward path of binary notifications. Similarly to the ECN approach [17], we assume that the CI bits marked by the routers in the forward path are echoed back to sources by some means (for instance, through transport acknowledgment packets). In addition, assume that the current values of source rates are available to routers by some means. This is necessary for most router algorithms that do not keep per-flow information (i.e. *stateless* algorithms) and, in our case, it is necessary because the marking probability depends also on the current value of source rates. For an IP-based implementation, we suggest the forward rate encoding of [4], which is a scheme that keeps encoded flow rates inside the regular IP headers, at cost of a small encoding error. In any case, the objective is to transmit to sources the values of p_k using a stream of binary indications (instead of multi-byte feedback messages). In the description that follows, we use the notation p_k and $p(k)$ interchangeably, while referring to the router fair share.

16.2.1 Router Marking Algorithm

Denote by r_i the rate of flow i encoded in each forward packet and define the following function:

$$q_i(k) = 1 - \alpha e^{-\beta \Delta_k(r_i)}, \quad \Delta_k(r_i) \geq 0, \tag{16.1}$$

where $\Delta_k(r_i)$ is given by

$$\Delta_k(r_i) = \frac{r_i - p(k)}{p(k)},$$

and α, β are positive parameters. The router proceeds as follows for each incoming packet:

- If $r_i > p_k$ then mark the packet with probability $q_i(k)$;
- If $r_i \leq p_k$, leave the packet untouched.

Packets are marked by setting the CI bit to 1 in the packet header. Thus, the marking probability depends on the difference between the encoded rate r_i and the current parameter $p(k)$, normalized to $p(k)$. Given that the marking decision is combined with the selective notification mechanism, only packets whose encoded rates are above $p(k)$ are entitled for marking, which means that the marking probability is calculated only when $r_i > p(k)$, being zero otherwise. This procedure has important consequences in terms of the estimation algorithm at the sources, as we shall discuss below. The target equilibrium is reached for flow i when, for each incoming packet, the difference $(r_i - p(k))$ is very small and the function (16.1) takes values close to the *equilibrium marking probability*

$$q(k) = 1 - \alpha.$$

16.2.2 Source Algorithm

Sources collect the feedback indications and estimate the corresponding marking probability. In basic terms, for each period n corresponding to a round-trip delay (rtt_i), the sources collect the feedback indications and estimate the marking probability through

$$\widehat{q}_i(n) = \frac{N_c(n)}{N_t(n)}, \tag{16.2}$$

where N_c is the number of decrease indications (i.e. indications with $CI = 1$) and N_t the total number of binary feedback indications. Through $\widehat{q}_i$, the source i adjusts the flow rate at the end of each interval rtt$_i$ using the inverse function of (16.1). This is equivalent to estimating the value of the parameter $p(k)$ and setting the flow rate to this value. Denoting by $r_i(n)$ the flow rate calculated by source i at the end of interval n, the rate is adjusted as follows:

$$r_i(n+1) = r_i(n) \, \frac{\beta}{\beta - \ln\left(\frac{1-\widehat{q}_i}{\alpha}\right)} \, . \qquad (16.3)$$

By considering in (16.1) the normalized difference between r_i and $p(k)$, instead of the absolute value of that difference, we have smaller encoding errors. This is due to the fact that the amplitude of the normalized values is much smaller than the amplitude of the absolute difference, which leads to smaller encoding errors for the same number of marked bits.

The procedure just described relies on the availability of enough samples per round-trip delay. This way, sources can rely on the values of estimated marking probability and adjust their rates. For scenarios where the round-trip delays are small and packet length is large, we may have a very small sampling rate. For these extreme cases, the direct application of (16.2) and (16.3) would be useless. To deal with this problem (as well as others related to several practical issues) we describe below an implementation where we have filtering procedures for both the rate adjustment process and estimation of marking probability. We tested several scenarios where the sampling frequency is very small (e.g. reaching average values between 1 and 2 samples per RTT), and still reached encouraging results. Below, we present simulation results (in experiment 2) where some flows exhibit an average sampling frequency of 2.4 packets per RTT, during some periods (flow rates around 400 Kbit/s, round-trip delays close to 50 ms and packets with 1024 bytes).

16.2.3 The Role of the Selective Binary Mechanism at the Routers

According to the previous procedure, the router only entitles packets to be marked, with a certain probability, when $r_i > p(k)$. The main reason for having this selective notification mechanism associated to the probabilistic marking function is to avoid the interference of feedback signals coming from different routers in the flow's path. Since we want to preserve the communication between each source and its bottleneck router (i.e. preserving the control loop of the algorithm with each source), the sources must not receive binary indications originating from the routers in their path, except their bottleneck.

The interference of signals would be severe if, for example, two routers in the flow's path had close values of fair share $p(k)$. Having approximate values for $p(k)$, the marks received at the sources and sent by these two routers would certainly lead to an erroneous estimation of the fair share at the sources. This would be a similar problem to that observed with proposals based on a fluid feedback assumption providing aggregate feedback (as mentioned in the Introduction).

In terms of implementation, the selective marking mechanism requires additional care in the design of the source algorithm, as discussed below. Since the sources sense feedback intervals where packets are marked by the bottleneck routers according to (16.1), alternating with intervals where the packets are not marked at all, sources have to probe for bandwidth in the absence of congestion marks. That is, sources have to estimate *implicitly* a marking probability while increasing their rates (a procedure described in the following section).

16.2.4 Implementation Issues

The function defined in (16.1) is depicted in Figure 16.1. Due to the selective notification mechanism, the marking procedure is valid for positive $\Delta_k(r_i)$, i.e. for $r_i > p(k)$. Thus, the curve of interest is defined only for $\Delta_k(r_i) > 0$, i.e. when a packet is entitled for probabilistic marking. When the flow rates are below the link parameter, i.e. $r_i \leq p(k)$, the packets are never marked and therefore sources have to probe for bandwidth. When this happens, the objective is to drive the source algorithm to a region as close as possible to the dotted segment of Figure 16.1 such that the values of $\Delta_k(r_i)$ are close to zero.

When $r_i > p(k)$ sources can always estimate the fair share through (16.2) and (16.3) or, at least, filtered versions of these procedures, because the stream of packets is marked according to the probability function (16.1). This is a very useful characteristic of this mechanism, because when sources are invited to decrease they are able to react fast and track accurately the decreasing values of $p(k)$.

Adaptive increasing of flow rates. During the lifetime of the flows, the sources are subject to probing for bandwidth phases, possibly with long periods where feedback indications have $CI = 0$. This is the case e.g. when the bottleneck capacity increases in a significant way or a large group of flows leaves the network. During the probing for bandwidth phases the sources

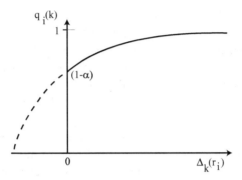

Figure 16.1 Marking probability function. The implementation at the routers does not include the dotted curve segment due to the selective notification mechanism.

increase their rates at the end of each round-trip interval, by decreasing the estimated probability iteratively, with adaptive step sizes. This way the flow rates are increased by variable amounts, until r_i reaches the value of $p(k)$ and eventually triggers a binary mark at the router. The probing for bandwidth phase stops when the first notification with $CI = 1$ arrives to the source.

This adaptive process proceeds as follows. When the first period (equivalent to a round-trip interval) without decrease notifications is detected, sources start with the target probability $(1 - \alpha)$ and a counter $N_i(n)$ set to 1. Sources update $\widehat{q}_i$, at the end of each interval, as follows:

$$\widehat{q}_i(n + 1) = \max\left(\widehat{q}_{\min}, (1 - \alpha) - \Lambda_q N_i(n)\right), \qquad (16.4)$$

where $N_i(n)$ is increased by 1 at the end of each consecutive interval and Λ_q is a small positive factor. The procedure stops when at least one decrease indication is detected (at this point, N_i is reset). If the probing for bandwidth is long enough, in particular, when a significant increase period is detected, the estimated probability reaches its lower limit, $\widehat{q}_{\min}$, which is a parameter obtained heuristically (in the experiments we have been using $\widehat{q}_{\min} = 0.05$). The parameter Λ_q is chosen taking into account the fine-tuning objectives of the adjustment procedure when the network conditions are varying slowly, and also the need for fast reactions when significant periods of implicit invitations are detected. We found, through simulation, that setting Λ_q to values between 0.05 and 0.06 leads to a good compromise among the two objectives. Also, from the experiments it is apparent that the procedure is insensitive to the number of flows and link capacities. Note that sources use (16.4) in the update formula (16.3) (or, alternatively, in the filtered version (16.6), described below) only when their flow rates increase. Otherwise, sources switch to the procedures described next.

Fine-tuning estimation of flow rates. When there are not wild variations on the bottleneck fair share (which is, in practice, the prevalent scenario for a large population of flows) the sources sense periods of binary notifications containing mixed decrease and increase invitations. That is, sources do not sense long periods with either $CI = 0$ or $CI = 1$. In this case, $\widehat{q}_i$ can be calculated using (16.2). However, the estimation procedures are subject to errors that might be non-negligible when e.g. the frequency of notifications is very small. The basic approach to limit these errors is to filter both the estimated probability and the flow rates. Taking the relations (16.2) and (16.3) into account, the sources update the estimated probability through

$$\widehat{q}_i(n+1) = \beta_d \, \widehat{q}_i(n) + (1 - \beta_d) \frac{N_c(n)}{N_t(n)}. \tag{16.5}$$

During the probing for bandwidth phases, the estimated marking probability switches to (16.4). In addition, the source rates are subject to the following filtering procedure:

$$r_i(n+1) = \beta_r \, r_i(n) + (1 - \beta_r) \, r_i(n) \frac{\beta}{\beta - \ln\left(\frac{1-\widehat{q}_i}{\alpha}\right)}. \tag{16.6}$$

The filter parameter β_d in (16.5) is changed according to the relation

$$\gamma_d = \frac{N_c(n)}{N_t(n)} \cdot \frac{1}{\widehat{q}_i(n)},$$

i.e. it depends on the variation of the most recent measured probability relative to the current estimated probability $\widehat{q}_i(n)$. We have been considering values for β_d ranging in the interval [0, 0.5], with $\beta_d = 0$ when $\gamma_d > 2$ and $\beta_d = 0.5$ when γ_d is close to 1. The parameter β_r in (16.6) is set to 0 during the probing for bandwidth phases, and varies in the interval [0, 0.5] for the remaining periods, according to the same thresholds of γ_d defined above. Note that with $\beta_r = 0$ the flow rates are updated without any filtering restrictions and sources can react fast to a significant increase in the measured marking probability. Also, when $\beta_d = 0$, sources are giving full confidence to the most recent value of $N_c(n)/N_t(n)$ in order to estimate the marking probability. This way, for $\beta_r = 0$ and $\beta_d = 0$, a critical event that is likely to occur when there is a significant decrease on the bottleneck fair share, the control loop between the bottleneck router and the sources is preserved as much as possible (without additional delays introduced by filtering procedures).

The implementation may also include ceiling restrictions on the flow rates. Considering $R_{\min_i}$ and $R_{\max_i}$ as the minimum and maximum flow rates, respectively, the flow rates may be subject to these restrictions as follows:

$$r_i(n+1) = \max\left(R_{\min_i}, \min(R_{\max_i}, r_i(n+1))\right).$$

In the experiments presented below, the initial flow rate is set to $R_{\min_i}$, and the maximum flow rates are always set to sufficiently high values such that flows are never restricted by $R_{\max_i}$.

16.3 Simulation Results

We now present simulation results obtained with the proposed marking scheme. The simulations were carried out at the packet level, using a discrete-event simulator. We test the scheme with the stateless router algorithm presented in [2], although other explicit-rate algorithms proposed in the literature could be considered as well. For the current purposes, we summarize next the relevant characteristics of the router algorithm. The router updates the fair share $p(k)$, according to the *max-min* fairness criterion, at the end of evaluation intervals I_k, taking into account the available link capacity and the aggregated input rate (measured during each interval). In the following experiments we set $I_k = 100$ ms. The adjustments of $p(k)$ are made through dynamic gain functions that are set according to the robust stability regions of the algorithm [2]. At any time, the value of $p(k)$ represents the fair share to be communicated to sources.

The marking function (16.1) is set with $\alpha = 0.7$ and $\beta = 3.0$. The probing for bandwidth phases follow the procedure in (16.4) with $\widehat{q}_{\min} = 0.05$ and $\Lambda_q = 0.06$. All flows start with $R_{\min_i} = 100$ Kbit/s. The remaining parameters are set as explained in Section 16.2.4.

We conduct two experiments. In the first experiment, we introduce the discussion by testing a simple configuration of three routers in a chain, with co-located sources and destinations subject to moderate round-trip delays. In the second experiment, we test a configuration with diverse source and destination nodes, subject to a larger number of flows and significant round-trip delays. Also, we test harder transients caused by flows entering and leaving the network in a synchronized way, and sudden changes in the bottleneck link capacity.

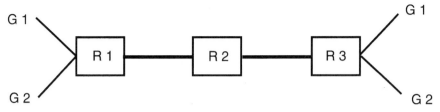

Figure 16.2 Simulated network configuration for **experiment 1**.

Since we want to assess the performance of the probabilistic marking scheme we also show, for comparison, simulation results for both experiments obtained with exactly the same settings, but having explicit-rate sources subject to multi-byte feedback messages (following an approach similar to e.g. the ABR protocol). Thus, in this case sources receive the exact values of $p(k)$ encoded in feedback messages. As a side effect, the results show also the merits of the router algorithm, in particular its stability and allocation efficiency. However, we keep the discussion around the performance of the proposed marking scheme, leaving aside as much as possible the specific aspects related to the router algorithm.

16.3.1 Experiment 1

The first experiment is based on the network configuration of Figure 16.2, with three routers in a chain, denoted by R1, R2 and R3. We have two groups of sources, denoted by **G1** and **G2** having, respectively, 30 and 70 rate-based greedy sources. We simulate flows with round-trip propagation delays uniformly distributed in the interval **[50–150] ms**. Given that the propagation delays between routers is the same for all flows, we set different delays for the flows by adding random propagation delays to the direct connection to router R1.

The groups of flows G1 and G2 enter the network at different times. Group G1 initiates at $t = 0.0$ s and G2 initiates at $t = 20.0$ s. Inside each group the flows start in a non-synchronized way: in G1, flows start every 0.02 s; in G2, every 0.3 s. All flows enter the network through router R1 and have destinations at R3. The simulation lasts 110 s and sources generate packets with length 1024 bytes. The total link capacity $\widehat{C}_k$ for **router R1** is set as follows:

$$\widehat{C}_k = \widehat{C} = 150 \text{ Mbit/s,} \quad \text{while } t < 40.0 \text{ s;}$$

$$\widehat{C}_k = \widehat{C} - 10 \times 10^6 (t - t_0) \text{ Mbit/s,} \quad \text{with } t_0 = 40.0 \text{ s}$$
$$\text{and } 40.0 \leq t \leq 48.0 \text{ s;}$$

$$\widehat{C}_k = 70 \text{ Mbit/s,} \quad \text{for } 48.0 < t < 80.0 \text{ s;}$$

$$\widehat{C}_k = 70 + 20 \times 10^6 (t - t_0) \text{ Mbit/s,} \quad \text{with } t_0 = 80.0 \text{ s}$$
$$\text{and } 80.0 \leq t \leq 110.0 \text{ s.}$$

For **router R2**, $\widehat{C}_k = 250$ Mbit/s at all times. The capacity of router R3 is irrelevant in this case, as all connections terminate at the output ports of R3. The available capacity, at each router, is set to $\rho \widehat{C}_k$ with $\rho = 0.9$.

We simulate a variation on the capacity of router R1 in the period $40.0 \leq t \leq 48.0$, for a rate variation of -10 Mbit/s^2, bringing its value to 70 Mbit/s at $t = 48.0$ s. For $t > 80.0$ s we simulate a severe increasing on the capacity of R1 so that when its value goes above 250 Mbit/s, the bottleneck changes to router R2 (an event occurring roughly at $t = 89.0$ s). This leads to a large period where all sources are invited to increase their rates.

Figure 16.3 shows the instantaneous flow rates of groups G1 and G2, using the probabilistic binary scheme. Figure 16.4 shows the instantaneous flow rates of groups G1 and G2, obtained with sources subject to explicit-rate marking (this time, the curves relative to the flows of both groups are super-imposed, as expected). Figure 16.5 shows the fair share $p(k)$ calculated by routers R1 and R2 over time, using explicit-rate feedback and binary sources subject to probabilistic marking. In Figure 16.5, at any time, the lower curves represent the bottleneck fair share. After roughly $t = 89.0$ s the curves invert positions due to the change of bottleneck routers (from R1 to R2). The upper curves in Figure 16.5 represent the values of $p(k)$ maintained internally by a non-congested router. In the implementation of the algorithm, non-congested routers always track the maximum observed rates passing through them, setting $p(k)$ roughly 30% above this maximum. This procedure avoids the unbound growing of $p(k)$ during non-congested periods.

Comparing the results of Figures 16.3, 16.4 and 16.5, we observe that the binary marking scheme shows a close performance in tracking the variations of the bottleneck fair share $p(k)$ to that obtained with explicit-rate sources. The estimation errors of the source rates lead to small oscillations around the values of $p(k)$ calculated by the bottleneck router. In particular, we observe in Figure 16.5 that the probabilistic marking scheme does not have significant influence on the stability of the algorithm, preserving the control loop in an

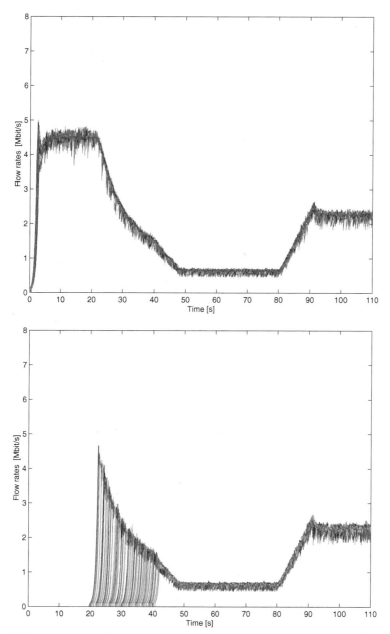

Figure 16.3 Instantaneous flow rates obtained in **experiment 1** (samples every 50 ms), using *probabilistic marking*. Top, group G1; bottom, group G2. To be compared to Figure 16.4.

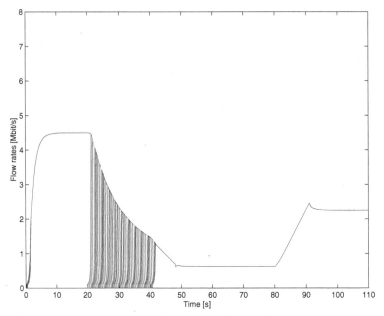

Figure 16.4 Instantaneous flow rates for groups G1 and G2 obtained in **experiment 1** (samples every 150 ms), using *explicit-rate marking*. To be compared to Figure 16.3.

effective way. On the other hand, the upper curves in Figure 16.5 exhibit differences because the link parameter of non-congested routers always track the maximum observed rates (setting $p(k)$ roughly 30% above these). Since the instantaneous flow rates of the binary sources oscillate around the equilibrium fair share of their bottleneck, the same happens to the values of $p(k)$ of non-congested routers.

From these results, we conclude that the deviations around the advertised fair share for binary sources is reasonably controlled and do not affect the stability of the router algorithm. In average, the behavior of the flow rates is close to that of explicit-rate sources which exhibit, as expected, a perfect matching in terms of instantaneous flow rates (Figure 16.4).

In the period $0 < t < 20$ s, the fair share converges to the target value of $(150 * 0.9)/30 = 4.5$ Mbit/s, while flows enter the network within a very short interval (each flow starts every 0.02 s). Thus, the set of 30 curves is practically superimposed in the graph. On the other hand, although explicit-rate sources also enter the network in a non-synchronized way, after a short

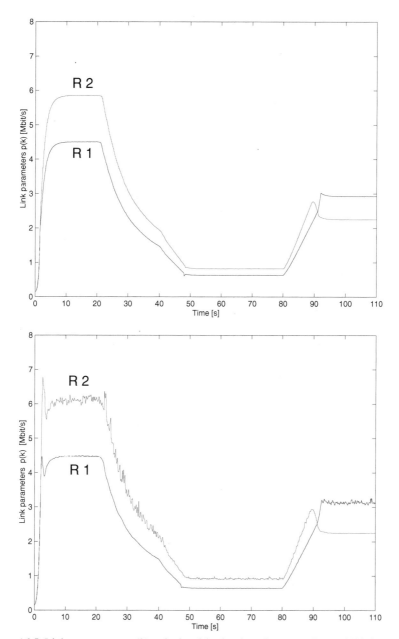

Figure 16.5 Link parameters $p(k)$ calculated by bottleneck routers R1 and R2 in **experiment 1**. Top: with *explicit-rate marking*; bottom: with *probabilistic marking*.

period they synchronize perfectly their rates all together since these sources receive a common multi-byte *explicit* value of $p(k)$.

After $t = 20.0$ s, the value of $p(k)$ decreases due to the new flows of G2 entering the network. In this period, each flow enters the network every 0.3 s and we can see all curves in the graphs. The probabilistic marking scheme is very effective in driving the flow rates currently in the network to lower levels, always in a controlled way and following the decreasing values of $p(k)$. It is apparent that, when sources decrease their rates, the rate oscillations around $p(k)$ are smaller, as compared to the periods where $p(k)$ varies slowly. This is due to the fact that sources have more consecutive intervals where feedback messages are generated under the condition $r_i > p(k)$ (with bits with $CI = 1$ and $CI = 0$ mixed up). In practice, the resulting feedback stream carries more information to the sources and these have the chance of calculating the marking probability with smaller estimation errors.

After $t = 40.0$ s, the fair share continues to decrease, this time due to the variation in the capacity of R1 (until $t = 48$ s). In the period $48 < t < 80$ the target fair share is given by $(70 * 0.9)/100 = 0.63$ Mbit/s, with 100 flows competing for the bandwidth of R1. After $t = 80.0$ s, the capacity of R1 increases, the flow rates increase and after a while R2 becomes the bottleneck, with a fair share of $(250 * 0.9)/100 = 2.25$ Mbit/s. The transition of bottlenecks, from router R1 to R2, happens in a transparent way to the sources.

Finally, Figures 16.6 and 16.7 show the bottleneck queue occupancy obtained with the binary marking scheme and explicit-rate sources, respectively. For this scenario, the occupancy of the routers, with both types of feedback, remains small at all times. We observe some activity on the queue of router R2, after $t = 90$ s, when R2 becomes the bottleneck.

16.3.2 Experiment 2

In the second experiment, we test the network configuration depicted in Figure 16.8 with five routers in a chain, named **R1–R5**. This scenario involves a total of 240 flows, generating from rate-based greedy sources distributed among five different groups, named **G1–G5**. Figure 16.8 depicts the source and destination for the flows of the different groups. The groups have **70, 70, 40, 20,** and **40** flows, respectively. In this scenario we simulate flows with round-trip propagation delays uniformly distributed in the interval **[50–300] ms**. The experiment lasts 110 s and sources generate packets with length 1024 bytes.

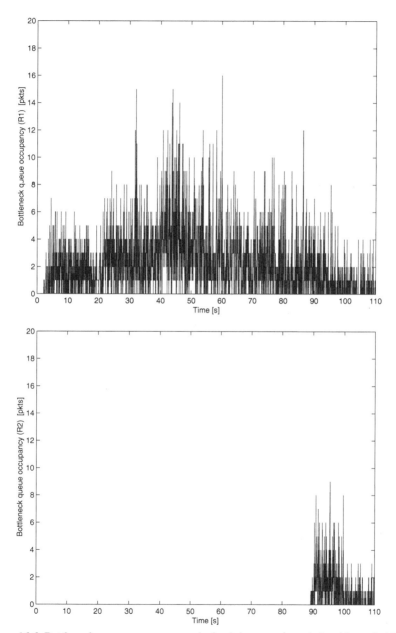

Figure 16.6 Bottleneck queue occupancy obtained in **experiment 1** with *probabilistic marking*. Top, router R1; at bottom, router R2.

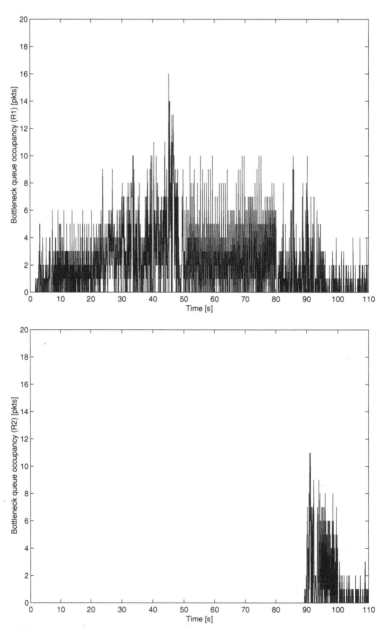

Figure 16.7 Bottleneck queue occupancy obtained in **experiment 1** with *explicit-rate marking*. Top, router R1; bottom, router R2.

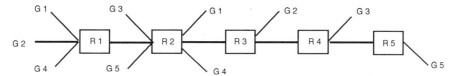

Figure 16.8 Simulated network configuration for **experiment 2**.

The groups enter the network at times **[0.0, 0.0, 15.0, 30.0, 95.0]** s. Inside each group flows start in a synchronized way. In addition, all the flows from group **G4** leave the network at time $t = 60$ s. The remaining groups stay in the network until the end of the simulation.

The total link capacity $\widehat{C}_k$ for **router R1** is set to 200 Mbit/s, while $t < 45.0$ s, and to 175 Mbit/s for $t \geq 45.0$ s. For **router R2**, $\widehat{C}_k$ is set to 50 Mbit/s, while $t < 75.0$ s, and to 75 Mbit/s for $t \geq 75.0$ s.

Routers R3 and R4 have link capacity of 100 Mbit/s, but they do not have the role of bottleneck routers at any time. Also, the link capacity of router R5 is irrelevant, as flows from G5 terminate at the output ports of R5. Similarly to the previous experiment, the available capacity for best-effort traffic, at each router, is set to $\rho\widehat{C}_k$ with $\rho = 0.9$.

We simulate hard transients caused by new groups of flows entering the network in a synchronized way, as well as sudden changes in the capacity of the bottleneck routers (R2 drops its capacity at $t = 45.0$ s and R2 doubles the capacity at $t = 75.0$ s). In addition, as compared to the scenario of experiment 1, we have now round-trip delays spanning over a larger interval.

Figure 16.9 shows the instantaneous flow rates obtained with the explicit-rate and binary feedback strategies. The graphs represent the set of super-imposed curves from the flows of each group. Figure 16.10 depicts the link parameters $p(k)$ calculated by the bottleneck routers R1 and R2 over time. Comparing the results of both graphs in Figure 16.9, we observe an encouraging performance for the flow rates with sources subject to probabilistic binary marking. The flow rates of binary sources exhibit oscillations, as expected, but these are significantly controlled in spite of the round-trip propagation delays affecting the flows. Also, observing the graphs in Figure 16.10, it seems that the stability of the algorithm is not affected by the tracking errors of the binary sources, which is a good indicator about the preservation of control loops between routers and sources. Thus, the proposed marking scheme may be a suitable alternative to explicit-rate feedback.

We now describe the different transients. In the period $0 < t < 15$ s, two groups of flows, G1 and G2, enter at $t = 0$ with flow rates that converge

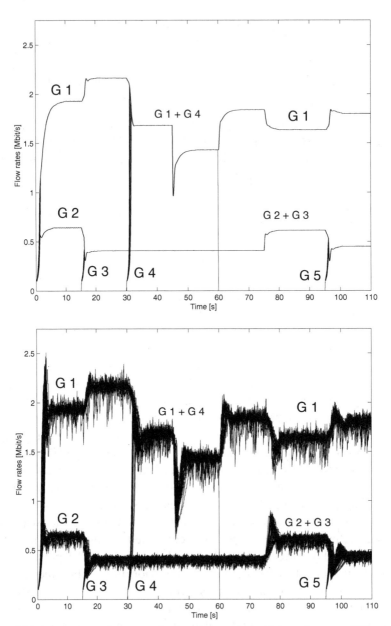

Figure 16.9 Instantaneous flow rates obtained in **experiment 2** (samples every 100 ms). Top, using *explicit-rate marking*; bottom, using *probabilistic marking*.

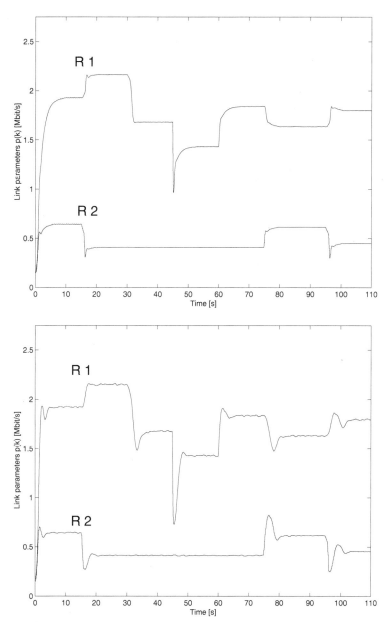

Figure 16.10 Link parameters $p(k)$ calculated by bottleneck routers R1 and R2 in **experiment 2**. Top, with *explicit-rate marking*; bottom, with *probabilistic marking*.

to different levels of fair share, since they are subject to different bottleneck routers. For the flows in G1 the bottleneck is R1, while for the flows in G2 the bottleneck is R2. Note that G1 (and later, also G4) will not be restricted at router R2, since these flows terminate at local ports of R2. In this period, the network converges to a situation where router R1 is serving 135 Mbit/s of locally restricted traffic (from the 70 flows of G1) and 45 Mbit/s of traffic restricted elsewhere (from the 70 flows of G2). The fair share for the flows in G1 converges to $p = (180 - 45)/70 \simeq 1.93$ Mbit/s (note that, in this period, router R1 has an available capacity of $\rho\widehat{C} = 180$ Mbit/s). The fair share for flows in G2 converges to $p = (45/70) \simeq 0.64$ Mbit/s. As expected, in Figure 16.9 we observe an initial over-shooting of flow rates for the binary sources. Given that in the initial probing for bandwidth phase these sources do not receive an explicit stream of binary notifications, procedure (16.5) cannot be applied. Instead, sources have to estimate their rates according to (16.4). This over-shooting (also observed in a lesser extent in experiment 1), is amplified here by the larger round-trip propagation delays of the flows.

The second transient starts at $t = 15.0$ s, the time when the 40 flows from G3 enter the network. The effect of this event is the re-allocation of fair share in R2, among the flows of G2 and G3. The new fair share for the flows of these groups is now given by $p = (45/110) \simeq 0.409$ Mbit/s. Since flows from G2 pass through R1 and their rates decreased, there are now more resources to re-distribute among the flows from G1, leading to a new fair share of $p = (180 - 0.409 * 70)/70 \simeq 2.162$ Mbit/s.

In the third transient, starting at $t = 30.0$ s, the 20 flows from G4 enter the network, which leads to a re-allocation of flow rates among G1 and G4. The new fair share for these flows is given by $p = (180 - 0.409 * 70)/90 \simeq 1.682$ Mbit/s. The flows restricted at R2 do not sense any perturbation, as expected.

At time 45.0 s, the capacity in R1 drops to 175 Mbit/s, which causes a negative transient. The algorithm at R1 reacts drastically by reducing the link parameter p_k, first to a level below the new fair share, and then converging to $p = (157.5 - 0.409 * 70)/90 \simeq 1.432$ Mbit/s (we have now, for router R1, an available capacity of $\rho\widehat{C} = 0.9 * 175 = 157.5$ Mbit/s). We see in Figure 16.9 that binary sources decrease their rates to a lower level, as compared to explicit-rate sources. This is due, basically, to the additional delay incurred by binary sources while estimating the marking probability, which obliges the router to decrease the link parameter p_k by a larger amount (as observed in Figure 16.10) than in the case of explicit-rate sources. As we show below, the side effect of this additional delay from binary sources leads to a larger queue occupancy of router R1 than in the case of explicit-rate sources. The rates from the groups restricted at R2 remain untouched.

At time $t = 60.0$ s, all flows from G4 leave the network, freeing a significant amount of resources at R1, readily distributed to the remaining flows of G1. The new fair share converges to $p = (157.5 - 0.409 * 70)/70) \simeq 1.841$ Mbit/s.

The transient triggered at $t = 75.0$ s is due to the sudden increase in the capacity of R2 (from 50 to 75 Mbit/s). This results in the distributed *max-min* re-allocation of fair share in R1 and R2. The new fair share at R2 converges to $p = (75 * 0.9/110) \simeq 0.614$ Mbit/s. Router R1 serves an amount of traffic restricted at R2 (the flows from G2) which is now given roughly by $(0.614 * 70) = 42.98$ Mbit/s. Thus, at R1, the fair share for locally restricted flows converges to $p = (157.5 - 0.614 * 70)/70 = 1.636$ Mbit/s.

Finally, at $t = 95$ s, the flows from group G5 enter the network. The new fair share at R2 converges to $p = (75 * 0.9/150) = 0.45$ Mbit/s, while at R1 converges to $p = (157.5 - 0.45 * 70)/70 = 1.8$ Mbit/s.

Figures 16.11 and 16.12 show the bottleneck queue occupancy for routers R1 and R2, with explicit-rate and binary feedback sources, respectively. Concerning the occupancy of R1, we observe that the negative transient at time 45 s, caused by the dropping of capacity in R1, leads to a higher occupancy peak for the case of binary sources. This is expected, since these sources introduce an additional delay (apart from the propagation delay) while estimating the marking probability. Thus, binary sources have a slower reaction while decreasing their rates, when compared to explicit-rate sources. In the remaining periods, no significant differences were observed.

For router R2, looking at the results obtained with explicit-rate sources, we observe occupancy peaks while the flows from groups G3 (at $t = 15$ s) and G5 (at $t = 95$ s) entered the network. With binary sources, the queue occupancy of R2 only exhibits a peak while flows from G5 entered the network (which is smaller than that observed with explicit-rate sources). During the probing for bandwidth phases, binary sources update their rates according to (16.4), which is a more conservative procedure than the rate updating process carried out by explicit-rate sources.

16.4 Conclusions

In this paper we present a new approach for congestion control based on binary probabilistic marking. Given the lack of support in the Internet for explicit-rate congestion signaling, the implementation of schemes based on

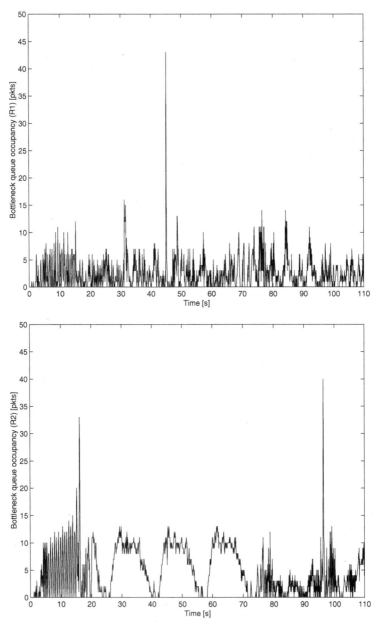

Figure 16.11 Bottleneck queue occupancy obtained in **experiment 2** with *explicit-rate marking*. Top, router R1; bottom, router R2.

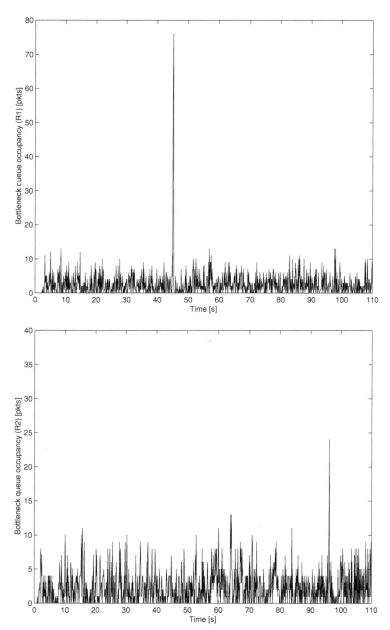

Figure 16.12 Bottleneck queue occupancy obtained in **experiment 2** with *probabilistic marking*. Top, router R1; bottom, router R2.

binary feedback is attractive. On the other hand, many studies have shown that explicit-rate algorithms can exercise congestion control efficiently, with improved performance when compared to binary control algorithms. The scheme proposed here, that encompasses a source algorithm and a router marking procedure, is devised to interact with router algorithms that were originally devised for explicit-rate feedback. The objective is to maintain close levels of performance when these algorithms exercise congestion control through binary signaling.

We compare, through simulation, the results obtained with two settings involving sources subject to the proposed binary marking scheme and sources subject to explicit-rate feedback. The results show that, for the case of probabilistic binary feedback, the router algorithm is able to exercise control over sources with similar performance to that obtained with explicit-rate feedback. We have considered a stateless router algorithm, where control is exercised without keeping per-flow information. Even though the command signals are now subject to estimation errors, the control loop between the bottleneck router and each source is preserved and the stability of the algorithm is not affected.

With this approach we take advantage of the best of two complementary strategies: (i) exercise control over sources through binary congestion signaling, which is a flexible and simple solution for the notification of sources; (ii) implement routers with explicit-rate algorithms, by choosing among many contributions that have been made and that show efficiency in congestion control.

References

[1] A. Almeida and C. Belo, A proposal for binary congestion control with rate-based sources and stateless routers, in *Proc. IEEE International Conference on Networks*, Singapore, September 2000.

[2] A. Almeida and C. Belo, Analysis and implementation of a stateless congestion control algorithm, in *Proc. International Symposium on Performance Evaluation of Computer and Telecommunication Systems (SPECTS'05)*, Philadelphia, PA, USA, July 2005.

[3] J. Borreicho and A. Almeida, Fairness and burstiness evaluation of router-assisted congestion control with window-based sources, in *Proc. Wireless and Optical Communications*, Montreal, Canada, May 2007.

[4] A. Almeida and C. Belo, Source rate encoding mechanisms for binary congestion control. *Performance Evaluation*, vol. 48, nos. 1–4, pp. 25–46, 2002.

[5] A. Almeida and C. Belo, Explicit rate congestion control with binary notifications, in *Proc. 10th IEEE Workshop on Local and Metropolitan Area Networks (LANMAN)*, Sydney, Australia, November 1999. An extended version appears in: D. Skellern, A.

Guha and F. Neri, (Eds.), *Evolving Access and Networking Techniques* (Selection of Papers from the 10th IEEE LANMAN Workshop), IEEE Press, 2001.

[6] K. Chen, K. Nahrstedt and N. Vaidya, The utility of explicit rate-based flow control in mobile ad hoc networks, in *Proc. IEEE Wireless Communications and Networking Conference*, Atlanta, Georgia, USA, March 2004.

[7] N. Dukkipati, M. Kobayashi, R. Zhang-Shen and N. McKeown, Processor sharing flows in the internet, in *Proc. International Workshop on Quality of Service (IwQoS)*, Passau, Germany, June 2005.

[8] A. Karnik and A. Kumar, Performance of TCP congestion control with explicit rate feedback, *IEEE/ACM Transactions on Networking*, vol. 13, no. 1, pp. 108–120, 2005.

[9] Y. Zhang, D. Leonard and D. Loguinov, JetMax: Scalable max-min congestion control for high-speed heterogeneous networks, in *Proc. IEEE INFOCOM*, Barcelona, Spain, April 2006.

[10] M. Lestas, A. Pitsillides, P. Ioannou and G. Hadjipollas, Adaptive congestion control: A congestion control protocol with learning capability, *Computer Networks*, vol. 51, no. 13, pp. 3773–3798, 2007.

[11] J. Kenney (Ed.), *Traffic Management Specification, Version 4.1*, ATM Forum Technical Committee, AF-TM-121.000, March 1999.

[12] A. Charny, An algorithm for rate allocation in a packet-switching network with feedback, Master Thesis in Electrical Engineering and Computer Science, Massachusetts Institute of Technology, May 1994.

[13] L. Benmohamed and S. Meerkov, Feedback control of congestion in packet switching networks: The case of a single congested node, *IEEE/ACM Transactions on Networking*, vol. 1, no. 6, pp. 693–708, 1993.

[14] R. Jain, S. Kalyanaraman and R. Viswanathan, *The OSU Scheme for Congestion Avoidance Using Explicit Rate Indication*, ATM-Forum Contribution 94-0883, September 1994.

[15] V. Jacobson, Congestion avoidance and control, in *Proc. ACM SIGCOMM*, Stanford, CA, pp. 314–329, August 1988.

[16] S. Floyd, TCP and explicit congestion notification, *ACM Computer Communication Review*, vol. 24, no. 5, pp. 8–23, 1994.

[17] K. Ramakrishnan, S. Floyd and D. Black, *The Addition of Explicit Congestion Notification (ECN) to IP*, September 2001. IETF RFC 3168.

[18] ECN Web page, URL http://www.icir.org/floyd/ecn.html.

[19] F. Bonomi, D. Mitra and J. Seery, Adaptive algorithms for feedback-based flow control in high-speed, wide-area ATM networks, *IEEE Journal on Selected Areas in Communications*, vol. 13, no. 7, pp. 1267–1283, 1995.

[20] S. Floyd, Connections with multiple congested gateways in packet-switched networks, Part 1: One-way traffic, *ACM Computer Communication Review*, vol. 21, no. 5, pp. 30–47, 1991.

[21] S. Floyd and V. Jacobson, Random early detection gateways for congestion avoidance, *IEEE/ACM Transactions on Networking*, vol. 1, no. 4, pp. 397–413, 1993.

[22] Z. Wang and J. Crowcroft, Eliminating periodic packet losses in 4.3-Tahoe BSD TCP congestion control algorithm, *ACM Computer Communication Review*, vol. 22, no. 2, pp. 9–16, 1992.

[23] L. Brakmo and L. Peterson, TCP Vegas: End to end congestion avoidance on a global internet, *IEEE Journal on Selected Areas in Communications*, vol. 13, no. 8, pp. 1465–1480, 1995.

[24] D. Sisalem and H. Schulzrinne, Congestion control in TCP: Performance of binary congestion notification enhanced TCP compared to Reno and Tahoe TCP, in *Proc. International Conference on Network Protocols*, Columbus, Ohio, pp. 268–275, October 1996.

[25] T. Hamann and J. Walrand, A new fair window algorithm for ECN-capable TCP (New-ECN), in *Proc. IEEE INFOCOM*, Tel Aviv, Israel, March 2000.

[26] R. Satyavolu, K. Duvedi and S. Kalyanaraman, *Explicit Rate Control of TCP Applications*, ATM-Forum Contribution 98-0152R1, February 1998.

[27] Y. Afek, Y. Mansour and Z. Ostfeld, Phantom: A simple and effective flow control scheme. in *Proc. ACM SIGCOMM*, Stanford, CA, pp. 169–182, August 1996.

[28] D. Chiu and R. Jain, Analysis of the increase and decrease algorithms for congestion avoidance in computer networks, *Computer Networks and ISDN Systems*, vol. 17, no. 1, pp. 1–14, 1989.

[29] R. Jain, A delay-based approach for congestion avoidance in interconnected heterogeneous computer networks, *ACM Computer Communication Review*, vol. 19, no. 5, pp. 56–71, 1989.

[30] K. Ramakrishnan and R. Jain, A binary feedback scheme for congestion avoidance in computer networks with a connectionless network layer, in *Proc. ACM SIGCOMM*, Stanford, CA, pp. 303–314, August 1988.

[31] K. Ramakrishnan, D. Chiu and R. Jain, Congestion avoidance in computer networks with a connectionless network layer. Part IV: A selective binary feedback scheme for general topologies, Technical Report DEC-TR-510, Digital Equipment Corporation, August 1987.

[32] M. Vojnović and J.-Y. Le Boudec, How (un)fair are the ABR binary schemes, actually?, in *Proc. SoftCOM'99*, Split, Rijeka (Croatia), Trieste, Venice (Italy), October 1999.

[33] M. Ritter, Network buffer requirements of the rate-based control mechanism for ABR services, in *Proc. IEEE INFOCOM*, San Francisco, CA, pp. 1190–1197, March 1996.

[34] O. Ait-Hellal and E. Altman, Performance evaluation of congestion phenomena in the rate based flow control mechanism for ABR, in *Proc. IEEE INFOCOM*, New York City, NY, March 1999.

[35] H. Ohsaki, M. Murata and H. Miyahara, Robustness of rate-based congestion control algorithm for ABR service class in ATM networks, in *Proc. IEEE GLOBECOMM*, London, pp. 1097–1101, November 1996.

[36] Y. Yang and S. Lam, General AIMD congestion control, in *Proc. International Conference on Network Protocols*, Osaka, Japan, November 2000.

[37] C. Rohrs, R. Berry and S. O'Halek, A control engineer's look at ATM congestion avoidance, in *Proc. IEEE GLOBECOMM*, Singapore, pp. 1089–1094, November 1995.

[38] S. Kunniyur and R. Srikant, A time scale decomposition approach to adaptive ECN marking, in *Proc. IEEE INFOCOM*, Anchorage, Alaska, April 2001.

[39] S. Athuraliya, V. Li, S. Low and Q. Yin, REM: Active queue management, *IEEE Network*, vol. 15, no. 3, pp. 48–53, 2001.

[40] S. Athuraliya, D. Laspsley and S. Low, An enhanced random early marking algorithm

for internet flow control, in *Proc. IEEE INFOCOM*, Tel Aviv, Israel, March 2000.

[41] S. Low and D. Lapsley, Optimization flow control I: Basic algorithm and convergence, *IEEE/ACM Transactions on Networking*, vol. 7, no. 6, pp. 861–874, 1999.

[42] K. Kar, S. Sarkar and L. Tassiulas, A simple rate control algorithm for maximizing total user utility, in *Proc. IEEE INFOCOM*, Anchorage, Alaska, April 2001.

17

Analysis of a RED Queue: A Singular Perturbation Approach

E. Altman[1], K. Avrachenkov[1] and B. J. Prabhu[2]

[1]*INRIA Sophia-Antipolis, 2004 route des Lucioles, 06902 Sophia-Antipolis, France; e-mail: {altman, k.avrachenkov}@sophia.inria.fr*
[2]*LAAS-CNRS, 7 Avenue du Colonel Roche, 31077 Toulouse Cedex 4, France; e-mail:bjprabhu@laas.fr*

Abstract

Several Active Queue Management (AQM) techniques for routers in the Internet have been proposed and studied during the past few years. One of the widely studied proposals, Random Early Detection (RED), involves dropping an incoming packet with some probability based on the estimated average queue length at the router. The analytical approaches to obtaining average drop probabilities in a RED enabled queue have been either based on using the instantaneous queue size for calculating the drop probability or have considered averaging with a fluid approximation. In this paper, we use a singular perturbation based approach to analyze a RED enabled queue with drop probabilities based on the estimated average queue size as has been proposed in the standard RED. The singular perturbation approach is motivated by the fact that the instantaneous and the estimated average queue lengths evolve at two different time scales. We present an analytical method to calculate the average queue size and the average drop probability for the non-responsive flows. We also provide analytical expressions for the Poisson arrivals and exponential service times case. Our model is derived under several approximations, and is validated through simulations.

Keywords: Singular perturbation, RED.

D. D. Kouvatsos (ed.), Traffic and Performance Engineering for Heterogeneous Networks, 371–397.

17.1 Introduction

Congestion control algorithms are end-to-end algorithms designed to fully utilize the available bandwidth on the links. When used in conjunction with drop tail queues in the routers, oscillatory behaviour and synchronized packet drops have been observed at the queues in the routers [1]. Synchronized increase and decrease of the window of different sources leads to inefficient use of the outgoing link. Active Queue Management (AQM) schemes seek to improve the link utilization by sending congestion signals in anticipation of congestion [2–4].

To overcome the synchronized window evolution, the Random Early Detection (RED) algorithm proposes to drop packets at random before congestion actually sets in [2]. Two of the design aims of RED are to accept occasional bursts and to maintain a reasonable average queue length when the system is heavily loaded with the objective of efficient use of the system capacity. Towards this end, the routers maintain a variable corresponding to the exponentially weighted moving average of the instantaneous queue length. An incoming packet is dropped with some probability which is a function of this estimated average queue length. The performance of the RED algorithm depends on the setting of parameters and the traffic characteristics [5]. Several alternatives and modifications have been proposed [4, 6, 7].

The aim of our study is to use singular perturbation techniques [8, 9] to provide an analytical expression for computing the drop probabilities and average queue length in a RED enabled queue with averaging as has been proposed in RED. Furthermore, we make use of the transition probability matrices in order to preserve the stochastic nature of the system as opposed to a fluid approximation which is deterministic. The dynamics of the average and the instantaneous queue lengths is modelled as a two dimensional Markov chain, which is then approximated by a non-homogeneous (or level dependent) Quasi-Birth-Death (QBD) process [10]. This model then permits us to obtain the joint steady state distribution of the average and instantaneous queue lengths.

The use of singular perturbation technique is motivated by the fact that two different time scales are involved in the evolution of the instantaneous and the average queue length. Indeed, for very small values of the averaging parameter, the average queue length can be assumed to vary much more slowly compared to the instantaneous queue length. The use of this decomposition allows us to compute the steady state distribution and the desired performance metrics at a reduced complexity. Next, we mention the techniques which have been previously used to analyze RED, and mention the differences with our work.

17.1.1 Related Literature

Various authors have studied the behaviour of RED, both through analysis and through simulations. In [11], the authors analyze the performance of a RED queue using the instantaneous queue length, instead of the average queue length, to compute the drop probabilities. In [12, 13], a fluid approximation is used to study the effect of the exponential averaging. In [14], the authors compute the joint distribution of the instantaneous and the averaged queue length of an $M/M/1/K$ queue as a solution of a set of differential equations. An analytical expression for the case of buffer size 1 and 2 is also provided. However, they consider packet drops only due to buffer overflow. The study of [15] uses a time scale decomposition to obtain in the limit a differential equation for the exponentially averaged queue length. They also provide a diffusion approximation to quantify the error due to the ODE approximation. Our method is also based on noting that average and the instantaneous queue lengths evolve at different time scales. However, unlike the ODE approximation obtained in [15], we use a discretized version of that fluid dynamics which has a simple probabilistic interpretation, which allows us then to obtain the joint distribution of the average and the instantaneous queue length (using non-homogeneous QBD). We note that this type of discretization is very frequent in numerical solutions of fluid models and of diffusions, including controlled ones and there is a huge literature that provides conditions under which the stochastic processes related to the discrete model converge (in various senses) to the original fluid model as the discretization step goes to zero, see e.g. theorems 2.7 and 4.2 in [16] (or the appendix of [17], as well as [18]).

The rest of the paper is organized as follows. In Section 17.2 we describe the RED algorithm and present the system model. In Section 17.3 we present key elements of singular perturbation approach to Markov chains. In Section 17.4 we apply the singular perturbation approach to compute the joint distribution of the average and instantaneous queue length, and to compute other performance metrics. In Section 17.5 we compare the results obtained using the analysis and with simulations. Finally, in Section 17.6 we present the conclusions.

17.2 System Model

17.2.1 The RED Algorithm

The basic algorithm of RED was proposed in [2] and can be summarized as follows. The router updates an exponentially averaged queue length variable (q_n, where the subscript n denotes the nth packet arrival) at every packet arrival. In addition to q_n, the router also keeps a count (c_n) of the number of packets accepted since the last dropped packet. The incoming packet is dropped with a probability, p_d, which is computed according to the following algorithm.

```
if qn < minth then
    pd = 0, cn = −1,
else if minth ≤ qn ≤ maxth then
    pb = maxp(qn − minth)/(maxth − minth),
    pd = min(1, pb/(1 − cn pb)),
else
    pd = 1,
```

where $\max_p$ (maximum drop probability), $\min_{th}$ (minimum threshold for queue), and $\max_{th}$ (maximum threshold for queue) are given constants. The variable c_n is set to zero each time an incoming packet is dropped. The estimated average queue length, q_n, is an exponentially weighted moving average of the instantaneous queue length, Q_n, and is computed as follows:

```
if Qn == 0 then
    qn+1 = (1 − α)ʷqn ,
else
    qn+1 = (1 − α)qn + α Qn+1,                    (17.1)
```

where α is the queue weight, and w is the average number of packets the router could have transmitted during the previous idle period. Figure 17.1 shows the drop function associated with the basic RED algorithm.

17.2.2 The System Model

Packets arrive to a RED enabled router with a buffer of size K packets. Let T_n denote the sequence of the interarrival arrival time between the $(n-1)$th and nth arrival. The sequence $\{T_n, n \geq 1\}$ is assumed to be independent and identically distributed with mean λ^{-1}, distribution function

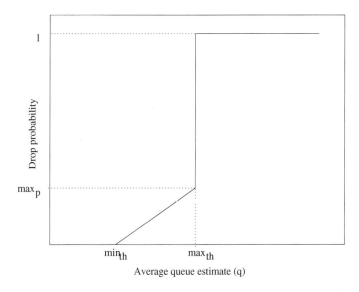

Figure 17.1 RED drop probability function.

$T(x)$, and Laplace–Stieltjes transform $T^*(s)$. The packet service times are assumed to be exponentially distributed with mean μ^{-1}. In the notation of queueing theory, the model corresponds to a $G/M/1/K$ queue with RED queue management scheme.

Assumption 1. *For* $\min_{th} \leq q_n \leq \max_{th}$, *an incoming packet is dropped with a probability* p_d *equal to* p_b.

Remark. In [2], the authors compare two methods of calculating the drop probability, p_d. In the first method, p_d is taken to be equal to p_b, and in the second method p_d is computed as in the algorithm described in Section 17.2.1. They conclude that, for the same number of average dropped packet, the losses are more spread out when the second method is used. Therefore, by assuming $p_d = p_b$ we can expect to obtain an upper bound on the probability of more than j successive packet losses.

Let Q_n and q_n be the instantaneous and estimated average queue lengths, respectively, just before the nth arrival. The recursive equations for the evolution of the state (q_n, Q_n) are given by

$$Q_{n+1} = \begin{cases} \max(Q_n - D_n, 0) & \text{w.p. } h(q_n) \quad \text{or if } Q_n = B; \\ \max(Q_n + 1 - D_n, 0) & \text{w.p. } 1 - h(q_n), \end{cases} \qquad (17.2)$$

$$q_{n+1} = (1 - \alpha)q_n + \alpha Q_{n+1}, \qquad (17.3)$$

where D_n is the number of packet departures during the interarrival time T_{n+1}, and $h(\cdot)$ denotes the RED packet drop probability function, i.e., $p_d = h(q_n)$.

With this formulation, the process $\{(q_n, Q_n), n \geq 0\}$ is a Markov chain. Its analysis is difficult, partly because the support of q is a countable subset of the interval $[0, B]$ where B is the maximum queue size.

We propose an approximation that reduces the support of q to a finite set denoted by $V_m = \{0, \ldots, mB\}$, where m is some integer. In the following discussion, $\hat{q}_n$ will refer to the discretized estimated average queue length. We assume below that α is sufficiently small such that $\alpha mB < 1$. The recursive equation for $\hat{q}_n$ is given by

$$\hat{q}_{n+1} = (1 - \alpha)\hat{q}_n + \alpha m Q_{n+1}, \quad \hat{q}_n \in \{0, 1, \ldots, mB\}. \qquad (17.4)$$

Then we replace equation (17.4) by the following approximation

$$\hat{q}_{n+1} = \hat{q}_n + C_n, \qquad (17.5)$$

where

$$C_n = \begin{cases} 1 & \text{w.p. } \alpha(m Q_{n+1} - \hat{q}_n)^+; \\ -1 & \text{w.p. } \alpha(\hat{q}_n - m Q_{n+1})^+; \\ 0 & \text{otherwise,} \end{cases}$$

with $x^+ = \max(x, 0)$.

We note that with this definition, we obtain

$$E[\hat{q}_{n+1} - \hat{q}_n | \hat{q}_n, Q_{n+1}] = 1 \cdot \alpha(m Q_{n+1} - \hat{q}_n)^+ + (-1) \cdot \alpha(\hat{q}_n - m Q_{n+1})^+$$
$$= \alpha(m Q_{n+1} - \hat{q}_n), \qquad (17.6)$$

so that (17.4) becomes the mean field of (17.5). With this approximation the transitions of $\hat{q}_n$ conditioned on Q_n become stochastic as opposed to the deterministic transitions in the original system.

The motivation for such an approximation is that we are interested in obtaining the steady state distribution of the couple $(\hat{q}_n, Q_n)$ which will allow us to study the various performance measures (e.g., average queue length, average drop probability, etc.) of the system. With this approximation we can model the $(\hat{q}_n, Q_n)$ process as a non-homogeneous QBD process for which algorithms are available to efficiently compute the steady state distributions.

In order to model the process $(\hat{q}_n, Q_n)$ as a QBD process, we first obtain the transition matrix of Q_n for a given value of $\hat{q}_n$. That is,

$$A_i = \{a_{jk}\}_i = P(Q_{n+1} = k | Q_n = j, \hat{q}_n = i), \qquad \forall i \in 0, 1, \dots, mB. \tag{17.7}$$

The matrix A_i is a $(B+1) \times (B+1)$ matrix. Using equation (17.2), the probability a_{jk} can be obtained as follows.

$$a_{jk} = \begin{cases} (1 - h(i))d_0, & k = \min(j+1, B); \\ (1 - h(i))d_{j-k+1} + h(i)d_{j-k}, & 0 < k \le j; \\ (1 - h(i))\sum_{l=\min(j+1,B)}^{\infty} d_l + h(i)\sum_{l=\min(j,B)}^{\infty} d_l, & k = 0, \end{cases} \tag{17.8}$$

where

$$d_l = \int_0^{\infty} \exp(-\mu x)\frac{(\mu x)^l}{l!} dT(x) \tag{17.9}$$

is the average probability of l departures between two arrivals.

Next, we obtain the transition matrix for $\hat{q}_n$ conditioned on given values of Q_{n+1} and $\hat{q}_n$. Let C_i^+, C_i^-, C_i^0 denote matrices given by

$$\begin{aligned}
C_i^+ := \{c_{jj}\} &= p(\hat{q}_{n+1} = i + 1 | Q_{n+1} = j, \hat{q}_n = i) \\
&= \alpha(m \cdot j - i)^+, \\
C_i^- := \{c_{jj}\} &= p(\hat{q}_{n+1} = i - 1 | Q_{n+1} = j, \hat{q}_n = i) \\
&= \alpha(i - m \cdot j)^+, \\
C_i^0 := \{c_{jj}\} &= p(\hat{q}_{n+1} = i | Q_{n+1} = j, \hat{q}_n = i) \\
&= 1 - \alpha(m \cdot j - i)^+ - \alpha(i - m \cdot j)^+, \\
&\qquad \forall i \in \{0, 1, \dots, mB\} \qquad \forall j \in \{0, 1, \dots, B\}.
\end{aligned}$$

C_i^+i, C_i^-, and C_i^0 are $(B+1) \times (B+1)$ matrices which denote the transition probability of $\hat{q}_{n+1} = \hat{q}_n + 1$, $\hat{q}_{n+1} = \hat{q}_n - 1$, and $\hat{q}_{n+1} = \hat{q}_n$, respectively, when $Q_{n+1} = j$. We note that C_i's are diagonal matrices which satisfy the equality

$$C_i^+ + C_i^- + C_i^0 = I. \tag{17.10}$$

Let $\tilde{P}$ denote the transition probability matrix for the two dimensional Markov chain $(\hat{q}_n, Q_n)$ with $\tilde{\pi}(\alpha)$ as its stationary distribution. Using the identity

$$p(\hat{q}_{n+1}, Q_{n+1}|\hat{q}_n, Q_n) = p(\hat{q}_{n+1}|\hat{q}_n, Q_n, Q_{n+1}) \cdot p(Q_{n+1}|\hat{q}_n, Q_n)$$
$$= p(\hat{q}_{n+1}|\hat{q}_n, Q_{n+1}) \cdot p(Q_{n+1}|\hat{q}_n, Q_n),$$

we can write $\tilde{P}$ as

$$\tilde{P} = \begin{bmatrix} A_0(C_0^- + C_0^0) & A_0C_0^+ & 0 & \cdots & & \cdots \\ A_1C_1^- & A_1C_1^0 & A_1C_1^+ & 0 & & \cdots \\ 0 & A_2C_2^- & A_2C_2^0 & A_2C_2^+ & & \cdots \\ \vdots & \vdots & \ddots & \vdots & & \vdots \\ \vdots & \vdots & \vdots & \ddots & & \vdots \\ 0 & \cdots & \cdots & & A_{mB}C_{mB}^- & A_{mB}(C_{mB}^0 + C_{mB}^+) \end{bmatrix}.$$
(17.11)

The diagonal rows entries of $\tilde{P}$ are matrices which give the transition probability matrix of Q for a given value of $\hat{q}_n$, and $\hat{q}_{n+1} = \hat{q}_n$ whereas the off-diagonal entries give the transition probability matrix of Q for a given value of $\hat{q}_n$ and $\hat{q}_{n+1} = \hat{q}_n + 1$, or $\hat{q}_{n+1} = \hat{q}_n - 1$. Using equation (17.10), we can rewrite $\tilde{P}$ as

$$\tilde{P} = P + PC,$$
(17.12)

where P is given by

$$P = \begin{bmatrix} A_0 & 0 & \cdots & 0 \\ 0 & A_1 & \cdots & 0 \\ \vdots & \vdots & \ddots & \vdots \\ 0 & \cdots & 0 & A_{mB} \end{bmatrix},$$

and C is given by

$$C = \begin{bmatrix} -C_0^+ & C_0^+ & 0 & \cdots \\ C_1^- & -(C_1^- + C_1^+) & C_1^+ & \cdots \\ \vdots & \vdots & \ddots & \vdots \\ 0 & \cdots & C_{mB}^- & -C_{mB}^- \end{bmatrix}.$$

We note that $\tilde{P}$ is a function of the averaging parameter α. In [19], the recommended value of α is of the order of 0.002. This suggests that the averaging parameter is, in general, small. This assumption on α allows us to obtain the steady state probabilities of $\tilde{P}$ using a singular perturbation approach which we outline below and which is computationally less expensive than algorithms to compute the invariant vectors of matrices.

17.3 Review of Singular Perturbation Approach

Let $P \in R^{n \times n}$ be a transition stochastic matrix representing transition probabilities in a Markov chain. Suppose that the structure of the underlying Markov chain is aperiodic. Let $P^* = \lim_{t \to \infty} P^t$ which is well-known to exist for aperiodic process. The matrix P^* is called ergodic projection. In the case when the process is also ergodic, P^* has identical rows, each of which is the stationary distribution of P, denoted by π. Let Y be the deviation matrix of P which is defined by $Y = (I - P + P^*)^{-1} - P^*$. It is well known that Y exists and it is the unique matrix satisfying

$$Y(I - P) = (I - P)Y = I - P^*,$$

and

$$P^*Y = YE = 0,$$

(where E is a matrix full of 1's) making it the group inverse of $I - P$. Finally, $Y = \lim_{T \to \infty} \sum_{t=0}^{T} (P^t - P^*)$.

We consider (linear) perturbations of the matrix P and their impact on the structure of the process. Specifically, for a scalar ε, $0 < \varepsilon < \varepsilon_{\max}$, and for some zero rowsum matrix Q, we look at the set of perturbed stochastic matrices $P(\varepsilon) = P + \varepsilon Q$, which are assumed to be unichain for any ε in the above mention region. A unichain Markov chain is a Markov chain which has a unique stationary distribution. We emphasize that unichain assumption is not assumed with regard to $P = P(0)$. Actually, the case when $P(0)$ contains some unrelated chains (with or without transient states) is our main focus. Here we survey some results on series expansions for $\pi(\varepsilon)$ and $Y(\varepsilon)$, which denote the stationary distribution and the deviation matrix, respectively, of $P(\varepsilon)$ for $0 < \varepsilon < \varepsilon_{\max}$ and consider their relationship to the corresponding entities in the unperturbed Markov chain for $P = P(0)$.

Let us concentrate on so-called Nearly Completely Decomposable (NCD) case, in which the state space under the unperturbed process is decomposed into a number of unichain Markov chains. Specifically, let $P(0) \in R^{n \times n}$ be a stochastic matrix representing transition probabilities in a completely decomposable Markov chain. By the latter we mean that there exists a partition Ω

of te state space into p, $p \geq 2$, subsets $\Omega = \{I_1, \ldots, I_p\}$ each of which being an ergodic class. We assume that the order of the rows and of the columns of P is compatible with Ω, i.e., for p stochastic matrices, $P_{I_1}, \ldots, P_{I_p}$,

$$P = \begin{pmatrix} P_{I_1} & 0 & \cdots & 0 \\ 0 & P_{I_2} & \cdots & 0 \\ \vdots & \vdots & \ddots & 0 \\ 0 & 0 & \cdots & P_{I_p} \end{pmatrix}.$$

Note that we assume above that P_{I_k} represents a unichain Markov chain.

For small values of ε, $P(\varepsilon) = P + \varepsilon Q$ is called nearly completely decomposable or sometimes nearly uncoupled.

Let us define the limiting stationary distribution $\pi^{(0)}$ as

$$\pi^{(0)} = \lim_{\varepsilon \to 0} \pi(\varepsilon).$$

For any subset $I \subset \Omega$, let

$$\kappa_I = \sum_{i \in I} \pi_i^{(0)}.$$

Also, let γ_I be the subvector of $\pi^{(0)}$ corresponding to subset I rescaled so as its entry-sum is now one. Then, γ_I is the unique stationary distribution of P_I. Note that computing γ_I is relatively easy as only the knowledge of P_I is needed.

Next we define the matrix $S \in R^{p \times p}$ which is usually referred to as the aggregate transition matrix. Each row, and likewise each column in S corresponds to a subset in Ω. Then, for subsets I and J, $I \neq J$, let

$$S_{IJ} = \sum_{i \in I} (\gamma_I)_i \sum_{j \in J} Q_{ij},$$

and let

$$S_{II} = 1 + \sum_{i \in I} (\gamma_I)_i \sum_{j \in J} Q_{ij} = 1 - \sum_{J \neq I} S_{IJ}.$$

Without loss of generality, assume that S_{II} is non-negative for all subsets I and hence S is easily seen to be a stochastic matrix. Note that the matrix S can be divided by any constant and ε can be multiplied by this constant leading to the same $n \times n$ transition matrix. Taking this constant small enough guarantees the stochasticity of S and hence this is assumed without loss of generality.

In particular, the stationary distribution of S is invariant with respect to the choice of this constant. Moreover, by assumption S is a unichain matrix.

Often it is convenient to express the aggregate transition matrix S in matrix terms. Namely, let $V \in R^{p \times n}$ be such that its ith row is full of zeros except for γ_{I_i} at the entries corresponding to subset I_i, and where $W \in R^{n \times p}$ is such that its jth column is full of zeros except for 1's in the entries corresponding to subset I_j. Note that $VW \in R^{p \times p}$ is the identity matrix. Moreover, V and W correspond to bi-orthonormal sets of eigenvectors of $P(0)$ belonging to the eigenvalue 1, V as left eigenvectors and W as right eigenvectors. Then, we can write

$$S - I + VQW.$$

The aggregate stochastic matrix S represents transition probabilities between subsets which in this context are sometimes referrred to as macro-states. However, although the original process among states is Markovian, this is not necessarily the case with the process among macro-states. Yet, as the following theorem indicates, much can be learned about the original process from the analysis of the aggregate matrix.

Theorem 1. *Let the perturbed Markov chain be nearly completely decomposable. The stationary distribution $\pi(\varepsilon)$ admits a Maclaurin series expansion in a punctured neighborhood of zero. Namely, for some vectors $\{\pi^{(m)}\}_{m=0}^{\infty}$ with $\pi^{(0)}$ being a probability vector satisfying $\pi^{(0)} = \pi^{(0)} P(0)$, and for some zerosum vectors $\pi^{(m)}$, $m \geq 1$,*

$$\pi(\varepsilon) = \sum_{m=0}^{\infty} \varepsilon^m \pi^{(m)}.$$

Moreover, for $I \in \Omega$, $\pi_I^{(0)} = \kappa_I \gamma_I$. Also, the sequence $\{\pi^{(m)}\}_{m=0}^{\infty}$ is geometric, i.e., for some square matrix U, $\pi^{(m)} = \pi^{(0)} U$ for any $m \geq 0$. Actually,

$$U = QY(0)(I + QWDV),$$

where D is the deviation matrix of the aggregate transition matrix S. Alternatively,

$$U = QY^{(0)},$$

where $Y^{(0)}$ is the first regular term of the Laurent series expansion for $Y(\varepsilon)$ (see also the next theorem). Finally, the validity of the series expansion holds for any ε, $0 \leq \varepsilon < \min\{\varepsilon_{\max}, \rho^{-1}(U)\}$ where $\rho(U)$ is the spectral radius of U.

Recall that for ε, $0 \leq \varepsilon < \varepsilon_{\max}$, we denote by $Y(\varepsilon)$ the deviation matrix of $P(\varepsilon)$. This matrix is uniquely defined and the case $\varepsilon = 0$ is no exception. Yet, as we see shortly, there is no continuity of $Y(\varepsilon)$ at $\varepsilon = 0$. We note that $Y(0)$ has the same shape as P has, namely

$$
Y(0) = \begin{pmatrix} Y_{I_1} & 0 & \cdots & 0 \\ 0 & Y_{I_2} & \cdots & 0 \\ \vdots & \vdots & \ddots & 0 \\ 0 & 0 & \cdots & Y_{I_p} \end{pmatrix},
$$

where Y_{I_i} is the deviation matrix of P_{I_i}, $1 \leq i \leq p$.

Theorem 2. *In the case of NCD Markov chains, the matrix $Y(\varepsilon)$ admits a Laurent series expansion in a punctured neighborhood of zero with the order of the pole being exactly one. Namely, for some matrices $\{Y^{(m)}\}_{m=-1}^{\infty}$ with $Y^{-1} \neq 0$, we have*

$$
Y(\varepsilon) = \frac{1}{\varepsilon} Y^{(-1)} + Y^{(0)} + \varepsilon Y^{(1)} + \varepsilon^2 Y^{(2)} + \cdots
$$

for $0 < \varepsilon < \varepsilon_{\max}$. Moreover,

$$
Y^{-1} = W D V,
$$

and in a component form,

$$
Y_{ij}^{(-1)} = D_{IJ}(\gamma_J)_j, \quad i \in I, \quad j \in J.
$$

The interested reader can find more details and references on singularly perturbed Markov chains in the survey [9].

17.4 Analysis

To use the singular perturbation approach outlined in the previous section, we first note that P is the matrix containing the transition probability matrices of the instantaneous queue length for a given value of the estimated average queue length when the averaging parameter $\alpha = 0$. Thus, P has mB ergodic classes each corresponding to a particular value of $\hat{q}$. The stationary distribution of each of these classes can be computed separately. This system of decomposable Markov chains is the unperturbed system which can be analyzed separately for different values of $\hat{q}$. As noted in the previous section, in general, the stationary distribution of the unperturbed system (i.e., $\alpha = 0$) is not the same as the stationary distribution of the original system when $\alpha \to 0$ [20]. This motivates us to use singular perturbation technique to determine the stationary distribution of our system when $\alpha \to 0$.

Later, we also show that the distributions of the instantaneous queue length and the average queue length can be computed for the general case (i.e. $\alpha \nrightarrow 0$). We note that the use of the approximation given in (17.5) has resulted in a level dependent Quasi-Birth-Death (QBD) type of transition matrix which is as given in (17.11). We can now use the algorithm, given in [21], for computing the steady state probability vector for level dependent QBD.

17.4.1 Limiting Case

In this subsection we obtain the steady state probability vector of $\tilde{P}$ when $\alpha \to 0$. We note that all the elements of the matrices C_i^+ and C_i^- have α as a common multiple and thus we can replace C_i^+ by αD_i^+ and C_i^- by αD_i^- where D_i^+ and D_i^- have elements independent of α. Thus we can rewrite equation (17.12) as

$$\tilde{P} = P + \alpha P D. \tag{17.13}$$

With $\varepsilon = \alpha$ and $Q = PD$ we recover the standard formulation for the singularly perturbed Markov chains. The singular perturbation approach allows us to find the stationary distribution of $\tilde{P}$ as $\lim_{\alpha \to 0} \tilde{\pi}(\alpha)$.

Let π_i denote the stationary distribution of A_i, and let V be a $(mB+1) \times (mB+1)(B+1)$ matrix such that in the ith row the entries, corresponding to the columns $\hat{q}_n = i$, are given by the rows of π_i, and is zero elsewhere. π_i is the stationary distribution vector of the unperturbed Markov chain corresponding to the instantaneous queue length when packets are dropped with a probability depending on $h(i)$. Let W be a $(mB+1)(B+1) \times (mB+1)$ matrix such that in the ith column the entries corresponding to the rows $\hat{q}_n = i$ are 1, and is zero elsewhere.

$$V = \begin{bmatrix} \pi_0 & \underline{0} & \cdots & \underline{0} \\ \underline{0} & \pi_1 & \cdots & \underline{0} \\ \vdots & \vdots & \ddots & \vdots \\ \underline{0} & \cdots & \cdots & \pi_{mB} \end{bmatrix}, \quad W = \begin{bmatrix} \underline{1}' & \underline{0} & \cdots & \underline{0} \\ \underline{0} & \underline{1}' & \cdots & \underline{0} \\ \vdots & \vdots & \ddots & \vdots \\ \underline{0} & \cdots & \cdots & \underline{1}' \end{bmatrix},$$

where $\underline{0}$ and $\underline{1}$ are $(B+1) \times 1$ vectors of 0 and 1, respectively.

Let S denote the generator matrix of the aggregated Markov chain. When α goes to 0, the stochastic sequence $\hat{q}_n$ converges weakly to a Markov chain induced by the aggregated transition matrix. Then S is given by

$$S = VPDW.$$

Since π_i is the stationary distribution of A_i, VP reduces to V, i.e.,

$$S = VDW. \tag{17.14}$$

Denote

$$f_i^+ = \pi_i D_i^+ \underline{1}' \tag{17.15}$$

$$= \sum_{j=\lceil \frac{i}{m} \rceil}^{B} \pi_i(j)(mj - i). \tag{17.16}$$

$$f_i^- = \pi_i D_i^- \underline{1}' \tag{17.17}$$

$$= \sum_{j=0}^{\lfloor \frac{i}{m} \rfloor} \pi_i(j)(i - mj), \tag{17.18}$$

where $\pi_i(j)$ is the jth element of π_i.

Using equations (17.15) and (17.17) we can rewrite S as

$$S = \begin{bmatrix} -f_0^+ & f_0^+ & 0 & \cdots \\ f_1^- & -(f_1^- + f_1^+) & f_1^+ & \cdots \\ \vdots & \vdots & \ddots & \vdots \\ 0 & \cdots & f_{mB}^- & -f_{mB}^- \end{bmatrix}.$$

The stationary distribution of this generator matrix can be obtained by solving $\gamma S = 0$. The stationary distribution, γ, is given by

$$\gamma := \{\gamma(i)\} = \gamma(0) \frac{\prod_{j=0}^{i-1} f_j^+}{\prod_{j=1}^{i} f_j^-}, \qquad i \in \{0, 1, \ldots, mB\}, \tag{17.19}$$

where

$$\gamma(0) = \left(\sum_{i=0}^{mB} \frac{\prod_{j=0}^{i-1} f_j^+}{\prod_{j=1}^{i} f_j^-} \right)^{-1}.$$

Proposition 1. *Let $\tilde{\pi}(\alpha)$ be the stationary distribution of $\tilde{P}$ for a given value of α, and let π_i denote the stationary distribution of A_i. Then, the limiting stationary distribution of $\tilde{P}$, $\tilde{\pi}(0)$ is given by*

$$\tilde{\pi}(0) = \begin{bmatrix} \gamma_1 \pi_1 & \gamma_2 \pi_2 & \cdots & \gamma_{mB} \pi_{mB} \end{bmatrix}, \tag{17.20}$$

where γ_i is given in (17.19).

Proof. Follows from Theorem 1. □

We note that the states $q > m \cdot \max_{th}$ are transient and the steady state probability of $q > m \cdot \max_{th}$ is 0. This can be shown as follows. From equation (17.16) we have

$$f_i^+ = \sum_{j=\lceil \frac{i}{m} \rceil}^{B} \pi_i(j)(mj - i).$$

Since the packet drop probability is 1 for $q \geq m \cdot \max_{th}$, the steady state probability, $\pi_i(j)$, becomes

$$\pi_i(j) = \begin{cases} 1 & j = 0; \\ 0 & j > 0; \end{cases} \quad \forall i \geq m \cdot \max_{th}. \tag{17.21}$$

This is true since all the packets arriving at the queue would be dropped and the queue would be empty. Since, $\pi_i(j)$ is as given in equation (17.21), and $i \geq m \cdot \max_{th} > 0$, we get that $f_i^+ = 0$, $\forall i \geq m \cdot \max_{th}$. From equation (17.18) we get $f_i^- > 0$, $\forall i$. From equation (17.19) it follows that $\gamma(i) = 0$, $\forall i \geq m \cdot \max_{th}$.

Remark. The stationary probability $\pi_i(j)$ is the steady state probability of an arrival finding j customers in a $G/M/1/K$ queue. These probabilities can be computed recursively. We note that in the limit $\alpha \to 0$, the instantaneous queue moves on a much faster time scale than the estimated average queue length, and hence approaches stationarity. Therefore, if we had exponential interarrival times and general service times (i.e., $M/G/1/K$) instead of general interarrival times and exponential interarrival times (i.e., $G/M/1/K$) we could use the steady state distribution of the $M/G/1/K$ system to compute the performance measures. In the limiting case, the performance measures of both $G/M/1/K$ and $M/G/1/K$ queues with RED queue management scheme could be obtained using this approach. However, for the general case ($\alpha \nrightarrow 0$), using our method, we can only obtain the performance measures of the $G/M/1/K$ queue with RED queue management scheme.

If the arrival process to the queue is Poisson with rate λ and the service times are exponentially distributed with mean μ^{-1}, then the stationary distribution of A_i, π_i, is given by

$$\pi_i := \{\pi_i(j)\} = \pi_i(0)\rho_i^j, \qquad (17.22)$$

where

$$\pi_i(0) = \left(\sum_{j=0}^{B} \rho_i^j\right)^{-1},$$

and $\rho_i = \lambda \cdot (1 - h(\hat{q}_n = i)) \cdot \mu^{-1}$. Here $h(\cdot)$ is the given drop function.

From the above analysis, using equation (17.19), we are able to obtain the stationary distribution of the average queue length variable, q. We use the PASTA property which states that the arrivals see this average queue length distribution. Hence, we can calculate the average packet drop probability as

$$P_{avg} = \sum_{i=0}^{mB} \gamma(i)h(i), \qquad (17.23)$$

For batch arrivals, we make the assumption that every packet in a batch is dropped with same drop probability. With this assumption, we can write the probability of dropping at least one packet in batch as

$$P_{avg} = \sum_{i=0}^{m(B-N)} (1 - (1 - \gamma(i))^N) \cdot h(i) + \sum_{i=m(B-N)+1}^{mB} h(i), \qquad (17.24)$$

17.4.2 Approximate Analysis

We present an approximate analysis which allows us to obtain the mean value of q without having to find the distribution.

The equilibrium value of q, q^*, can be obtained by solving

$$q^* = mE(Q(q^*)), \qquad (17.25)$$

where $E(Q(q^*))$ is the expected queue length when we drop with constant probability corresponding to q^*.

Using equations (17.16) and (17.18) we get

$$f_i^+ - f_i^- = m\sum_{j=0}^{B} j\pi_i(j) - i = mE(Q_i) - i, \qquad (17.26)$$

where $E(Q_i)$ is the expected queue length in the unperturbed Markov chain with $\hat{q}_n = i$. The term $f_i^+ - f_i^-$ can be thought of as an indicator for transitions of the average queue length, q. We note that in states where $mE(Q_i) > i$, $f_i^+ - f_i^-$ is positive, hence indicating a tendency to increase q. Similarly, when $mE(Q_i) > i$, $f_i^+ - f_i^-$ is negative, hence indicating a tendency to decrease q. The equilibrium value of q will then occur at the point where $f_i^+ - f_i^- = 0$. Hence, q^* can be obtained by solving (17.25). For the existence and uniqueness of the solution, we assume that the drop probability function (this is the case in the gentle variant of RED [22]) is a continuous non-decreasing function, $h(\hat{q})$, on the estimated average queue length. The effective offered load, ρ^*, for a given equilibrium value of q^* is then given by

$$\rho^* = \rho(1 - h(q^*)). \tag{17.27}$$

Since ρ^* is a continuous non-increasing function on q^*, we can express q^* as

$$q^* = h^{-1}\left(1 - \frac{\rho^*}{\rho}\right) := g(\rho^*). \tag{17.28}$$

We note that g is continuous non-increasing function on ρ^*. Since the average queue length, $E(Q(\rho^*))$, in the stationary regime is a continuous non-decreasing function of the offered load, then we can say that $E(Q(q^*))$, is a continuous non-increasing function of q^*. As q^* and $EQ(\cdot)$ the map to the same interval i.e., $[0, B]$, we can say that the solution to equation (17.25) exists and is unique.

Remark. The fixed point equation was also obtained in [15] as the stationary solution of an ordinary differential equation.

17.4.3 General Case

In Section 17.4.1 we presented an analysis for the limiting case $\alpha \to 0$. In this section we outline an algorithm with which we can compute the steady state probabilities when $\alpha \nrightarrow 0$. However, α still has to satisfy the condition $\alpha m B < 1$.

The transition probability matrix of the two dimensional Markov chain $(\hat{q}_n, Q_n)$ given in equation (17.11) can be written as

$$\tilde{P} = \begin{bmatrix} L_0 & F_0 & 0 & \cdots & \cdots \\ B_1 & L_1 & F_1 & 0 & \cdots \\ 0 & B_2 & L_2 & F_2 & \cdots \\ \vdots & \vdots & \vdots & \ddots & \vdots \\ 0 & \cdots & \cdots & B_{mB} & L_{mB} \end{bmatrix}, \qquad (17.29)$$

where B_i, L_i, F_i denote the backward, local, and forward transition matrices, respectively. This is the transition matrix for a level dependent QBD process. Let the $\tilde{\pi}$ be stationary distribution of $\tilde{P}$. Let $\tilde{\pi} = [\tilde{\pi}_0 \tilde{\pi}_1 \ldots \tilde{\pi}_{mB}]$, where $\tilde{\pi}_i$'s are vectors of size $1 \times B$. $\tilde{\pi}_i$ is the steady state probability vector of Q for $q = i$. $\tilde{\pi}$ can be found using the following algorithm [21].

1. Compute the S_i matrices using the following recursion

$$S_0 = L_0$$
$$S_i = L_i + B_i(I - S_{i-1})^{-1}F_{i-1}, \qquad 1 \le i \le mB.$$

2. $\tilde{\pi}_{mB}$ is computed by solving

$$\tilde{\pi}_{mB} = \tilde{\pi}_{mB}S_{mB},$$
$$\tilde{\pi}_{mB} \cdot \underline{1} = 1.$$

3. $\tilde{\pi}_i$, $0 \le i \le mB$ is computed using the recursion

$$\tilde{\pi}_i = \tilde{\pi}_{i+1}B_{i+1}(I - S_n)^{-1},$$

4. $\tilde{\pi}$ is found by normalization

$$\pi = \frac{\tilde{\pi}}{\tilde{\pi} \cdot \underline{1}}$$

17.5 Numerical Results

In this section we present the results obtained through the analysis as described in the previous section and the results obtained through simulations. Our aim to see the effect of the approximations that we made in the model and to verify that these effects are indeed negligible. The arrival process is assumed to be Poisson batch arrival process with rate λ and fixed batch sizes, N. The service times are assumed to be exponential with mean $1/\mu$. The offered load, ρ, is defined as $\rho = \lambda N / \mu$.

Remark. The validity of exponential interarrival time has been widely debated. Various authors have shown that for different scenarios the traffic is self-similar and not Poissonian [23]-[25]. However, in [26], the authors argue that under heavy load conditions the traffic tends towards a Poisson process. In a recent article [27], the main findings suggest that, when the level of multiplexing is large, packet arrivals appear Poisson at small time scales, non stationary at medium time scales, and long-range dependent at large time scale. Thus, as a preliminary study, we present some results related to the performance of a RED queue when the interarrival time is exponentially distributed.

The analytical results were obtained by numerically calculating the values using MATLAB. The buffer length, B, is taken to be 100 whereas the discretization parameter, m, is taken to be 5. The drop probability function is taken to be of the form

$$P_{\text{drop}} = \begin{cases} 0, & \hat{q} < \min_{\text{th}}, \\ \frac{\hat{q} - \min_{\text{th}}}{\max_{\text{th}} - \min_{\text{th}}}, & \min_{\text{th}} \le \hat{q} \le \max_{\text{th}}, \\ 1, & \hat{q} > \max_{\text{th}}. \end{cases} \tag{17.30}$$

We note that all the values of $\min_{\text{th}}$, $\max_{\text{th}}$, and $\hat{q}$ are scaled by a factor m in the analysis in order to discretize the estimated average queue size. The unscaled values for $\min_{\text{th}}$ and $\max_{\text{th}}$ are 15 and 40, respectively, and the drop probability function versus the unscaled estimated average queue length is as shown in Figure 17.2. We shall use the average drop probability and the probability of at least one drop in a burst as performance measures. The simulations were performed using C, and approximately 10^8 packets were generated during the simulations.

17.5.1 Limiting Case

First, we present the results for the limiting case. In the simulations, the averaging parameter, α, was taken to be 0.0001. Figures 17.3 shows the analytical as well as the simulation results for average packet drop probability as a function of the offered load for different values of burst sizes. There is a fairly good correspondence between the analytical and simulation results.

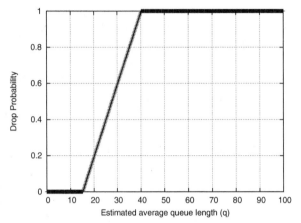

Figure 17.2 RED drop probability function.

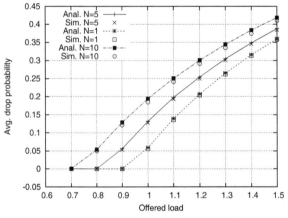

Figure 17.3 Average packet drop probability versus offered load.

In [11] it has been argued that the number of drops becomes infinite with positive probability as $\alpha \to 0$. However, for the drop function as shown in Figure 17.1, the number of consecutive drops appears to follow a geometric behaviour for small values of the number of consecutive drops. We plot the probability that there are greater than or equal to n consecutive drops as a function of n for a drop tail and a RED queue in Figure 17.4. The decay rate of the tail is much faster in the case of a RED queue when compared to that of a drop tail queue. For small values of n, the decay rate is close to p^n, where p is the average packet drop probability. Most of the mass can

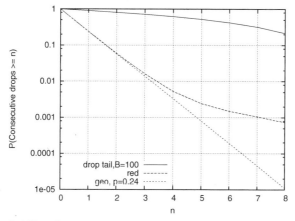

Figure 17.4 Probability of greater than or equal to n consecutive drops. $\rho = 1.1$. Burst size $= 10$.

be seen to be concentrated within small values of n and, so the geometric approximation appears to be a reasonable one. A continuous drop function ensures that the drops are independent and, thus, reduces the synchronisation of drops for different flows. This argument supports using gentle RED. The tail drop probability during this simulation scenario was 2×10^{-4} which also suggests that synchronized losses are reduced.

As mentioned above, the probability of dropping consecutive packets becomes independent when the drop function is continuous. This suggests that the probability of drops in a burst is binomially distributed. The probability of at least one drop in a burst can thus be approximated by $1 - (1 - p)^N$, where p is the average drop probability and N is the burst size. This behaviour is seen in Figure 17.5 in which the probability of dropping at least one packet in a burst is plotted as a function of the offered load.

Next, we validate the approximate analysis for obtaining the average queue length. In order to obtain the average queue length we need to solve the fixed point equation as given in equation (17.25). For example, the average queue length as observed by a incoming packet in a $M/M/1$ queue with finite buffer B is given by

$$E(Q_{q*}) = \frac{\rho_{q*}}{(1 - \rho_{q*})} - (B + 1)\frac{\rho_{q*}^{B+1}}{(1 - \rho_{q*}^{B+1})}, \qquad (17.31)$$

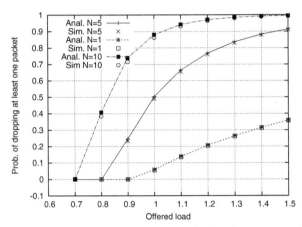

Figure 17.5 Probability of drop of at least one packet in a burst versus offered load.

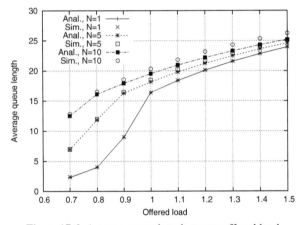

Figure 17.6 Average queue length versus offered load.

where

$$\rho_{q^*} = \rho \left(1 - \frac{q^* - \min_{th}}{\max_{th} - \min_{th}} \right).$$

We can now obtain the equilibrium value of average queue length by solving

$$q^* = \frac{\rho_{q^*}}{(1 - \rho_{q^*})} - (B + 1) \frac{\rho_{q^*}^{B+1}}{(1 - \rho_{q^*}^{B+1})}. \qquad (17.32)$$

Similarly, for the batch arrivals we use the expression for the infinite buffer case as an approximation. The equilibrium value for batch arrivals is obtained

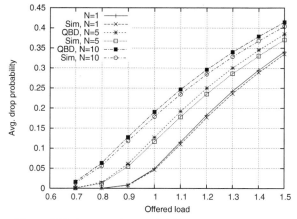

Figure 17.7 Average drop probability versus offered load.

by solving

$$q^* = \frac{\rho_{q^*}}{(1 - \rho_{q^*})} \frac{N+1}{2}. \qquad (17.33)$$

For the values of $\min_{th}$, $\max_{th}$, and B assumed in this section, we plot in Figure 17.6 the average queue as obtained from solving equations (17.32) and (17.33) numerically using MATLAB and the results obtained through simulations.

17.5.2 General Case

Next, we consider the case when $\alpha \nrightarrow 0$. In the rest of this section we assume that the discretization parameter, m, is 1. The drop function is the same as in the previous section. We need that $\alpha m B$ be less than 1. Thus, α has to be chosen such that $\alpha < 0.01$ in order to use the QBD algorithm. We choose $\alpha = 0.009$ which is close to 0.01. In Figures 17.7 and 17.8 we plot the average drop probability and the average queue length, respectively, as a function of the offered load. We note that the behaviour is similar to that observed in the limiting case.

In order to see the effect of choosing an averaging parameter which is not necessarily small, we plot in Figure 17.9 the distribution function of the average queue length, q, using the QBD analysis, simulations, and the singular perturbation (SP) analysis. The offered load was 1.2 and the burst size was taken to be 10. As can be observed from the figure, the effect of increasing a is to increase the variance of q. We cannot, however, reduce the

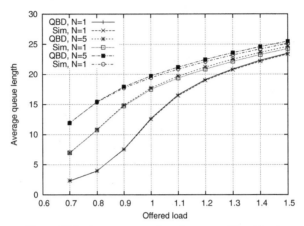

Figure 17.8 Average queue length versus offered load.

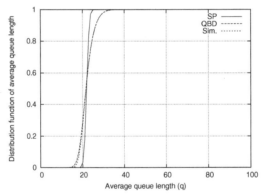

Figure 17.9 Distribution function of the average queue length; $N = 10$ and offered load is 1.2, $a = 0.009$.

variance of q by taking the limit. There is a limiting distribution of q (as was seen from the limiting case) and hence a limiting variance which is different from 0.

17.6 Conclusions

We used a singular perturbation based approach to find the limiting stationary distribution of the average queue length and the packet drop probability in a RED enabled queue. For Poisson arrival and exponential service times the limiting stationary distribution was given as closed form expressions. The equilibrium point of the average queue length was found to be the fixed point solution of a function depending on the average queue length of a system with no averaging. The analytical results were observed to match fairly accurately with results obtained through simulations. We also showed that by using a continuous drop function we can reduce the number of consecutive losses thereby avoiding synchronized losses. We showed that for the case when the averaging parameter is not small, the stationary distribution can be computed using an algorithm for level dependent QBDs. However, for small values of the averaging parameter, the computational complexity of using this algorithm was much greater than that by using the singular perturbation approach. In this work, we considered scenario in which packet generation process is independent of packet drops. Future work will involve the study of a RED queue with closed loop TCP traffic sources.

Acknowledgements

This work was partially supported by the EURO NGI network of Excellence, by the INRIA's TCP ARC collaboration project and by the *Indo-French Center for Promotion of Advanced Research (IFCPAR)* under research contract number 2900-IT. The work of the third author was carried out while he was with INRIA Sophia-Antipolis.

References

[1] L. Zhang and D. Clark, Oscillating behavior of network traffic: A case study simulation, *Internetworking: Research and Experience*, vol. 1, no. 2, pp. 101–112, 1990.

[2] S. Floyd and V. Jacobson, Random early detection gateways for congestion avoidance, *IEEE/ACM Transactions on Networking*, vol. 1, no. 4, pp. 397–413, August 1993.

[3] R. Pan, B. Prabhakar and K. Psounis, CHOKe: A stateless active queue management scheme for approximating fair bandwidth allocation, in *Proc. of the IEEE INFOCOM*, 2000.

[4] W. Feng, D. Kandlur, D. Saha and K. Shin, The blue queue management algorithms, *IEEE/ACM Transactions on Networking*, vol. 10, no. 4, pp. 249–264, August 2002.

[5] M. Christiansen, K. Jeffay, D. Ott and F. D. Smith, Tuning RED for web traffic, *IEEE/ACM Transactions on Networking*, vol. 9, no. 3, pp. 249–264, June 2001.

[6] S. Athuraliya, V. H. Li, S. H. Low and Q. Yin, REM: Active Queue Management, *IEEE Network*, vol. 15, no. 3, pp. 48–53, May/June 2001.

[7] T. J. Ott, T. V. Lakshman and L. Wong, Sred: Stabilized red, in *Proc. of IEEE INFOCOM*, pp. 1346–1355, 1999.

[8] P. J. Schweitzer, Perturbation series expansion of nearly completely-decomposable Markov chains, in *Teletraffic Analysis and Computer Performance Evaluation*, O. J.

Boxma, J. W. Cohen and H. C. Tijms (Eds.), Elsevier Science Publishers, (North Holland), pp. 319–328, 1986.

[9] K. E. Avrachenkov, J. A. Filar and M. Haviv, Singular perturbations of Markov chains and decision processes. A survey, in *Handbook of Markov Decision Processes: Methods and Applications*, E. A. Feinberg and A. Shwartz (Eds.), International Series in Operations Research and Management Science, Kluwer Academic Publishers, pp. 113–153, 2002.

[10] G. Latouche and V. Ramaswamim, *Introduction to Matrix Analytic Methods in Stochastic Modeling*, ASA-SIAM Series on Statistics and Applied Probability, SIAM, 1999.

[11] T. Bonald, M. May and J. Bolot, Analytic evaluation of RED performance, in *Proc. of the IEEE INFOCOM*, 2000.

[12] V. Mishra, W. Gong and D. Towsley, Fluid based analysis of a network of AQM routers supporting TCP flows with an application to RED, in *Proc. of the ACM SIGCOMM*, 2000.

[13] C. Hollot, V. Misra, D. Towsley and W. Gong, A control theoretic analysis of RED, in *Proc. of the IEEE INFOCOM*, 2001.

[14] E. Kuumola, J. A. C. Resing and J. Virtamo, Joint distribution of instantaneous and averaged queue length in an M/M/1/K system, in *Proc. of the 15th ITC Specialist Seminar "Internet Traffic Engineering and Traffic Management"*, P. Tran-Gia and J. Roberts (Eds.), Elsevier, pp. 58–67, 2002.

[15] V. Sharma, J. Virtamo and P. Lassila, Performance analysis of the RED algorithm, *Probability in the Engineering and Information Sciences*, vol. 16, pp. 367–388, 2002.

[16] F. Camilli and M. Falcone, Approximation of optimal control problems with state constraints: Estimates and applications, in *Nonsmooth Analysis and Geometric Methods in Deterministic Optimal Control*, B. S. Mordukhovic and H. J. Sussman (Eds.), IMA Volumes in Applied Mathematics, vol. 78, Springer Verlag, pp. 23–57, 1996.

[17] M. Bardi and I. Capuzzo-Dolcetta, *Optimal Control and Viscosity Solutions of Hamilton–Jacobi–Bellman Equations*, Birkhäuser, 1997.

[18] H. J. Kushner and P. G. Dupuis, *Numerical Methods for Stochastic Control Problems in Continuous Time*, Springer, 1992.

[19] S. Floyd, RED: Discussions of setting parameters, November 1997. Available at http://www.icir.org/floyd/red.html.

[20] P. J. Schweitzer, Perturbation theory and finite Markov chains, *Journal of Applied Probability*, vol. 5, pp. 410–413, 1968.

[21] D. P. Gaver, P. A. Jacobs and G. Latouche, Finite birth-and-death models in randomly changing environments, *Advances in Applied Probability*, vol. 16, pp. 715–731, 1984.

[22] S. Floyd, Recommendation on using the gentle variant of RED, Available at http://www.icir.org/floyd/red.html, March 2000.

[23] W. E. Leland, M. S. Taqqu, W. Willinger and D. V. Wilson, On the self-similar nature of ethernet traffic, *IEEE/ACM Transactions Networking*, vol. 2, no. 1, pp. 1–15, 1994.

[24] V. Paxson and S. Floyd, Wide-area traffic: The failure of Poisson modeling, *IEEE/ACM Transactions on Networking*, vol. 3, no. 3, pp. 226–244, June 1995.

[25] M. E. Crovella and A. Bestavros, Self-similarity in World Wide Web traffic: Evidence and possible causes, *IEEE/ACM Transactions on Networking*, vol. 5, no. 6, pp. 835–846, December 1997.

[26] J. Cao, W. S. Cleveland, D. Lin and D. X. Sun, Internet traffic trends toward Poisson and independent as the load increases, chapter in *Nonlinear Estimation and Classification*, Lecture Notes in Statistics, vol. 171, Springer, pp. 83–109. 2003.

[27] T. Karagiannis, M. Molle, M. Faloutsos and A. Broido, A nonstationary Poisson view of internet traffic, in *Proc. of IEEE INFOCOM*, March 2004.

PART SIX

APPLICATION LAYER MULTICAST

18

Peer-to-Peer Multimedia Conferencing System Based on SIP Signaling

Andre Rios, Alberto J. Gonzalez and Jesus Alcober

*Department of Telematics Engineering, Universitat Politecnica de Catalunya/
i2CAT Foundation, Barcelona, Spain;
e-mail: {andre.rios, alberto.jose.gonzalez, jesus.alcober}@upc.edu*

Abstract

Currently, collaborative tools such as multimedia conferencing are using the Peer-to-Peer (P2P) paradigm and, thus, they are becoming a hot topic in Internet research. Conferencing systems allow an interactive communication and facilitate the joint work among users regardless of their physical location. These systems conventionally work in a centralized manner using a high performance device called Multipoint Conference Unit (MCU), following a client-server scheme. Mainly, corporative environments and telcos are using this last solution. On the other hand, in Internet, P2P conferencing solutions are still in early stages of development. In this sense, most of the existing conferencing systems support few participants and use pure Application Layer Multicast (ALM) techniques that do not properly solve interactivity among users and heterogeneity requirements. In addition, some of them have a lack of mechanisms for media mixing and do not allow future integration with standard IP Telephony and Instant Messaging systems (SIP-compliant). A video conferencing system should be easy to setup, maintain and any participant should be able to manage the conference. The SIP conferencing framework offers these last facilities but commonly it is not applied to P2P environment. Therefore, building a flexible and scalable P2P multimedia conferencing system is still an open issue. In this work, we present a prototype

*D. D. Kouvatsos (ed.), Traffic and Performance Engineering for Heterogeneous
Networks,* 401–416.

that attempts to solve the issues described above. This prototype is based on a hybrid mechanism that uses ALM algorithm and standard SIP messaging. The preliminary testing has demonstrated that this solution is a valid platform for building a flexible multipoint conferencing system.

Keywords: Application Layer Multicast (ALM), conferencing, mixing, multipoint, peer-to-peer (P2P), Session Initiation Protocol (SIP).

Abbrevations

ALM	Application Layer Multicast
CPU	Central Processing Unit
MCU	Multipoint Conference Unit
MPEG-TS	Moving Picture Experts Group – Transport Stream
P2P	Peer-to-Peer
RAM	Random Access Memory
SDP	Session Description Protocol
SIP	Session Initiation Protocol
TV	Television
VLC	VideoLan Client

18.1 Introduction

Recently, P2P networks have gained popularity thanks to the deployment of file-sharing applications. However, nowadays real-time media streaming applications arouse a great interest both to commercial level and academic research. Live streaming introduces new challenging problems [1, 2] different to ordinary file-sharing. In general, P2P media streaming solutions have different features that determine the operation of the applications; i.e., large volume of media data along with stringent timing constraints, the dynamic and heterogeneous nature of P2P networks and the unpredictable behavior of peers. Additionally, if the media streaming service is a conference, then exists the "big challenge" to maintain the interactive experience, especially when there are many participants (large-audience).

Conference services are generally deployed for open group applications in order to distribute public events like lectures, conferences, concerts or Internet TV. Multiparty conferencing systems with a reduced and controlled membership group (close groups) are better suited to set up meetings, discussions, seminars or consultations. We are interested to design a system that supports both schemes in order to have a more generic and flexible platform.

Mainly, a conference is a real-time application that in its point-to-point basic scheme involves bidirectional communication with stringent requirements for end-to-end delay. On the other hand, the development of a multipoint conference in a P2P scheme has additionally other critical challenges such as bandwidth and processing limitations of peers and the delay associated with multicast among peers.

There are several existing solutions handling P2P conferencing. A first traditional approach is IP multicast, which is based on network layer. Currently, its deployment is limited due to its complexity and scalability and security problems. A second approach is to use a full-meshed scheme but clearly its main problem is the scalability and applicability, especially when the number of users of the system is high. Finally, there is a third approach, which currently is the most widely used: Multicast at Application Layer (ALM) [3]. Its main advantage is to reduce the network usage and to balance the links in the underlying network.

Besides, there are several works that try to mix P2P techniques with the Session Initiation Protocol (SIP) [4]. Mainly, there are two approaches [15]: SIP using P2P (replace SIP location service by a P2P protocol) and P2P over SIP (implementation of P2P using SIP signaling). Currently there is an IETF working group, with efforts to standardize P2P-SIP [12].

Taking into account these above mentioned antecedents, we are interested in exploring ALM techniques that allow to improve the existing solutions in order to offer a system with real-time support and adding features like media mixing (centralized or distributed) depending on the scheme selected: pure conference or multiparty.

In a parallel way, we also understand that a good conferencing service should besides meet certain basic requirements such as the following ones: it should be easy to setup and maintain and it should be able to dynamically add and drop participants to the conference. In this sense, SIP-based conference control framework (SIPPING) offers a suitable environment to work while allowing interoperability with other SIP-compliant systems.

Finally, this work proposes a P2P multipoint conference prototype based on ALM and SIP, which is a first approach to overcome some problems associated to P2P conferencing.

This work is organized as follows. In Section 18.2, an overview of P2P media conferencing approaches applied is presented. Section 18.3 analyzes how SIP allows to develop multiparty conferencing and discusses their implications in a P2P scheme. Section 18.4 introduces the proposed conferencing solution based on hybrid ALM+SIP. In Section 18.5 a developed prototype and its development environment are described in order to check the different programmed functionalities and to validate the concepts on which the application is based. Finally, Section 18.6 concludes this work and discusses the future work.

18.2 Related Works

In general, conferencing systems allow two or more participants to communicate with each other in real-time using both audio and video. The conventional solution in these cases is to use a Multipoint Conference Unit (MCU). This solution allows a centralized communication among users, but it lacks of scalability due to the large amount of resources (bandwidth, CPU, etc) required in the central point where the MCU is located. Currently, this solution is commonly utilized by telcos and enterprises, but in Internet its applicability is limited.

As it was mentioned above, the second solution is IP Multicast. At this moment, there are many MBONE applications [5] for conferencing, but its deployment is at local level in a controlled environment, such as universities or research institutions.

Most of the recent techniques for P2P media streaming over Internet utilize multicast model at application layer. Usually these techniques are more known by applications like P2P-TV or P2P-VoD, however, several papers have already proposed to use ALM [6–9] to deliver decentralized conferencing over Internet.

Additionally, there are not many documented solutions handling P2P conferencing using SIP. Commercial products such as Damaka [16] and Openwengo [17] only offer either point-to-point or multipoint based on MCU concept with support for few users.

To sum up, despite the great number of ALM systems, the majority of them only put effort in optimizing the end-to-end delay and failure recovery mechanisms.

Our approach differs from the prior work in operation aspects. Mainly, we use the SIP protocol to setup the conference and this choice would enable us to eventually offer new value-added services such as instant messaging. Besides, we add mixing capabilities to all peers and, consequently, it is possible to reduce the media traffic and to improve the display of the video stream.

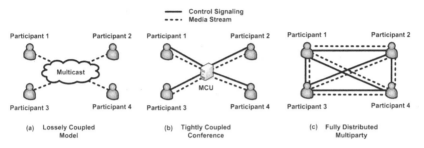

Figure 18.1 Models of multiparty communication.

18.3 Conferencing Using SIP

SIP is a protocol that provides mechanisms based on dialogs (SIP messages) for the initiation, modification and termination of media sessions between user agents. While the original SIP scheme was developed for a two party communication, SIP communication sessions with multiple participants present other technical complications no resolved in the initial specification. In order to overcome these issues, different conferencing frameworks have been proposed: SIPPING [10] and XCON [11]. This work is based on the former because does not define new conference control protocol to be used and provide provisions for inviting, expelling and configuring media streams during the conference. Figure 18.1 depicts different models of multiparty communications that SIP can support.

SIPPING workgroup has defined some new functional entities: FOCUS, Conference Notification Service, Conference Policy Server and Mixer. FOCUS is a central controller entity, which receive all of INVITE requests to a conference. Moreover, it maintains a SIP signaling relations with each and every participant and takes the responsibility of sending media to all the participants in the conference. The Conference Notification Service maintains a Subscription/Notification service with the participants for the modification of the conference primitives. On the other hand, every conference is governed by a certain set of rules, which are enforced by Conference Policy Server. Finally, media transmission and distribution is done by the Mixer. It receives a set of streams of the same type and combines their media in a type specific manner and redistributes the resultant media to every participant in the conference. Besides, it includes transport protocols such as RTP for communications. Figure 18.2 shows a basic SIPPING scheme.

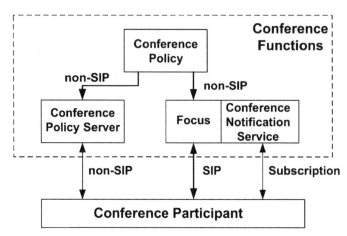

Figure 18.2 Sipping conference framework.

SIPPING framework is open to define the physical realization. For instance, it is possible to mention some models: centralized server, endpoint server, media server component and distributed mixing. In this work we try to exploit all of these functionalities into a P2P network in order to allow large-audience conferencing.

18.4 Proposed P2P Conferencing System

This section explains how the proposed system allows to create a large-audience conference by taking advantage of P2P techniques for distributing the conference stream among all the participants, using SIP as standard signaling and messaging protocol.

18.4.1 Peer Roles

The system discerns between two specific roles of the peers that join the system: the leader and the participants.

The leader is the peer that creates the conference, the one in charge of creating the virtual conference room and announcing to the rest of the users the event. The peers that receive the announcement of the conference will become participants when joining the virtual room. Once the conference has started, a participant can adopt two behaviors: actively participate on the conference or passively participate on the conference. This yields two types of participants, the first one is called speaker, which has video capturing capabilities, and

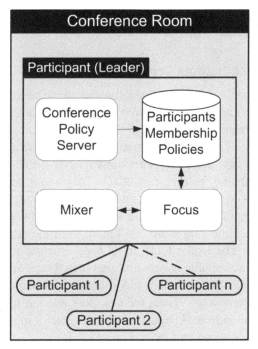

Figure 18.3 Participant SIP architecture.

the second one is known as listener, which only wants to be present on the conference and does not take part.

The proposed conference system allows a limited number of speakers, in this case we defined a limit of four simultaneous speakers on the conference. Note that this number cannot be very high, otherwise the conference would be not useful. In order to allow the participation of peers in an ordered way, the leader manages a list of participation events. The list consists on an ordered queue that allocates the queries of the different participants that want to become speakers. This list applies a (First-In-First-Out) FIFO behavior. The leader has also the ability to reject some participants of the conference.

18.4.2 Architecture

The proposed conferencing system borrows ideas from the Framework for conferencing with SIP which is currently standardized by the IETF SIPPING Working Group. In Figure 18.3, the general SIP conferencing architecture implemented by each peer within the system is shown. This architecture allows any peer in the system to become leader, speaker or listener.

18.4.3 Distribution Network Construction

The conference media stream will be delivered to the audience by constructing an Application Layer Multicast Tree (ALM) rooted at the leader peer. The leader peer will create a map of the conference connected peers by creating a logic tree structure. The tree is constructed using the data gathered from new joining peers. The new incoming peers must specify the number of supported children by calculating its own fanout (f). The fanout can be obtained by dividing the peer current available connection speed by the stream bitrate.

A new peer, which tries to join a conference room, sends an INVITE Join SIP message indicating which its fanout is in the attached SDP to the conference leader. The leader looks in its peer structure for the first peer in the system with an available connection (deputy). Once obtained the deputy peer, it sends its SIP URI to the new incoming peer through the SDP attached to the 200 OK SIP message. Then, the new peer just has to send an INVITE SIP message to the indicated deputy in order to start joining the conference. The described process can be seen in Figure 18.4.

18.4.4 Failure Mechanism

We define failure as the situation where a peer cannot receive the media stream because the communication with its parent peer is broken (i.e. when a peer accidentally leaves or there is a problem in the link between parent and children). A peer detecting a failure tries to establish a new connection with another peer.

If a peer fails in receiving data from its parent, then it will ask the leader for another peer with enough available fanout in order to become its new parent. The peer also communicates to the leader, which is the peer that has failed in order to update its peer-map. Once the peer establishes the new connection with its new parent, the communication can continue.

When the peer detects the failure, it uses an INFO message in order to inform the leader source that its parent is unavailable. Then, the leader peer deletes the leaving peer and looks for a new parent for the peer and notifies it to the peer on the 200 OK response SIP message (see Figure 18.5).

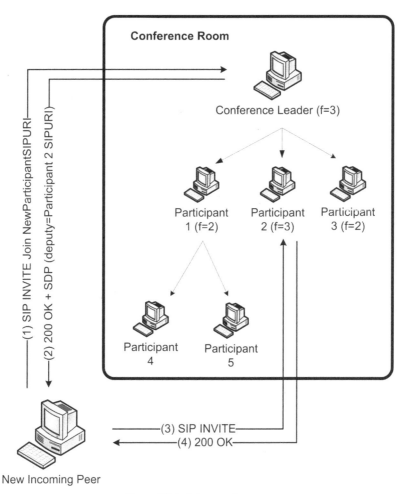

Figure 18.4 Join a conference.

18.4.5 Mixing Speakers

The leader peer is able to mix the different streams generated by the participants that want to become speakers. The participants send REFER SIP messages in order to add its stream to the conference. If the maximum number of simultaneous speakers is reached, then the candidate speaker is added to a queue and can wait till its turn arrives. Once the speaker can participate in the conference, it sends its video stream to the leader who mixes (see Figure 18.6) the streams generated by the speakers (S1, S2, S3 and S4) and creates a single

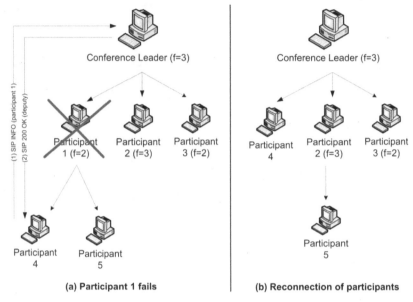

Figure 18.5 Peer reconnection.

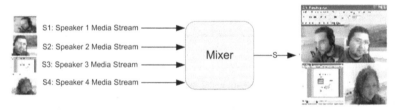

Figure 18.6 Mixing four speakers (S1, S2, S3 and S4) into a single stream S.

stream (S) that will be distributed (see Figure 18.7) all along the network from the leader to the participants.

18.5 Implementation and Testbed

In order to verify the proposed conferencing system, a testbed was prepared. The main goal was to determine the correctness of the different signaling and distribution techniques that were used. According to this objective, it was implemented a conferencing system prototype that runs over the presented testbed. This prototype allows to distribute the conference media stream among all the participants in a conference by using a P2P technique based on

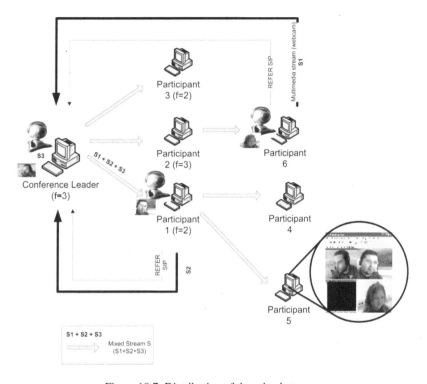

Figure 18.7 Distribution of the mixed stream.

ALM. The control plane is based on the SIP signaling protocol. In addition, the leader peer has the ability to compose the media stream that will be spread all along the conference thanks to its mixing capability. For this purpose, it has been developed a collection of tests. The main goals of them are the following ones. Initially, through a simple testing scenario, the idea is to verify the operation of the system in a P2P scheme with few nodes (PCs) and then to check the mechanisms to construct the logical tree structure and the recovery mechanisms when a node fails or leaves the conference. Finally the correct operation of the signaling plane will be verified. On the other hand, it must be pointed out that in the system the role of the leader peer is a key aspect that can suppose the bottleneck of the whole system because it needs extra processing capability of the video streams for mixing and forwarding tasks. The objective is to check the amount of CPU and RAM charge needed when mixing (with and without transcoding). At implementation level it is also important to see what the best choice to develop the forward mechanism

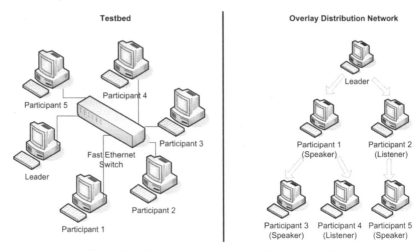

Figure 18.8 Testbed and overlay distribution network.

is, taking into account that the system will work with high quality video streams quality under real-time conditions. In this sense, several software implementations will be evaluated.

It has been developed a prototype application in order to test the distribution of four mixed media streams. The mixing function has been realized using VideoLan [13] Client (VLC) 0.8.5, which allows to create a mosaic of different video/audio streams. Basically, we have done a video mosaic mixing four video streams obtained by different video cameras connected to the PCs of the participants and then streaming the resulting video on the network by using a single MPEG-TS stream. Note that we are also streaming the four sound tracks from the four streams in the same MPEG-TS stream. The deployed scenario can be seen in Figure 18.8.

The testing scenario is composed by six PCs connected to a Fast Ethernet switch. In this case, three peers act as speakers while the others are acting as listeners. A binary tree in which each node is father of two sons can represent the distribution scheme.

Through this scenario, it was verified the proper operation of the construction of the P2P network and the recovery mechanism. At signaling level, we prove that SIP is a very versatile protocol that facilitates the start-up and maintenance of a multi conference.

With regard to the operation of the leader and its mixing capabilities, in Table 18.1, the CPU charge and the RAM consumption measured at the

Table 18.1 Resources consumption measurement at mixer node (leader peer).

Number of videos	Video bitrate	Outgoing bitrate	Resolution	CPU	RAM
4	500 Kbps	2 Mbps	200x200	10 %	12 KB
			400x400	52 %	30 KB
		1 Mbps	200x200	42 %	27 KB
			400x400	55 %	29 KB
4	1,5 Mbps	6 Mbps	200x200	67 %	75 KB
			400x400	77 %	80 KB
		3 Mbps	200x200	68 %	77 KB
			400x400	78 %	80 KB
4	25 Mbps	100 Mbps	200x200	100 %	70 KB
			400x400	99 %	74 KB
		50 Mbps	200x200	100 %	70 KB
			400x400	100 %	70 KB

leader node during the mixing process is represented. Different tests were carried out considering streams of several video qualities. The results show that the computational cost is very high in a desktop PC (Pentium IV, 3.2 GHz, 1GB RAM) when the considered multimedia flows have high bitrates (25 Mbps, High Definition media). Moreover, it is possible to observe that the computational cost grows when demanding transcoding is required, specially when the transcoding process involves a large decrease in the bitrate of the composed media stream.

The peer that acts as leader configures the transmission network interface to receive the desired stream from the speakers and also to send the stream to its children. The connected participants receive the media stream from their parent peer and forward it to their respective children.

The transmission layer of the developed prototype was implemented using different programming languages (Java and C), different socket implementations (java.io blocking socket, java.nio non-blocking socket, PacketReflector [14] C based) and different operating systems (Windows XP and Ubuntu Linux). When observing the gathered results, the best performance was achieved when using the PacketReflector implementation running under Linux. We achieved a transmission bitrate of 25 Mbps when sending four mixed streams. The test results are graphically represented in Figure 18.9.

The latest results confirm that a node, in order to fulfill the role of leader, needs extra capacity to carry out its work. By definition, any node can become

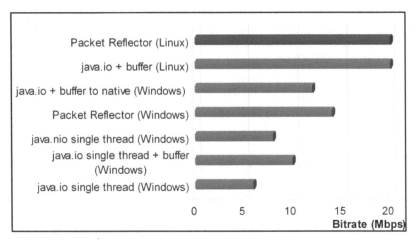

Figure 18.9 Maximum bitrate achieved with different implementations.

leader, but given the heterogeneity of the terminals, it should be set initially who may or may not be leader due to this constraint.

18.6 Conclusions and Future Work

In this work, we introduced a prototype system, which allows to create large-audience media conferences by applying ALM techniques for distributing the media stream over the network created by all the participants of the conference. The use of this P2P technique allows making an efficient use of the network resources. The system uses SIP as signaling protocol, which provides flexibility to the solution and allows the development of new value-added services such as instant messaging, etc.

Moreover, the developed prototype is based on the mixing of the different speaker streams in order to make able the resulting conference stream to all the participants in the conference. With this feature, it is possible to offer both a pure conference and multiparty conference. Finally, the study presented the test results gathered from the deployed testbed. The results obtained from the testbed allowed to verify the correct use of SIP as signaling protocol and the use of ALM as an efficient manner to distribute the media stream focusing on a conferencing system. Besides, we identify the transcoding process as a bottleneck, when demanding transcoding is required. We need to further revise the mechanisms (ALM+SIP) that optimize overall performance and increase the scalability. In the next future, we are also interested in improving the failure mechanisms and room management.

It is being studied how to implement a subscription mechanism based on SIP (SUBSCRIBE) in order to efficiently detect the failure events produced in the P2P network. It must be analyzed the resulting signaling overhead to maintain the subscription and the notification among peers.

Furthermore, some current studies on P2P video streaming are applying media coding techniques, such as Multiple Description Coding (MDC) and LC (Layered Coding), mainly in order to improve error resilience and media adaptation to the specific features of the end user terminal. In multimedia conferencing systems the interactive feeling and presence are key aspects, however, they are clearly affected when the number of users grows and exists heterogeneity among the end terminals. The use of these media coding techniques for P2P conferencing applications is still in an early stage due to real-time limitations. On the other hand, the main idea is not to have a supercomputer to work in an efficient manner, but conventional PCs. Nevertheless, this research topic is very interesting thinking in the capability to work with different video qualities and improve, for instance, the mixing and transcoding processes. Moreover, P2P for its own nature presents challenges associated to the high churn and free-riding so, theses techniques could give robustness to the system.

Acknowledgements

This work was partially funded by i2CAT Foundation and MCyT (Spanish Ministry of Science and Technology) under the project TSI2007-66637-C02-01, which is partially funded by FEDER. The authors would like to thank Antoni Oller and Juan Lopez from the Universitat Politecnica de Catalunya (UPC) for their help and support.

References

[1] D. Meddour, M. Mushtaqm and T. Ahmed, Open issues in P2P multimedia streaming, in *Proceedings MULTICOMM*, Istambul, June 2006.
[2] W. P. Ken, X. Jin and H. G. Chan, Challenges and approaches in large-scale P2P media streaming, *IEEE Multimedia*, vol. 14, no. 2, pp. 50–59, April 2007.
[3] M. Brogle, D. Milic and T. Braun, Supporting IP multicast streaming using overlay networks, in *Proceedings Q'Shine 07*, Vancouver, Canada, August 2007.
[4] J. Rosenberg, H. Schulzrinne, G. Camarillo, A. Johnston, J. Peterson, R. Sparks, M. Handley and E. Schooler, SIP: Session Initiation Protocol, IETF RFC 3261, June 2002.

[5] S. Petrik and V. Skala, Report on videoconferencing systems, Technical Report No. DCSE/TR-2006-04, University of West Bohemian Pilsen, December 2006.

[6] X. Wu, K. Kishore and V. Krishnaswamy, Enhancing application-layer multicast for P2P conferencing, in *Proceedings IEEE Consumer Communications & Networking Conference (CCNC)*, Las Vegas, NV, January 2007.

[7] M. Dowlatshahi and F. Safei, Multipoint interactive communication for peer to peer environments, in *Proceedings IEEE International Conference on Communications (ICC)*, Istambul, June 2006.

[8] Ch. Luo, J. Li and S. Li, DigiMetro: An application-layer multicast system for multiparty video conferencing, in *Proceedings Globecom 2004*, Dallas, TX, November 2004.

[9] H. Horiuchu, N. Wakamiya and M. Murata, A network construction method for a scalable P2P video conferencing system, in *Proceedings IASTED European Conference Internet and Multimedia Systems and Applications*, Chamonix, France, March 2007.

[10] J. Rosenberg, A framework for conferencing with the Session Initiation Protocol (SIP), IETF RFC 4353, February 2006.

[11] R. Even and N. Ismail, Conferencing Scenarios, IETF 4597, July 2006.

[12] P2PSIP WG IETF, http://www.p2psip.org/

[13] VideoLan Project, http://www.videolan.org/

[14] J. Highfield, UDP packet reflector rum, http://spirit.lboro.ac.uk/mug/mug.html

[15] K. Singh and H. Schulzrinne, Columbia University, New York, June 2005, http://www.cs.columbia.edu/IRT/p2p-sip

[16] Damaka. http://www.damaka.com/

[17] Openwengo. http://www.openwengo.org/

PART SEVEN

NUMERICAL AND SOFTWARE TOOLS

19

Automated Formulation and Solution of Markov Modulated Queues with Geometric Processes

David J. Thornley, Harf Zatschler and Peter G. Harrison

*Department of Computing, Imperial College London, London SW7 2AZ, U.K.;
e-mail: {djt,hz3,pgh}@doc.ic.ac.uk*

Abstract

We present an automated formulation mechanism that generates systems of
equations for efficiently solving for the equilibria of finite or infinite Markov
modulated queues with geometrically batched Poisson arrivals and departures. In common with earlier work, the method benefits from its ability to
generate equations which describe a homogeneous linear matrix recurrence
relation of finite domain over the majority of the system. The new approach
can do this for queues with arbitrary numbers of arrival processes of positive
and negative customers of mixed types and service completion processes.
Moreover, geometric distributions can be scaled and superimposed to produce
a range of convex probability mass functions. We thereby provide a fully operational practical approach for the analytical modelling of many present day
communication and computer systems, e.g. the Internet and mobile networks.

The geometric distribution produces an infinite range of batch sizes,
which creates unbounded transitions in queue length, leading to Kolmogorov
balance equations with an unbounded, possibly infinite, number of terms.
Our method constructs an equivalent ensemble of transformed Kolmogorov
balance equations of finite range. The key contribution is the mechanical
derivation of these transformed equations, for which we have a working implementation used to provide a range of examples. Previously, such equations
had to be derived from first principles for every variant of every applicable

*D. D. Kouvatsos (ed.), Traffic and Performance Engineering for Heterogeneous
Networks,* 419–457.

queueing system in a manual procedure, which is highly error prone. An additional, crucial benefit is the ability to formulate and solve systems in which there may be inactive processes, or two or more processes with similar burstiness characteristics in a given modulation state. Such degenerate systems cannot be solved using these previous methods.

Keywords: Queueing theory, spectral expansion, batched transitions.

19.1 Introduction

The performance evaluation community is constantly on the lookout for novel, practical methods for modelling real systems. The basic requirement on all such methods is sufficient accuracy to enable performance planning and/or troubleshooting. Emphasis is also commonly placed on the efficiency of the solution methods to be used, as it is not uncommon to require iterative refinement over large network structures. The work we present is a new generation of a track of research into queues based on Markov modulated Poisson occurrences of job transactions in batches of geometric size. It can therefore account for traffic which is both bursty and correlated over time. Other work using geometric batches focuses either on the parameter matching problem when using aggregated traffic [8], or on generating generalized formulae for balance equations [11]. Kouvatsos et al. have achieved considerable success in network modelling using aggregated geometrically batched traffic. Chakka and Do [29] follow closely the procedures introduced in [4], extending them to accommodate multiple arrival processes of positive customers.

In this paper, we distill the core algebraic premise from the work in [4] and use this to synthesize an automated approach to model formulation [17] which is not susceptible to the sort of errors inevitably experienced when performing manipulations by hand. The practicality of our approach is demonstrated, e.g. in [12] and [10], where queues with multiple, independent arrival processes were solved using our developing methods.

In performing this synthesis, we have identified a constraint on the class of problems solvable by methods following the approach in [4, 11]. Such methods cannot be used to solve systems in which any arrival or service process either has a zero batch size parameter (giving unit batches) in any given modulation state, or has a batch size distribution parameter identical to another process in the same class and modulation state. For example, if two arrival processes are identically batched in a given modulation state, these formulations will fail. Our new method is not subject to these constraints because our explicit, recursive definition of the transformed ensemble of constraints takes such situations into account.

The general background to the present contribution is summarized in Section 19.2 and in Section 19.3 we briefly review the compound Poisson process, together with the generalized exponential distribution, considering their significance in a Markov modulated environment. In Section 19.4 we define our notation and terminology, specifying the class of queueing systems we address. The notation is introduced constructively, to provide a complete description of a very large class of queue models. The crux of the paper is the presentation of the elimination function described in Section 19.5. This provides the basis for the equation transformation mechanism which enables us to render any geometrically batched queueing behaviour into a form solvable by matrix geometric techniques, such as highly efficient direct solvers [3, 13] or spectral expansion [1]. We use Mathematica® [5] as a proofing tool for our methods, as it allows us to perform operations both symbolically and numerically. We include an example symbolic result for localized balance equations to demonstrate the relative simplicity of function in comparison with earlier work [4].

19.2 Background

Work on the construction of geometrically batched queues for exact solution (e.g. not using the maximum entropy approach of Kouvatsos [19]) took off with an ATM model by Harrison and Chakka [20], with exact solutions using spectral expansion as espoused by Mitrani and Chakka [21]. This was later developed to incorporate a negative customer arrival process [22]. This important step introduced the concept of superposition of departure processes in the processing completion and customer removal due to negative customers. Chakka and Harrison then provided proofs of the correctness of localized equations for a canonical geometrically batched queue [23], and with positive and negative customer arrivals [4]. In the same year, Harrison's MEGAN research project to investigate the use of geometrically batched queues in networks began. Harrison's concept of a geometrically batched node with superposed arrival streams and link traffic approximated as a compatible process led to a successful implementation of a processor-farming network model with multiple queues and feedback [12]. As part of this process, the formulation approach for the queue itself underwent a radical change, resulting in an automated process which will formulate finite balance equations for a geometrically batched queue with an arbitrarily complex description [17].

Concurrently with our dissemination of this automated approach, Chakka et al. provided important example applications, with the addition of a more complex processing description, using the original approach of hand-crafting the equations [11, 24, 25]. We successfully implemented an important enhancement to the queue formulation – suggested to us by Chakka in person – in which queue length dependency is approximated by defining processes in terms of bands of queue lengths, with a clear demonstration of the improvement of accuracy by comparison with simulation [26] in a network with tight feedback to emphasize the value of the banding. Do et al. have used an *unbatched* banded queue to model a web service [27]. We believe this restriction may have been required as a result of the prohibitively high complexity of the approach to balance equation localization used. We present here a mechanical approach, which represents an addtional simplification of the recursive approach we first presented in [17].

We, in common with Chakka and his co-workers, have consistently favoured spectral expansion, as originally advocated by Mitrani and Chakka [21], due to its explicit presentation of the behaviour of the solution. The controversy surrounding this choice was resolved in [18]. The choice of solution mechanism is now open, and depends more on the precise form of the queue, and the context within which it is to be used. Using the spectral expansion approach has also enabled more freedom of expression in the construction of individual traffic streams [28]. In the present paper we detail the construction of queues with multiple independent traffic, service and customer removal processes.

19.3 The Geometric Batch Size Distribution

A geometric distribution of batch sizes, in which a batch of size s occurs with probability $(1-\theta)\theta^{s-1}$ for some parameter $0 \le \theta < 1$, provides a monotonic distribution where small batches are more likely than large batches. This clearly cannot be used to model in close detail, for example, the distribution of packet sizes in Ethernet traffic, which displays very clear peaks. It does, however, provide a range of batch sizes which enables the modelling of the principal effect on performance metrics of burstiness, which is to increase queue lengths, and hence degrade response times. Batches also allow us to examine the effect burstiness has in creating stronger correlation in traffic in networks with loops [10, 12].

The compound Poisson process (CPP) describes point arrivals of customers at exponentially distributed intervals in geometrically sized batches.

The generalized exponential (GE) distribution of processing times describes a similar batched behaviour for service completions. The probability distribution function F_{GE} of the inter-arrival time random variable A of a CPP is a "generalized exponential" distribution [8] with rate parameter λ, and batch size parameter θ, say $F_{GE}(a) = P(A \leq a) = 1 - (1 - \theta)e^{-\lambda a}$. The $(1 - \theta)$ term gives an impulse at the origin ($F_{GE}(0) = \theta$) giving a non-zero probability to a null inter-arrival time. This allows a sequence of one or more zero inter-arrival times, and hence non-unit (geometric) batches. A deterministic unit batch size is given by setting the geometric distribution parameter to zero.

Batched processes extend the domain of application of queueing models, for example modelling the bursty traffic observed in present day communication systems. Markov modulation can represent alternate modes of operation of a queueing system, for example the operational or broken status of a server, different types of traffic and certain types of correlation. The simple Markov modulated MM M/M/c queue has been well understood for many years, usually being solved by considering a two dimensional Markov chain (modulation phase × queue length, giving a finite width lattice strip).

We derive an automated formulation mechanism that facilitates the inclusion of batches of geometrically distributed size in Markov modulated, multiprocessor queues of finite or infinite capacity with an arbitrary number of arrival and service completion processes, to yield a complete set of transformed balance equations solvable using spectral expansion or matrix geometric methods. For more general applications, geometric distributions – e.g. associated with the arrival processes – can be scaled and superimposed to produce a class of monotonic, unbounded, convex probability mass functions.

The geometric distribution produces an infinite range of batch sizes, which creates unbounded transitions in the queue length, leading to Kolmogorov equations with an unbounded, possibly infinite, number of terms. Previous analyses such as [4] derive equivalent transformed Kolmogorov equations with a finite number of terms, solved by the spectral expansion method [1]. In these, the transformed equations are derived by hand from first principles, for every model and model-variant. The key contribution in this report is the derivation of the transformed Kolmogorov equations. We present a formulation and solution methodology that allows the piecewise, automated construction of equilibrium solutions (when they exist) for the state occupation probabilities of these Markovian queues. This methodology allows the addition of arbitrarily many batched arrival processes of either positive or negative customers and various queueing disciplines in a Markov modulated environment.

19.4 Queue Formulation

We begin with the Kolmogorov balance equations for the simplest type of queue – $M/M/1/L$. For each new concept included in our paradigm, we successively augment this representation until the full balance equations for all the modelled concepts are synthesized.

In all the following, the balance equations are written in the form $r_j = 0$ for $0 \le j \le L$. The left-hand side (LHS) of the jth Kolmogorov balance equation r_j for the $M/M/1/L$ queue is:

$$r_j^{M/M/1/L} = \pi_{j-1}\lambda f_{j>0} - \pi_j\big[\lambda f_{j<L} + \mu f_{j>0}\big] + \pi_{j+1}\mu f_{j<L} \qquad (19.1)$$

This expression gives the net incoming probability flux to level j, where π_j is the equilibrium probability of occupying state j and f is the indicator function defined by $f_{true} = 1$ and $f_{false} = 0$. The terms λ and μ are the arrival and service rates respectively.

19.4.1 Modulation

We allow the current state of an independent underlying continuous time, discrete state Markov process to select the parameters for the interval distributions of the arrivals and departures. This N-state process is fully defined by its instantaneous transition rate (generator) matrix Q:

$$Q = \begin{pmatrix} -\sum_{i\neq 1} q_{1i} & q_{12} & \cdots & q_{1N} \\ q_{21} & -\sum_{i\neq 2} q_{2i} & & q_{2N} \\ \vdots & & \ddots & \vdots \\ q_{N1} & \cdots\cdots\cdots & & -\sum_{i\neq N} q_{Ni} \end{pmatrix}$$

The joint queue length/modulation states of the resulting queue forms a 2D lattice strip. The dimension describing the state of the modulator is finite, and the other dimension – the queue-length – may be finite or infinite. The equilibrium probability of being in any particular column of the lattice is given by the steady state phase probability of the modulation process π defined by the two equations $\pi.Q = \mathbf{0}$, and $\|\pi\|_1 = 1$.[1] Here $\mathbf{0}$ is a row vector of N zeros.

---

[1] We express the sum of elements in a vector $\mathbf{x}$ as its L_1 norm $\|\mathbf{x}\|_1 = \sum_i x_i$.

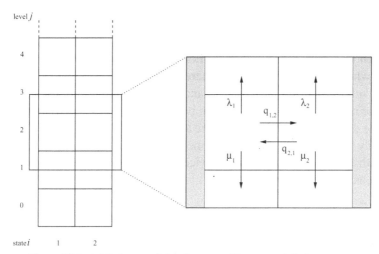

Figure 19.1 A Markov modulated queue with two modulation states.

To allow for modulation in our model, we treat queue levels (lengths) as single entities by representing the state occupation probability (SOP) at level j as the N-component row vector $\mathbf{v}_j$. This vector consists of the SOPs of the N modulation states at that level. In addition, the representation of arrivals and services is augmented to give an analogous vector equation.

Consider a two state modulated queue, whose instantaneous arrival process of positive customers is a Poisson process whose rate is selected from λ_1, λ_2, and service process has rates selected from μ_1, μ_2. This is defined by:

$$\Lambda = \begin{pmatrix} \lambda_1 & 0 \\ 0 & \lambda_2 \end{pmatrix}, \quad M = \begin{pmatrix} \mu_1 & 0 \\ 0 & \mu_2 \end{pmatrix}$$

The rates are selected based on the state of an independent continuous time Markov chain; the "modulator". This notation allows us to associate SOPs on the lattice rows with rate parameter matrices by use of row vectors of SOPs pre-multiplying standard-looking rate terms, with the substitution of the identity matrix I for 1, and the appropriate diagonal matrix for a rate term.

$$\mathbf{r}_j^{MM/MM/1/L} = \mathbf{v}_{j-1}\Lambda f_{j>0} + \mathbf{v}_j\left[Q - \Lambda f_{j<L} - M f_{j>0}\right] + \mathbf{v}_{j+1}M f_{j<L}$$

Compare this expression with that for the unmodulated queue shown in expression (19.1) to see the substitution of vectors and matrices as appropriate for SOPs and rates respectively. The inclusion of Q provides the modulation transitions.

19.4.2 Geometric Batches and the GE Distribution

The queue can accommodate geometrically batched occurrences of both arrivals and processing completions. Formally, we have transitions within the queue whose vertical component (increasing or decreasing queue length) has size s with probability $(1 - \theta)\theta^{s-1}$, and the horizontal component is zero. As with the rates Λ and M, we write the batch parameters in matrix form, where Θ is the diagonal matrix of geometric arrival batch size parameters (θ) and Φ similarly for service batch size parameters (ϕ).

With geometric arrival and service batches included, our LHS balance equations become:

$$
\mathbf{r}_j^{\text{MM CPP/GE/1/L}} = \sum_{i=0}^{j-1} \mathbf{v}_i \Lambda (I - \Theta)^{f_{j<L}} \Theta^{j-i-1}
$$

$$
+ \mathbf{v}_j \left[Q - \Lambda f_{j<L} - M f_{j>0} \right]
$$

$$
+ \sum_{i=j+1}^{L} \mathbf{v}_i M (I - \Phi)^{f_{j>0}} \Phi^{i-j-1}
$$

19.4.3 Multiple Processors

So far our model has assumed the presence of exactly one processor ($c = 1$), with its processing rates given by the matrix M and batches described by the matrix Φ. To introduce multiple homogeneous processors, the processing rate matrix M is replaced by $C_j = \min(j, c)M$ at queue length j, with the batch matrix Φ remaining unchanged. If all processors are busy, then service batches can clear jobs down to level $c - 1$ inclusive; i.e. there is *unbounded* batch flow to levels c and above, and *bounded* (truncated) batch flow to $c - 1$. If any processors are idle, then the processing batch size is exactly 1, as there are no jobs in the waiting room, and a processor can only clear its own job in service. The LHS is thus:

$$
\mathbf{r}_j^{\text{MM CPP/GE/c/L}} = \sum_{i=0}^{j-1} \mathbf{v}_i \Lambda (I - \Theta)^{f_{j<L}} \Theta^{j-i-1}
$$

$$
+ \mathbf{v}_j \left[Q - \Lambda f_{j<L} - C_j f_{j>0} \right]
$$

$$
+ \sum_{i=j+1}^{L} \mathbf{v}_i C_i (I - \Phi)^{f_{j>c-1}} \Phi^{i-j-1} f_{\substack{(i=j+1) \\ \vee (j \geq c-1)}}
$$

The switch $f_{j>c-1}$ bounds downward flow at level $c - 1$. The term $f_{(i=j+1)\vee(j\geq c-1)}$ selects valid flows, which are batches from anywhere in the waiting room to just below c, or single jobs from a single processor.

19.4.4 Breakdowns and Repairs

We treat breakdowns and repairs (in the sense used by Mitrani and Chakka [1]) by allowing the number of processors to vary from 0 to c across the modulation phases. Thus the number of phases $N = c + 1$. This is introduced into the left-hand side expression by replacing references to c with the vector $(c_1, \ldots, c_N)$ of the numbers of operational processors in each modulation state. Our left-hand side expression thus becomes:

$$\mathbf{r}_j^{\text{MM CPP/GE/Mc/L}} = \sum_{i=0}^{j-1} \mathbf{v}_i \Lambda (I - \Theta)^{f_{j<L}} \Theta^{j-i-1}$$

$$+ \mathbf{v}_j \left[Q - \Lambda f_{j<L} - C_j f_{j>0} \right]$$

$$+ \sum_{i=j+1}^{L} \mathbf{v}_i C_i (I - \Phi)^{F_{j>c_m-1}} \Phi^{i-j-1} F_{\substack{(i=j+1) \\ \vee(j\geq c_m-1)}}$$

where $F_{P(m)}$ is a diagonal matrix of values whose ith diagonal element is $f_{P(i)}$. We define the result of raising a square matrix A to the power B with the same dimensions to be a similar matrix of elements $a_{i,j}^{b_{i,j}}$. (In fact all matrices operated upon here are diagonal.) Also, the mth element of C_j is now $\min(j, c_m)\mu$, where $c_m = (m - 1)$. We combine breakdowns and repairs with modulated arrivals by the standard technique of taking the Kronecker product of the independent modulation matrices.

An interesting extension, which will not be discussed further here, has been proposed by Thornley in [16], where jobs in service are lost when a processor breaks down. Similarly, repairs can be accompanied by a simultaneous arrival to model the recovery of previously lost customers. Inclusion of this behaviour involves the incorporation of diagonal transitions in the lattice, which are represented by non-diagonal rate matrices.

19.4.5 Negative Customers

Negative customers [2] create additional probability flux downward in the lattice, corresponding to queue length decrease due to customer loss. This is used to model network phenomena such as losses and load balancing [9]. As well as specifying the Poisson rate and batch size parameters in matrices K and R respectively, a killing mode has to be chosen. We consider three modes: t^v or "tail vulnerable" removes a job from the tail of the queue even if it is in service, t^s or "tail safe" removes a job from the tail of the queue but not when in service, and h^p or "head per" which removes a customer from the head of the queue (in service) at an independent but equal rate per processor, leading to a lower loss rate when some processors are inactive.

Killing mode t^v is the simplest, as it can kill any job in or out of service, and the batches are bounded only at level zero. Mode t^s is bounded at level c, as it cannot kill any job in service. Mode h^p causes flux identical to processing completions.

$$
\mathbf{r}_j = \sum_{i=0}^{j-1} \mathbf{v}_i \left[\Lambda (1 - \Theta)^{f_j < L} \Theta^{j-i-1} \right]
$$

$$
+ \mathbf{v}_j \left[Q - \Lambda\, f_{j<L} - K\, f_{((j>\kappa) \vee h^p)} \beta_j - C_j \right]
$$

$$
+ \sum_{i=j+1}^{L} \mathbf{v}_i \left[K(1-R)^{f_j > \kappa} R^{i-j-1} f_{\substack{(j \geq \kappa) \\ \vee (h^p \wedge i = j+1)}} \beta_i \right.
$$

$$
\left. + C_i (1 - \Phi)^{f_j > c-1} \Phi^{i-j-1} f_{\substack{(i=j+1) \\ \vee (j \geq c-1)}} \right]
$$

where κ is the lowest level reachable by killing, i.e. $\kappa_m = c_m f_{t^s} + (c_m - 1) f_{h^p}$, at which batch killing is truncated. We define the mth diagonal element of the killing factor matrix at level (queue length) j, $\beta_j = \min(j, c_m)/\max_m(c_m)$ for h^p killing and $c_m/\max_m(c_m)$ otherwise.

19.4.6 Multiple Processes

To add multiple processes, we augment the arrival term to reflect a sum of streams. For multiple arrivals this would take the form of

$$
\mathbf{r}_j^{\text{MM CPP}_{k\cdots}} = \sum_{i=0}^{j-1} \mathbf{v}_i \sum_{k=1}^{n^{\text{arr}}} \Lambda_k (I - \Theta_k)^{f_j < L} \Theta_k^{j-i-1} + \cdots \text{(similarly to above)}
$$

where n^{arr} is the number of arrival streams.

This introduction of sums of streams can also be performed for processing completions and negative customers to create a more general balance equation:

$$
\mathbf{r}_j^{\text{general}} = \sum_{i=0}^{j-1} \mathbf{v}_i \left[\sum_{k=1}^{n^{\text{arr}}} \Lambda_k (1 - \Theta_k)^{f_{j<L}} \Theta_k^{j-i-1} \right]
$$

$$
+ \mathbf{v}_j \left[Q - \sum_{k=1}^{n^{\text{arr}}} \Lambda_k \, f_{j<L} - \sum_{k=1}^{n^{\text{kill}}} K \, f_{((j>\kappa)\vee \mathrm{h}^p)} \beta_j - \sum_{k=1}^{n^{serv}} C_{k,j} \right]
$$

$$
+ \sum_{i=j+1}^{L} \mathbf{v}_i \left[\sum_{k=1}^{n^{\text{kill}}} K_k (1 - R_k)^{f_{j>\kappa}} R_k^{\,i-j-1} \, f_{\substack{(j\geq\kappa) \\ \vee(\mathrm{h}^p \wedge i-j+1)}} \beta_i \right.
$$

$$
\left. + \sum_{k=1}^{n^{serv}} C_{k,i} (1 - \Phi_k)^{f_{j>c-1}} \Phi_k^{\,i-j-1} \, f_{\substack{(i=j+1) \\ \vee(j\geq c-1)}} \right]
$$

19.5 Localized Balance Equation Ensemble

We have a function $\mathbf{r}_j$ which, for a given level j, provides the (symbolic) left-hand side of a Kolmogorov (balance) equation $\mathbf{r}_j = 0$. These balance equations generally include probability vectors from all levels of the queue due to the presence of batch arrivals and departures. We call these upward and downward streams respectively, referring to the direction of probability flux in the Kolmogorov balance equations. In order to efficiently solve for the steady-state solution of very large finite or infinite systems, we localize balance equations using a process we call *stream elimination*. Its aim shares a goal with Gaussian elimination – that of "block-diagonalizing" the system. A key difference is that stream elimination exploits the structure of the queue to enable it to solve infinite systems.

We note that the matrix Kolmogorov balance equations for individual levels can be taken as columns in a larger Matrix problem, as illustrated in expression (19.2). In this expression, $\Omega_{j,k}$ is the matrix coefficient of $\mathbf{v}_k$ in the balance equation for level j. With unbounded batch transitions, such as found with geometrically batched processes, all $\Omega_{j,k}$ are non-zero. Our automated system implicitly performs column operations on this system, using coefficients calculated using our characterization of the relationships between off diagonal block terms in the system as specified before manipulation.

$$
(\mathbf{v}_0 \ \mathbf{v}_1 \ \dots \ \mathbf{v}_L)
\begin{pmatrix}
\Omega_{0,0} & \Omega_{1,0} & \cdots & \\
\Omega_{0,1} & & & \\
\vdots & & \ddots & \\
& & & \Omega_{L,L}
\end{pmatrix}
= \mathbf{0}
\tag{19.2}
$$

Note that this expression establishes the relationship between the state occupation probabilities to within a common multiple. We add a normalization expression, summing the probabilities and equating to 1 to complete the solution. Once we have written down this complete set of equations, we can abstract away from the queueing problem itself, and manipulate the set of expressions directly. We do, however, use observations on the formulation of the problem to enable us to calculate the coefficients we use for the standard column operations. At least one of the pair of coefficient matrices used for the elimination is non-singular, so the system is not rendered singular by these operations.

If the system had included only unbatched streams, the matrix would be tri-diagonal. The wider off-diagonal terms are created by the geometric batching. Our goal is to eliminate these terms to create a system which we can represent more efficiently. This is achieved by reducing the system to a number of adjacent diagonals. For a finite system ($L \neq \infty$), this can be considered to map directly to a Gaussian elimination type of solution. When the system is infinite, we clearly cannot write down the bottom-right corner of the matrix in the expression, but we can represent it.

Each stream's set of geometric batching terms is eliminated by combining scalar multiples of two neighbouring left-hand side terms, which map to columns in the above expression. The *target* level (usually called j) is the level to which the resulting localized left-hand side pertains. The *eliminator* level provides the terms which will be subtracted to remove the stream.

19.5.1 Stream Elimination Example

We can eliminate any stream from a target level using any eliminator which has contributions of that same stream. For example, we can remove the summation terms associated with the stream of positive arrivals from any level using any other except level L (for a finite queue).

The closer the eliminator is to the target, the fewer terms for that stream remain. So, the choice is between using the level below, or the level above, and either would work for any given target level. There is in fact no choice, certainly for downward streams – and also for upward streams in a finite queue – if we consider the situation at the limits of the queue $j = 0$ and $j = L$, as we explain below.

The simplest process to consider is the arrival of batches of positive customers. These arrive at every level in the queue. The stream arriving at level L must be cancelled out using the level below, as there are none above. To cancel the positive arrival stream for target level $L - 1$, we could use either the level above or below. To see that we cannot use the level above to get an independent equation, we will show that the determinant of a matrix representation of the solution would have a zero determinant. The balance equations for our class of queues can be written as follows:

$$\mathbf{v}A = \mathbf{0}$$

where $\mathbf{v}$ is the row vector $(\mathbf{v}_1, \mathbf{v}_2, \ldots, \mathbf{v}_L)$, and the columns $(j - 1)N + 1$ through jN of A give the matrix coefficients of the vectors $\mathbf{v}_j$ for the target level j. The matrix A is as follows for an MM CPP/GE/1/4 queue with N modulation states:

$$A = \begin{pmatrix} (Q-\Lambda) & \Lambda(I-\Theta) & \Lambda(I-\Theta)\Theta & \Lambda(I-\Theta)\Theta^2 & \Lambda(I-\Theta)\Theta^3 \\ C & (Q-\Lambda-C) & \Lambda(I-\Theta) & \Lambda(I-\Theta)\Theta & \Lambda(I-\Theta)\Theta^2 \\ C\Phi & C(I-\Phi) & (Q-\Lambda-C) & \Lambda(I-\Theta) & \Lambda(I-\Theta)\Theta \\ C\Phi^2 & C(I-\Phi)\Phi & C(I-\Phi) & (Q-\Lambda-C) & \Lambda(I-\Theta) \\ C\Phi^3 & C(I-\Phi)\Phi^2 & C(I-\Phi)\Phi & C(I-\Phi) & (Q-C) \end{pmatrix}$$

Our transformation of the system involves column operations, which we can achieve by post-multiplication of A by an appropriate matrix, say B. The rank of A is $5N - 1$, i.e. one less than the number of states. When a column is removed and replaced by an expression reflecting the normalization constraint, the system is correctly specified, so we can solve it. If we postmultiply by a singular matrix, the rank of the system will be reduced, and will no longer be soluble with the normalization constraint. Thus, the determinant of B must be non-zero. For example, to remove the arrival stream terms in column five, we could post-multiply by a matrix B:

$$B = \begin{pmatrix} I & 0 & 0 & 0 & 0 \\ 0 & I & 0 & 0 & 0 \\ 0 & 0 & I & 0 & 0 \\ 0 & 0 & 0 & I & -\Theta \\ 0 & 0 & 0 & 0 & I \end{pmatrix}$$

The determinant of B is 1, so our solution is safe. This yields a modified A matrix:

$$\begin{pmatrix} (Q-\Lambda) & \Lambda(I-\Theta) & \Lambda(I-\Theta)\Theta & \Lambda(I-\Theta)\Theta^2 & 0 \\ C & (Q-\Lambda-C) & \Lambda(I-\Theta) & \Lambda(I-\Theta)\Theta & 0 \\ C\Phi & C(I-\Phi) & (Q-\Lambda-C) & \Lambda(I-\Theta) & 0 \\ C\Phi^2 & C(I-\Phi)\Phi & C(I-\Phi) & (Q-\Lambda-C) & \Lambda(I-\Theta)-(Q-\Lambda-C)\Theta \\ C\Phi^3 & C(I-\Phi)\Phi^2 & C(I-\Phi)\Phi & C(I-\Phi) & (Q-C)-C(I-\Phi)\Theta \end{pmatrix}$$

Now consider the removal of the positive arrival components from column 4 (queue level 3) *in addition*. We can achieve this by post multiplying A by one of the two following B matrices:

$$B_{\text{bad}} = \begin{pmatrix} I & 0 & 0 & 0 & 0 \\ 0 & I & 0 & 0 & 0 \\ 0 & 0 & I & 0 & 0 \\ 0 & 0 & 0 & \Theta & -\Theta \\ 0 & 0 & 0 & -I & I \end{pmatrix}, \quad B_{\text{ok}} = \begin{pmatrix} I & 0 & 0 & 0 & 0 \\ 0 & I & 0 & 0 & 0 \\ 0 & 0 & I & -\Theta & 0 \\ 0 & 0 & 0 & I & -\Theta \\ 0 & 0 & 0 & 0 & I \end{pmatrix}$$

These remove upward streams of positive arrivals from the balance equations for levels 3 and 4. B_{ok} removes the arrivals using the level above, and B_{bad} uses the level below as eliminator. Since columns four and five are linearly dependent, the determinant of B_{bad} is 0, so use of this matrix in multiplication would render the equation ensemble degenerate. The determinant of B_{ok} is 1, so it is safe to use.

A matrix B_{all} to remove all upward streams could be either of the following (since removing the upward stream at level 1 is not necessary):

$$B_{\text{all}} \begin{pmatrix} I & -\Theta & 0 & 0 & 0 \\ 0 & I & -\Theta & 0 & 0 \\ 0 & 0 & I & -\Theta & 0 \\ 0 & 0 & 0 & I & -\Theta \\ 0 & 0 & 0 & 0 & I \end{pmatrix} \quad \text{or} \quad \begin{pmatrix} I & 0 & 0 & 0 & 0 \\ 0 & I & -\Theta & 0 & 0 \\ 0 & 0 & I & -\Theta & 0 \\ 0 & 0 & 0 & I & -\Theta \\ 0 & 0 & 0 & 0 & I \end{pmatrix}$$

Thus, we are constrained to eliminating upward streams using the balance equation at one level lower than the target level. A similar argument demands removal of downward streams using the raw left-hand side from the level above.

The expression resulting from elimination of a stream includes terms from all the queue levels of the constituent left-hand sides. In each modulation state m, there are a number, say u_m, of distinct upward streams to cancel out, which requires the use of left-hand sides from levels $j - u_m \ldots j - 1$. Similarly in the general case,[2] there are d_m distinct downward streams to this level, which requires eliminators from $j + 1 \ldots j + d_m$. The (vector) balance equations cover the maximum range over the modulation states upward and downward. We define $\hat{u}$ as the maximum over m of u_m, and $\hat{d}$ similarly. A localized equation for level j therefore comprises levels $j - \hat{u}$ through $j + \hat{d}$.

19.5.2 The Recursive Elimination Operator

We define the transformation operator $E_{\tau,j}^{v,X}$, to be applied to a left-hand side vector-valued *function* $\mathbf{r}$ (mapping integers j to left-hand sides $\mathbf{r}_j$). This returns a left-hand side without batch sums of the stream given by the diagonal matrix of rates X, e.g. $X = \Lambda$:

$$E_{\tau,j}^{v,X}\mathbf{r} = \mathbf{r}_j \frac{\partial \mathbf{r}_{j+v}}{\partial(\mathbf{v}_\tau X)} - \mathbf{r}_{j+v}\frac{\partial \mathbf{r}_j}{\partial(\mathbf{v}_\tau X)} = \mathbf{r}_j A + \mathbf{r}_{j+v} B \qquad (19.3)$$

The index j is the target level for which the returned function is a localized Kolmogorov equation. In it, batches of the stream with matrix rate term X are removed by linear combination of the balance equation left-hand sides given by $\mathbf{r}$ at j and at the elimination level $j + v$. The weightings are determined by the partial derivatives which serve only to select the appropriate coefficients. We therefore eliminate downward streams (processing completions and the action of negative customers) by using terms above the target, so $v = 1$ for these. Upward streams (positive customer arrivals) are eliminated using terms below the target so $v = -1$ for these.

The index τ is the queue level at which the coefficients of X are calculated, and we call this the *test level*. This must be outside the range of levels whose left-hand sides are used in constructing the localized left-hand side for level j. We call this range the *elimination bracket*.

There may be a non-zero common factor between the two resulting elimination terms A and B in equation (19.3), and we can divide through by this because we ultimately equate the resulting left-hand side to zero. For example, eliminating the positive arrival stream in an MM CPP/MM/1 queue with only batched positive arrivals for level 2 involves the following terms:

---

[2] Within the processing region, tail-safe negative customers cannot operate, resulting in fewer downward streams within this region.

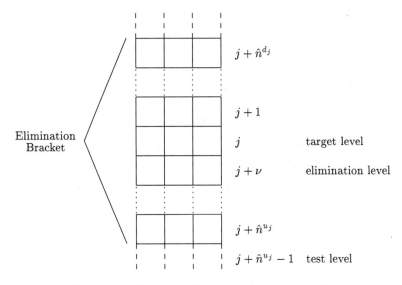

Figure 19.2 Elimination of an upward stream ($\nu = -1$).

$$\frac{\partial \mathbf{r}_2}{\partial (\mathbf{v}_0 \Lambda)} = \frac{\partial}{\partial (\mathbf{v}_0 \Lambda)} \left(\mathbf{v}_0 \Lambda (I - \Theta) \Theta + \mathbf{v}_1 \Lambda (I - \Theta) \right.$$

$$\left. + \mathbf{v}_2 [Q - \Lambda - C] + \mathbf{v}_3 C \right)$$

$$= (I - \Theta) \Theta$$

$$\frac{\partial \mathbf{r}_1}{\partial (\mathbf{v}_0 \Lambda)} = \frac{\partial}{\partial (\mathbf{v}_0 \Lambda)} \left(\mathbf{v}_0 \Lambda (I - \Theta) + \mathbf{v}_1 [Q - \Lambda - C] + \mathbf{v}_2 C \right)$$

$$= (I - \Theta)$$

We use these in the elimination function as follows:

$$E_{0,2}^{-1,\Lambda} \mathbf{r} = \mathbf{r}_2 \frac{\partial \mathbf{r}_1}{\partial (\mathbf{v}_0 \Lambda)} - \mathbf{r}_1 \frac{\partial \mathbf{r}_j}{\partial (\mathbf{v}_0 \Lambda)}$$

$$= \mathbf{r}_2 (I - \Theta) - \mathbf{r}_1 (I - \Theta) \Theta$$

$$= \mathbf{r}_2 - \mathbf{r}_1 \Theta, \text{ post-multiplying by the inverse of the non-zero}$$

$$\text{common matrix factor, all } \theta_m < 1$$

$$= \mathbf{v}_1 [\Lambda - (Q - C) \Theta] + \mathbf{v}_2 [Q - \Lambda - C (I + \Theta)] + \mathbf{v}_3 C$$

19.5.3 Degenerate Streams during Elimination

Apart from reliable automation, a key advantage our methods offer is the accommodation of different numbers of independently distributed batched streams across modulation states. If an entirely pre-derived matrix balance equation is used, as in [11] for example, solution the system can become degenerate if two streams in a modulation state have identical batch parameters, and unstable if they are very similar. These circumstances arise naturally during the process of solving for the steady state of a network. It is therefore essential that the queue solution mechanism can cope with such potential degeneracy. In this section, we explain our method's response to the circumstance of identical batching.

For each arrival or service completion process, it is possible to have a mixture of null (having zero rate, or a zero batch distribution parameter) and non-null batched streams across the modulating states. When performing the E-transformation, for a phase with zero rate or batch parameter, the corresponding component of the vector left-hand side returned is simply the argument's component at j. This is because there are no non-local terms to remove. For example, to eliminate a negative customer stream from the raw balance left-hand side for a finite t^v queue, if $\delta_1 = 1/2$ and $\delta_2 = 0$, we have (taking $\mathbf{r}_{j,m}$ to be the mth element of the vector returned by $\mathbf{r}_j$):

$$
E_{L,j}^{+1,R}\mathbf{r} = \left(\underbrace{\mathbf{r}_{j,1}\frac{\partial \mathbf{r}_{j+1,1}}{\partial(\mathbf{v}_L\delta_1)} - \mathbf{r}_{j+1,1}\frac{\partial \mathbf{r}_{j,1}}{\partial(\mathbf{v}_L\delta_1)}}_{\substack{\text{elimination was required} \\ \text{in this modulation state}}}, \quad \underbrace{\mathbf{r}_{j,2}}_{\text{no elimination}}\right)
$$

Examining the elements of matrices A and B in equation (19.3), we note that whenever it is not necessary to eliminate a stream in modulation state m, the mth elements of the diagonals of A and B are informative, as follows. Element $a_{m,m}$ is zero if the stream has already been removed. The value of $b_{m,m}$ is zero if either the stream has been removed, or the batch parameter of that stream is zero. The only case which requires care is when the stream has already been removed, wherein both $a_{m,m}$ and $b_{m,m}$ are zero, and we simply set $a_{m,m}$ to 1 if it is calculated in equation (19.3) as zero. This keeps the term from the source equation at the target level, this being $\mathbf{r}_{j,2}$ in the example above.

Equation (19.3) gives us a left-hand side *function of j* with stream (X, ν) removed. Elimination of further streams is simply the application of the same elimination procedure to the intermediate partially localized left-hand sides.

19.5.4 Worked Example

We now perform a dry run of generating the localized balance equation for level 11 of a $c = 10$, t^{ν} infinite waiting room MM CPP/GE/c/L G-queue. Equation (19.3) is used recursively. This creates branches which encompass r_{10} through r_{13}. Thus, the test level for upward streams must be lower than 10 and the test level for downward streams must be higher than 13. We choose $\tau^{u} = 14$ and $\tau^{d} = 9$ as these are the values used in normal operation of the system.

In our elimination terminology, this localized balance equation is $\mathbf{b}_{11} = E_{14,11}^{+1,K} E_{14}^{+1,C} E_{10}^{-1,\Lambda} \mathbf{r}$. The K (negative customer arrival stream) elimination uses equations at levels 11 and 12, which in turn involve application of the C (service completion) elimination term at, respectively, 11 and 12, and 12 and 13, and each of these involves an application of the Λ (positive arrivals) elimination term, which in total uses the raw balance equations for levels 10 through 13.

First, we calculate a leaf of this tree, eliminating the positive customer arrival summation terms by taking the raw balance equations (as developed in Section 19.4.6) for levels 10 and 11 and combining them appropriately:

$$
\begin{aligned}
\mathbf{r}_{10} = \cdots + \quad & \overbrace{\mathbf{v}_9[\Lambda\,(I - \Theta)]}^{\frac{\partial \mathbf{r}_{10}}{\partial(\mathbf{v}_9\Lambda)}} \\
+ \quad & \mathbf{v}_{10}[Q - \Lambda - K - C] \\
+ \quad & \mathbf{v}_{11}[K(I - R) + C(I - \Phi)] \\
+ \quad & \mathbf{v}_{12}[K(I - R)R + C(I - \Phi)\Phi] \\
+ \quad & \mathbf{v}_{13}[K(I - R)R^2 + C(I - \Phi)\Phi^2] \\
+ \quad & \mathbf{v}_{14}[K(I - R)R^2 + C(I - \Phi)\Phi^3] + \cdots
\end{aligned}
$$

$$\mathbf{r}_{11} = \cdots + \overbrace{\mathbf{v}_9[\Lambda\,(I - \Theta)\Theta]}^{\frac{\partial \mathbf{r}_{11}}{\partial(\mathbf{v}_9\Lambda)}}$$

$$+ \quad \mathbf{v}_{10}[\Lambda(I - \Theta)]$$

$$+ \quad \mathbf{v}_{11}[Q - \Lambda - K - C]$$

$$+ \quad \mathbf{v}_{12}[K(I - R) + C(I - \Phi)]$$

$$+ \quad \mathbf{v}_{13}[K(I - R)R + C(I - \Phi)\Phi]$$

$$+ \quad \mathbf{v}_{14}[K(I - R)R + C(I - \Phi)\Phi^2] + \cdots$$

The coefficients of Λ for level 9 are $\frac{\partial \mathbf{r}_{10}}{\partial(\mathbf{v}_9\Lambda)} = (I - \Theta)$ and $\frac{\partial \mathbf{r}_{11}}{\partial(\mathbf{v}_9\Lambda)} = (I - \Theta)\Theta$, hence in the ratio $I : \Theta$, so subtracting $\mathbf{r}_{10}\Theta$ from $\mathbf{r}_{11}$, we obtain $\mathbf{u}_{11} = E_{9,11}^{-1,\Lambda}\mathbf{r}$:

$$\mathbf{u}_{11} = \quad \mathbf{v}_{10}[\Lambda - (Q - K - C)\Theta]$$

$$+ \quad \mathbf{v}_{11}[Q - \Lambda - K(I + \Theta(I - R)) - C(I + \Theta(I - \Phi))]$$

$$+ \quad \mathbf{v}_{12}[K(I - R)(I - R\Theta) + C(I - \Phi)(I - \Phi\Theta)]$$

$$+ \quad \mathbf{v}_{13}[K(I - R)R(I - R\Theta) + C(I - \Phi)\Phi(I - \Phi\Theta)]$$

$$+ \quad \mathbf{v}_{14}[K(I - R)R^2(I - R\Theta) + C(I - \Phi)\Phi^2(I - \Phi\Theta)] \quad + \cdots$$

And we can write down the expressions for u_{12} and u_{13} directly from this, as the constituent behaviours are identical:

$$\mathbf{u}_{12} = \quad \mathbf{v}_{11}[\Lambda - (Q - K - C)\Theta]$$

$$+ \quad \mathbf{v}_{12}[Q - \Lambda - K(I + \Theta(I - R)) - C(I + \Theta(I - \Phi))]$$

$$+ \quad \mathbf{v}_{13}[K(I - R)(I - R\Theta) + C(I - \Phi)(I - \Phi\Theta)]$$

$$+ \quad \mathbf{v}_{14}[K(I - R)R(I - R\Theta) + C\underbrace{(I - \Phi)\Phi(I - \Phi\Theta)}_{\frac{\partial \mathbf{u}_{12}}{\partial(\mathbf{v}_{14}C)}}] \quad + \cdots$$

$$\mathbf{u}_{13} = \quad \mathbf{v}_{12}[\Lambda - (Q - K - C)\Theta]$$

$$+ \quad \mathbf{v}_{13}[Q - \Lambda - K(I + \Theta(I - R)) - C(I + \Theta(I - \Phi))]$$

$$+ \quad \mathbf{v}_{14}[K(I - R)(I - R\Theta) + C\underbrace{(I - \Phi)(I - \Phi\Theta)}_{\frac{\partial \mathbf{u}_{13}}{\partial(\mathbf{v}_{14}C)}}] \quad + \cdots$$

In all of these, only the level immediately below the target level appears in the localized expression, just as would have been the case without batches.

To eliminate the processing stream, with rate matrix C, we take pairs of these equations and provide them to the elimination operator $E_{14,j}^{+1,C}$ to be applied at levels $j = 11, 12, 13$. To calculate the expression for $E_{14,11}^{+1,C} \mathbf{u}$ we find the coefficients of C at the test level in $\mathbf{u}_{12}$ and $\mathbf{u}_{13}$, which are respectively $(I - \Phi)\Phi(I - \Phi\Theta)$ and $(I - \Phi)(I - \Phi\Theta)$. The ratio between these is $\Phi : I$, so we subtract $\mathbf{u}_{12}\Phi$ from $\mathbf{u}_{11}$ to yield the left-hand side $\mathbf{d}$ with only K-batch terms:

$$
\begin{aligned}
\mathbf{d}_{11} = \quad & \mathbf{v}_{10}[\Lambda - (Q - K - C)\Theta] \\
+ \quad & \mathbf{v}_{11}[Q(I + \Theta\Phi) - \Lambda(I + \Phi) \\
& \quad - K(I + \Theta(I - R + \Phi)) - C(I + \Theta)] \\
+ \quad & \mathbf{v}_{12}[-Q\Phi + \Lambda\Phi + K((I - R)(I - R\Phi) \\
& \quad + (I + \Theta)\Phi(I - R)) + C] \\
+ \quad & \mathbf{v}_{13}[K(I - R)(I - R\Theta)(R - \Phi)] \\
+ \quad & \mathbf{v}_{14}[K \underbrace{(I - R)R(I - R\Theta)(R - \Phi)}_{\frac{\partial \mathbf{d}_{11}}{\partial(\mathbf{v}_{14}K)}}] + \cdots \quad \text{only terms in } K
\end{aligned}
$$

Again, the behaviour for level 12 is identical, so we immediately have $\mathbf{d}_{12}$:

$$
\begin{aligned}
\mathbf{d}_{12} = \quad & \mathbf{v}_{11}[\Lambda - (Q - K - C)\Theta] \\
+ \quad & \mathbf{v}_{12}[Q(I + \Theta\Phi) - \Lambda(I + \Phi) \\
& \quad - K(I + \Theta(I - R + \Phi)) - C(I + \Theta)] \\
+ \quad & \mathbf{v}_{13}[-Q\Phi + \Lambda\Phi + K((I - R)(I - R\Phi) \\
& \quad + (I + \Theta)\Phi(I - R)) + C] \\
+ \quad & \mathbf{v}_{14}[K \underbrace{(I - R)(I - R\Theta)(R - \Phi)}_{\frac{\partial \mathbf{d}_{12}}{\partial(\mathbf{v}_{14}K)}}] + \cdots \quad \text{only terms in } K
\end{aligned}
$$

In these expressions, only the level immediately above the target contains terms associated with processing completions, as would have been the case with unbatched processing.

Assuming unequal batch terms (i.e. $R \neq \Phi$), these can be combined using $E_{14,j}^{+1,K} \mathbf{d}$ to eliminate the final batch term in K by subtracting $\mathbf{d}_{12}R$ from $\mathbf{d}_{11}$ to give a localized balance $\mathbf{b}_{11}$:

$$
\begin{aligned}
\mathbf{b}_{11} = \quad & \mathbf{v}_{10}[\Lambda - (Q - K - C)\Theta] \\
+ \quad & \mathbf{v}_{11}[Q(I + \Theta(\Phi + R)) - \Lambda(I + \Phi + R) \\
& \quad - K(I + \Theta(I + \Phi)) - C(I + \Theta(I + R))] \\
+ \quad & \mathbf{v}_{12}[-Q(\Phi + R(I + \Theta\Phi)) + \Lambda(\Phi + R(I + \Theta\Phi)) \\
& \quad + K(I + \Phi(I + \Theta)) + C(I + R(I + \Theta))] \\
+ \quad & \mathbf{v}_{13}[Q\Phi R - \Lambda\Phi R - K\Phi - CR]
\end{aligned}
$$

In this expression, we see two levels appear above the target level, one for each of negative customers and processing completions streams. This has the same form as the result found in [4].

Note that the coefficient of $\mathbf{v}_{13}$ is only strictly valid if it was necessary to remove all streams. The elimination ratio was determined from a term $K(I - R)(I - R\Theta)(R - \Phi)$, which contains zeros wherever the diagonal elements of R and Φ are equal. In these circumstances, the information required to formulate an elimination ratio is not present. This is an example of stream degeneracy treated in Section 19.5.3.

These coefficients in $\mathbf{b}_j$ pertain to balance equations for the repeating region of an MM CPP/GE/c/L G-queue. The repeating region is described in detail in Section 19.5.5 below, but in essence, there is a large region of the queue which is described by a common balance equation for all levels, which taks the form of a linear homogeneous matrix recurrence relation on the steady state SOP vectors. We can write this as $\mathbf{v}_{j-1}Q_0 + \mathbf{v}_j Q_1 + \mathbf{v}_{j+1}Q_2 + \mathbf{v}_{j+2}Q_3 = \mathbf{0}$, where

$$
\begin{aligned}
Q_0 = {} & \Lambda - (Q - K - M)\Theta \\
Q_1 = {} & Q(I + \Theta)(\Phi + R) - \Lambda(I + \Phi + R) - K(I + \Theta(I + \Phi)) \\
& - C(I + \Theta)(I + R) \\
Q_2 = {} & -Q(\Phi + R(I + \Theta\Phi)) + \Lambda(\Phi + R(I + \Theta\Phi)) + K(I + \Phi(I + \Theta)) \\
& + C(I + R(I + \Theta)) \\
Q_3 = {} & Q\Phi R - \Lambda\Phi R - K\Phi - CR
\end{aligned}
$$

In a more general queue with more arrival processes and/or departure processes, we require more terms, and the solution is written as follows:

$$\sum_{k=0}^{\hat{u}+\hat{d}} \mathbf{v}_{j+k-\hat{u}} Q_k = \mathbf{0} \tag{19.4}$$

19.5.5 Repeating Region

Queues commonly have a region of queue lengths for which the Kolmogorov balance equations form a recurrence relation on state occupation probabilities in a short contiguous set of queue lengths. The untransformed (or raw) Kolmogorov balance equations for a geometrically batched queue generally involves a large or infinite range of queue lengths in any given balance equation. Our process provides a transformed ensemble of balance equations providing the necessary locality of reference to show a repeating region.

A localized balance equation from within the repeating region produces the following linear homogeneous matrix recurrence relation of order $\hat{d}+1+\hat{u}$ which is valid for SOP vectors within the entire region, cf. equation (19.4).

$$\sum_{k=0}^{\hat{u}+\hat{d}} \mathbf{v}_{j+k-\hat{d}} Q_k = 0 \tag{19.5}$$

Localized balance equations for different levels are identical when they have been produced by eliminating identical batched streams, so this region stretches strictly between levels at which any batches are truncated. For example, in an MM/MM/1/L queue, this region stretches from level 1 to $L-1$, because departure batch transitions are truncated at level 0, and arrival batches are truncated at level L.

As another example, batch processing completions are truncated at level $c-1$ when there are c processors with geometrically batched sevice completions. This means that the raw balance equations for $j = c - 1$ and $\geq c$ have different batch contributions: transitions from level i are $C\Phi^{i-j-1}$ to level $j = c - 1$ but $C(I - \Phi)\Phi^{i-j}$ to $j \geq c$. This means that localized balance equations which use level $c - 1$ in their composition are different to those using only levels c and above.

If there are $\hat{u}$ upward streams, then the lowest level to which the repeating region equation pertains in this example is $c + \hat{u}$.

We label the bottom of the repeating region ϵ^b and the top as ϵ^t. This includes all levels i at which every stream transitioning from level i to level j has a rate of the form $X(I - Y)Y^{|j-i|-1}$, where X is a rate parameter matrix, and Y is a batch size parameter matrix. We find that ϵ^b is either equal to c, if the negative customers are either absent or of type h^p or t^v, or equal to $c + 1$ otherwise.

19.6 Solution Methods

We shall be concentrating on two methods for solving this recurrence relation – Matrix Geometric and Spectral Expansion. The differences lie largely in questions of efficiency and ease of analysis. We first consider the more popular matrix geometric approach, followed by spectral expansion.

19.6.1 Matrix Geometric Solution

The matrix geometric method aims to relate the SOP of adjacent vectors within the repeating region through a non-singular matrix F, i.e.

$$\mathbf{v}_{j+1} = \mathbf{v}_j F \Leftrightarrow \mathbf{v}_{j+i} = \mathbf{v}_j F^i \qquad \epsilon^b \le j \le j + i \le \epsilon^t \qquad (19.6)$$

or, by starting the recurrence at the bottom of the repeating region,

$$\mathbf{v}_{\epsilon^b + i} = \mathbf{v}_{\epsilon^b} F^i \qquad 0 \le i \le \epsilon^t - \epsilon^b \qquad (19.7)$$

By substituting (19.7) into (19.5) we see that a solution using this representation can only exist if F satisfies the matrix polynomial

$$\sum_{k=0}^{\hat{u}+\hat{d}} F^k Q_k = 0 \qquad (19.8)$$

Methods of solving for F are surveyed in [15], but see Section 19.6.2.1 for an explanation of a limitation of this method. The vector $\mathbf{v}_{\epsilon^b}$ is constrained by the boundary conditions imposed by the remaining queue lengths outside the repeating region, i.e. the "processor filling" region of queue lengths less than c, and in the case of a finite queue, $\mathbf{v}_{\epsilon^t}$ by the full queue region.

19.6.2 Spectral Expansion Solution

Modern matrix geometric methods are in general considerably more efficient in the practical context than spectral expansion methods, when they are appropriate. However, the Kolmogorov equations necessary for the solution of queues with several independent streams of geometrically batched processes give rise to eigensystems too large to be fully represented in the most efficient, iterative matrix geometric algorithms [12]. This is because these solution techniques provide at most two matrices to represent the solution. These provide $2N$ eigenvalues, and the solution may require more than this with non-zero coefficients. For this reason, we consider the spectral expansion solution, as it always provides the full eigensystem; see below.

The spectral expansion solution for the linear homogeneous matrix equation (19.5) is the vector of geometric series provided by the eigenvalues ξ_i of the characteristic function

$$Q(\xi) = \sum_{k=0}^{\hat{u}+\hat{d}} \xi^k Q_k$$

found by setting the determinant $\det |Q(\xi)| = 0$, and projecting onto the corresponding left eigenvectors ψ_i satisfying $\psi_i Q(\xi_i) = 0$.

Suppose there are ϵ^n eigenvalue/eigenvector pairs (or *eigenmodes*). Each eigenvector defines a basis function component, and we can write the SOP vector for queue level j for arbitrary α_i as follows:

$$\mathbf{v}_j = \sum_{i=1}^{\epsilon^n} \alpha_i \xi_i^j \psi_i \tag{19.9}$$

We can satisfy the boundary conditions defined by the balance equations of the levels neighbouring the repeating region, together with the normalization constraint, by suitable choice of α_i $(1 \leq i \leq \epsilon^n)$.

These balance equations include both *explicit* $\mathbf{v}$ vectors which give the state occupation probabilities outside the repeating region, and vectors constructed using the eigensystem, and hence effectively *implicit*.

When the queue is infinite, any eigenvalues of magnitude greater than or equal to 1 (i.e. lying on or outside the unit disk in the Argand plane) must take zero coefficients, as their infinite sum cannot converge, and hence cannot be normalized.

The choice of raising the eigenvalues to the power j, the queue level, is somewhat arbitrary: as long as the eigenvalue is non-zero, the scaling due to any constant offset from j can be absorbed into the associated coefficients α_i.

We exclude zero eigenvalues and their associated eigenvectors from our solutions for the sake of uniformity. When there are zero eigenvalues, their eigenvectors do provide a means for stitching the boundary conditions into the repeating region, but at the expense of complication of our automated representation. These would replace one vector of explicit unknowns, but since there are the same number of values to solve for, they do not improve efficiency.

The number of eigenvalues pertaining to this system ϵ^n is dictated by the polynomial order of the determinant of the characteristic equation. This is equal to the total number of distinct streams in the system by the following reasoning. The recurrence relation in an M/M/1 queue is $p_{j-1}\lambda - p_j(\mu + \lambda) + p_{j+1} = 0$. This is quadratic, so has two solutions, and there are two streams: arrivals of positive customers at rate λ with no batches, and processing completions at rate μ. If there are batches, the transformed equation covers the same levels (cf. Section 19.5.4). Addition of a distinct stream increments the order of the equation, hence adding a solution. Thus, there are as many solutions – eigenvalues – as distinct streams, i.e. $\epsilon^n = \sum_{m=1}^{N} u_m + d_m$. The form of modulation does not add or remove eigenvalues, as it does not affect transitions up or down in the queue, it simply couples the eigenmodes, creating non-trivial eigenvectors.

Eigenvalues occur as real numbers or as complex conjugate pairs. The geometric series of the latter traces a spiral in the Argand plane. Since the final solution must be real, any imaginary parts must cancel out. The corresponding coefficients are necessarily complex conjugates, as this yields a real result, and the coefficients' arguments therefore provide a phase shift in the resulting characteristic of the basis function provided by the *pair* of eigenvalues and associated (identical) vectors.

19.6.2.1 Relationship between Solution Methods
It is apparent that in general there are multiple[3] solutions for F in (19.8). Traditionally the minimal non-negative solution [3], $\hat{F}$, is sought. Matrix geometric methods solve the same recurrence relation (19.5) as the spectral expansion method and it is known that each eigenvalue/vector pair of $\hat{F}$ cor-

---

[3] There are exactly $\epsilon^n!/N!(\epsilon^n - N)!$ solutions, for a queue with N modulation states and $\epsilon^n = \sum_{m=1}^{N} u_m + d_m$ distinct streams.

responds to exactly one of the eigenvalue/vector pairs $(\xi_i, \boldsymbol{\psi}_i)$ found during spectral expansion.

We can express the vector $\mathbf{v}_{\epsilon^b}$ as a linear combination of these eigenvectors.

$$\mathbf{v}_{\epsilon^b} = \sum_{m=1}^{N} \gamma_m \boldsymbol{\psi}_m \tag{19.10}$$

Then from equation (19.7) we have

$$\mathbf{v}_{\epsilon^b+i} = \sum_{m=1}^{N} \gamma_m \xi_m^i \boldsymbol{\psi}_m \qquad 0 \le i \le \epsilon^t - \epsilon^b \tag{19.11}$$

Comparing (19.9) and (19.11) it becomes apparent that if N is smaller than the number of streams (and hence eigenvalues) ϵ^n, it is impossible to express the general solution using only one matrix $\hat{F}$. To gain access to more than N eigenmodes of (19.8) using the Matrix Geometric method, a sum of two or more independent matrix solutions F_i to (19.8) might be employed (as suggested in [4]). See [18] for an in-depth analysis.

19.6.3 Explicit Region

In general, in the region $0 \le j \le c-1$, there exists no specific structure in the balance equations that can be exploited and we represent the SOP vectors for these levels explicitly as $\mathbf{v}_j$. To constrain the SOP vectors we may generate the localized balance equations using the methodologies in Section 19.5 at the corresponding levels $0, \ldots, c-1$.

19.6.4 Boundary Regions

For each modulation state m, the Kolmogorov equations just above the processing region, $c \le j \le c+u_m$, and in the case of finite queues, the region just below the top, $L-d_m \le j \le L$, give the boundary conditions linking the explicit and repeating regions. Localization of balance equations operates as before, yet derived equations consist of mixtures of SOP vectors represented either explicitly as $\mathbf{v}_j$ (for $0 \le j \le \epsilon^b - 1$ and $j = L$ for finite queues) or using (19.9) or (19.7) for all other levels j.

Due to the vector representation used for SOPs it is in practice convenient to generate localized (vector) balance equations for all modulation states for all levels involved in the solution $c \le j \le c+\hat{u}$ and $L-\hat{d} \le j \le L$ and drop constituent (scalar) equations for those modulation states that contain degenerate streams and hence do not contribute to the solution.

19.7 Solving for the Steady State

Having automatically generated the most efficient, localized balance equations which capture the full relationship between probability flux flows in the queue, we need to solve them. To achieve this, we use a basic property of the homogeneous region(s) of the queue.

19.7.1 Repeating Region Equations

The set of repeating region equations $\epsilon^b \leq j \leq \epsilon^t$ have been transformed into a compact representation, and this allows us to use the solution to the underlying Matrix recurrence equation. Depending on the particular solution method used, it is either of the form (19.7) or (19.9). Both forms supply us with a set of $\epsilon^n = N \times (\hat{u} + \hat{d})$ free parameters. By construction, the localized balance equations for the repeating region are satisfied for any values of these parameters.

19.7.2 Boundary Equations

The free parameters supplied by the solution of the matrix recurrence now have to be constrained by a set of boundary conditions that link the explicit regions[4] with matrix representations pertaining to the repeating region. These boundary conditions are given by the boundary equations introduced in Section 19.6.4. The levels at which these are used are $0 \leq j \leq \epsilon^b + \hat{u} - 1$ and for a finite queue, $\epsilon^t - \hat{d} \leq j \leq L$. Localized balance equations at these levels then involve a mixture of explicit SOP variables $\mathbf{v}_j$, and matrix/eigensystem variables. Recall from Section 19.5.5 that $\epsilon^b \geq c$.

19.7.3 Accounting for Degenerate Streams

The equations gathered so far are algebraically equivalent to the ensemble of raw balance equations $\mathbf{r}_j$ for all levels $0 \leq j \leq L$ within the model, yet there is a significantly reduced number of them. This number needs to be further reduced should there be degenerate streams present. In our context, degenerate streams are either streams with batch parameter zero, or streams that are

---

[4] $0 \leq j \leq c - 1$ and $j = L$.

eliminated as a side-effect of another stream's elimination. This phenomenon occurs when there exist two streams running in the same direction with equal batch parameter. Every degenerate stream decreases the count ϵ^n of free parameters by one, causing the set of generated repeating region equations and boundary conditions to be over-specified. Solving such linear systems using floating-point values is not possible in general so we choose to drop a redundant equation from within the modulation state in which the degeneracy occurred. If the degenerate stream was in modulation state m of an arrival stream, we drop the (scalar) equation $(\mathbf{b}_{\epsilon^b+\hat{u}-1})_m$, for a negative customer or service stream in the same modulation state, the dropped equation is $(\mathbf{b}_{\epsilon^t+\hat{d}})_m$. For multiple degenerate streams, we successively drop adjacent equations in the same modulation state, e.g. $(\mathbf{b}_{\epsilon^b+\hat{u}-2})_m$ would be dropped if there were two degenerate arrival streams.

19.7.4 Normalization

After accounting for degenerate streams in the model and having dropped the required equations from the transformed ensemble, we still have a set of equations that are algebraically equivalent to the ensemble of raw balance equations. Just like these, they have a kernel of dimension 1. The resulting linear system can be made non-singular by replacing an arbitrarily chosen equation (which contains reference to explicit SOP variables) by the normalization constraint. For an infinite queue, this takes the form of either

$$\left(\sum_{j=0}^{\epsilon^b-1} \mathbf{v}_j + \sum_{i=1}^{\epsilon^n} \alpha_i \frac{\xi_i^{\epsilon^b}}{1-\xi_i} \boldsymbol{\psi}_i \right) \cdot \mathbf{e} = 1 \qquad (19.12)$$

for the spectral expansion method, or

$$\left(\sum_{j=0}^{\epsilon^b-1} \mathbf{v}_j + \mathbf{v}_{\epsilon^b}(I-R)^{-1} \right) \cdot \mathbf{e} = 1 \qquad (19.13)$$

for the Matrix Geometric Method. For a finite queue these expressions become

$$\left(\sum_{j=0}^{\epsilon^b-1} \mathbf{v}_j + \mathbf{v}_L + \sum_{i=1}^{\epsilon^n} \alpha_i \frac{\xi_i^{\epsilon^b} - \xi_i^{\epsilon^t+1}}{1-\xi_i} \boldsymbol{\psi}_i \right) \cdot \mathbf{e} = 1 \qquad (19.14)$$

and

$$\left(\sum_{j=0}^{\epsilon^b-1} \mathbf{v}_j + \mathbf{v}_L + \mathbf{v}_{\epsilon^b}(I - R^{\epsilon^t - \epsilon^b + 1})(I - R)^{-1} \right) \cdot \mathbf{e} = 1 \qquad (19.15)$$

respectively.

The resulting linear system can be solved by a variety of methods. We use LU decomposition in Mathematica® for convenience.

19.8 Multiple Negative Customer Streams

As we have explained, our scheme takes care of multiple streams of positive and negative customers and multiple service completion processes. To emphasize the versatility of application, we provide here an example balance equation automatically generated by our code for level 0 of a queue with a single arrival stream, a single service completion process, but crucially two independent streams of negative customers *of different killing modes*. As can be seen, equation (19.16) is enormous, despite our limiting the system to single positive arrival and processing completion streams. The queue is Markov modulated, so as always, we have matrices for the rates and batch distribution parameters. It has a single processor operating as (M, Φ), and a single positive arrival process (Λ, Θ). One of the negative customer streams operates as (K_1, Δ_1, h_p), i.e. in "head per" mode. The second operates as (K_2, Δ_2, t_s), i.e. in "tail safe" mode

Handling this huge expression manually would be a preposterous undertaking. Our automated methods are necessary to ensure that the results of the localization process can be trusted. Results of a manual process are subject to errors at every stage with each new queue analyzed. Our methods use a general expression for the unlocalized balance equations, and an elementary elimination expression. The correctness of our results requires only the correctness of the general expression, and the elementary elimination operator. This ensures that the localized balance equations for all queue lengths in finite or infinite queues with any number of processors and job transition processes will be correct.

$$\mathbf{v}_0 \cdot ((\Delta_2\Theta - I) * \Lambda * (\Delta_2 - \Phi)(\Theta * \Phi - I)(\Delta_1(\Delta_1 - \Delta_2 + \Delta_1\Delta_2\Theta) +$$
$$(\Delta_1 - \Delta_2)(\Delta_1\Theta - I)\Phi) +$$
$$aQ \cdot ((\Delta_2 - I)(\Delta_2 - \Phi)(\Phi - I)((-\Delta_2)\Phi -$$
$$\Delta_1^2(I + \Delta_2 + \Phi) + \Delta_1(\Delta_2 + \Phi + \Delta_2\Phi)))) +$$

$$\mathbf{v}_1 \cdot ((-\Delta_1 + \Delta_2)K_2(\Delta_1 - \Phi)(\Delta_2 - \Phi) +$$
$$(-\Delta_1 + \Delta_2)M(\Delta_1 - \Phi)(\Delta_2 - \Phi) +$$
$$(\Delta_1\Theta - I)\Lambda((-\Delta_1)\Delta_2^3 + \Delta_2^3(I + \Delta_1\Theta)\Phi - \Delta_2^3\Theta\Phi2 +$$
$$(-\Delta_1 + \Delta_2)(\Delta_2\Theta - I)\Phi3) +$$
$$aQ \cdot ((I - \Delta_1)((-\Delta_1)\Delta_2^3 + (I + \Delta_1)\Delta_2^3\Phi - \Delta_2^3\Phi2 +$$
$$(\Delta_2 - I)(-\Delta_1 + \Delta_2)\Phi3))) +$$

$$\mathbf{v}_2 \cdot ((-(\Delta_2 - \Phi))((-\Delta_2)\Phi((K_1 + K_2)\Phi + \Delta_2(K_1 + M + \Lambda\Phi)) +$$
$$\Delta_1(\Delta_2 K_1\Phi + (K_1 + K_2)\Phi2 +$$
$$\Delta_2^2(K_1 + M + \Theta\Lambda\Phi2)) + \Delta_1^2(\Delta_2^2\Lambda(1 - \Theta\Phi) + \Phi(M + \Lambda\Phi) +$$
$$\Delta_2(K_2 + \Lambda\Phi(1 - \Theta\Phi)))) +$$
$$aQ \cdot (\Delta_1^2\Delta_2^3 - \Delta_1^2\Delta_2^3\Phi + (\Delta_1 - I)\Delta_2^3\Phi2 +$$
$$(\Delta_1 - \Delta_2)(\Delta_1(\Delta_2 - I) - \Delta_2)\Phi3)) +$$

$$\mathbf{v}_3 \cdot ((\Delta_2 - \Phi)((-\Delta_2)\Phi + \Delta_1(\Delta_2 + \Phi))(\Delta_2 K_1\Phi +$$
$$\Delta_1(\Delta_2 M + (K_2 + \Delta_2\Lambda)\Phi)) +$$
$$aQ \cdot ((-\Delta_1)\Delta_2(\Delta_2 - \Phi)\Phi((-\Delta_2)\Phi + \Delta_1(\Delta_2 + \Phi))))$$
$$(19.16)$$

Expression (19.16) provides the left-hand side of the localized balance equation pertaining to queue length 0. For this particular queue, this is the only queue length at which the balance equation differs from that which would be found in a queue with three independent service streams with the same rates as this queue's service and two negative customer streams.

These algebraic representations of the balance equations serve to illustrate the result of application of some of the basic principles, but are not of use in practice. The necessity for isolation of matched or null streams has been explained in (19.5.3), which obviates reliance on algebraic matrix equations in application to practical situations. Our implementation work has two aspects. The algebraic mathematica prototype allows us to test our expressions, and a numerical version, following an identical fundamental procedure, but with the addition of the degeneracy consideration, is used in practical application.

19.9 Numerical Examples

We show some queue equilibrium solutions as plots of state occupation probability against modulation state and queue length. We begin by illustrating basic results of geometric job batching with simple examples.

19.9.1 Batched Arrivals Increase Mean Queue Length

We compare the mean queue length of a unit arrival process against batched arrival processes with the same mean rate. For simplicity, this example is not modulated. The mean rate $\bar{\lambda}$ of a compound Poisson process of rate λ and batch size distribution parameter θ is calculated as follows:

$$\bar{\lambda} = \sum_{s=1}^{\infty} s\lambda(1-\theta)\theta^{s-1} = \lambda(1-\theta)\sum_{s=1}^{\infty} s\theta^{s-1} = \frac{\lambda}{(1-\theta)}$$

So for our purposes, we can vary θ and produce a λ that provides any chosen equivalent mean arrival rate.

Here are some example results from our implementation. We consider an infinite single-server queue with service rate $\mu = 1.0$ and an arrival process with mean rate $\bar{\lambda} = 0.9$, a variable contribution being attributed to geometrically distributed batches, with parameter θ as above. We vary θ between 0 and 0.9 in steps of 0.1 and look at the resulting mean queue lengths. Figure 19.3 shows the output of our prototype tool with a single modulation state with these parameters.[5]

---

[5] These agree with the results of using the expression for mean queue length of the M/M/1 queue [14].

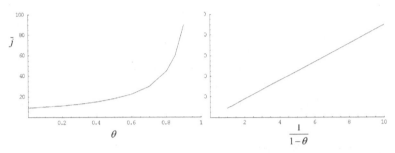

Figure 19.3 Mean queue length $\bar{j}$ scales linearly with $1/(1 - \theta)$ for constant $\bar{\lambda}$.

19.9.2 Fast Modulation

Consider a two state modulation process with the following Q matrix:

$$Q = \begin{pmatrix} -1 & 1 \\ 1 & -1 \end{pmatrix} k$$

Varying k changes the modulation rate leaving $\pi = (1/2, 1/2)$. Queue A has positive arrivals $(1/2, 1)$, processing rates $(3/4, 1)$ and a capacity of 20. Queue B is identical to A, except the arrival rates are swapped to the other modulation states, so its arrival rates are $(1, 1/2)$. Both queues are modulated by a Markov chain with transition matrix Q as defined above. Figure 19.4 shows the queue length distributions when we solve the two queues for $k = 1/100, 1$ and 10. This example shows two modulated queues with the same arrival and processing rate components, but allocated to different modulation states, converging to a common queue length distribution as the modulation rate increases.[6] If we consider a single arrival stream of positive customer batches, modulated by Qk, with rate parameters λ_i and batch parameters θ_i, we suggest that as $k \to \infty$, the process converges to inter-arrival delays with probability density function $\bar{\lambda}e^{-\bar{\lambda}t}$, where $\bar{\lambda} = \|\pi.\Lambda\|_1$ and the probability of batch size s is $\sum_{i=1}^{N} \pi_i \lambda_i (1 - \theta_i)\theta_i^{s-1} / \sum_{i=1}^{N} \pi_i \lambda_i$.

19.9.3 Multiple Unreliable Processors

Figure 19.5 shows the marginal queue length probabilities in a queue with 4 processors with breakdowns and repairs. The modulated system is described

---

[6] This particular comparison works when the swapped rates λ are allocated to modulation states with the same equillibrium probability.

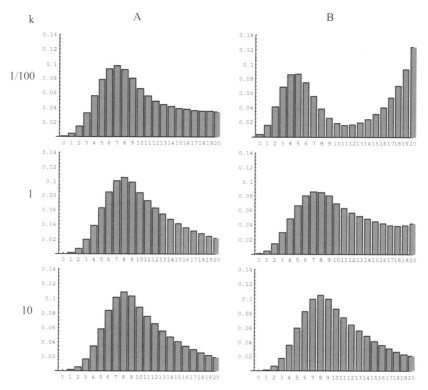

Figure 19.4 Graphs of queue length probability against queue length for queues A and B (which have arrival rates switched between modulation states) for a range of values of k. This illustrates the convergence to a limiting behaviour as $k \to \infty$.

as follows:

$$Q = \begin{pmatrix} -\Sigma \ldots & Nr & 0 & \cdots & 0 \\ b & -\Sigma \ldots & (N-1)r & \ddots & \vdots \\ 0 & 2b & -\Sigma \ldots & \ddots & 0 \\ \vdots & \ddots & \ddots & \ddots & r \\ 0 & \cdots & 0 & Nb & -\Sigma \ldots \end{pmatrix}$$

Here, $b = 3/100$ is the breakdown rate for individual processors and $r = 1/10$ is similarly the individual node repair rates. The queue has a capacity of $L = 40$.

We model 4 processors, so we have $N = 5$ modulation states. Each processor processes batches of parameter $1/2$ at rate $1/5$.

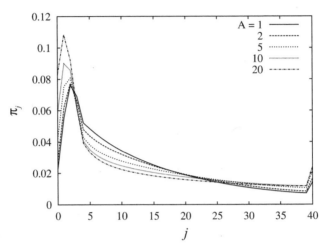

Figure 19.5 Marginal queue length distributions for a finite 4 processor breakdown and repair queue with A arrival streams of distinct batch size distribution.

The marginal queue length probabilities $\pi_j = \|\mathbf{v}_j\|_1$ are shown for distinct types of arrival traffic. This traffic has been chosen to have mean rate 1, but consists of a varying number of superimposed streams with distinct batch parameters.

For a given parameter A, we construct A arrival streams with rate parameter λ_k and batch parameter ϕ_k common to all modulation states, where $\phi_k = k/A + 1$ and $\lambda = \lambda_1 = \ldots = \lambda_k$ chosen such that the mean rate remains 1. This choice also causes the mean batch size to be 2 for all A. Other methods achieve traffic matching for the output of geometrically batched queues [8] by using aggregation to a single stream with the mean batch distribution. This provides sufficient accuracy for bounds on network performance to be calculated. By enabling exact representation of the individual batch behaviour of the separate traffic components, we improve the match of network traffic and queue length at equilibrium when matched against simulation. Correlation of traffic in the network worsens the match both for the aggregation and independent stream methods, but the latter fares better [12].

19.9.4 Individual and Global Processor Array Breakdown and Repair

Figure 19.6 illustrates an infinite queue where in addition to breakdowns and repairs of individual processors, there is the facility for the simultaneous breakdown and repair of all processors. The modulation generator is as follows:

$$
Q = \begin{pmatrix}
-\Sigma\ldots & Nr & & & r_N \\
b_0 + b & -\Sigma\ldots & (N-1)r & & r_N \\
b_0 & 2b & -\Sigma\ldots & \ddots & \vdots \\
\vdots & & & \ddots & \ddots & r + r_N \\
b_0 & & & & Nb & -\Sigma\ldots
\end{pmatrix}
$$

Parameters b and r are as before, with b_0 and r_0 the catastrophic failure and global repair rates respectively. The modulation structure for this example is taken from [1], and the same parameters $b_0 = b = 0.05$ and $r_0 = r = 0.1$ are used. The model incorporates $N = 20$ modulation states that track the activity states of 19 homogeneous processors.

There are three arrival and one service streams that are common to all modulation states. Arrivals occur with Poisson rates $1/3, 1/4, 3/14$ and batch parameters $1/3, 1/5, 1/7$, services have rate $1/20$ and batch parameter $1/2$.

The repeating region for this queue structure has 60 eigenvalues within the unit disc. Generating the balance equations susceptible to spectral expansion or matrix geometric methods with the approach used in [4] would involve an enormous investment for each new queue. Using our new representation and localization scheme, we simply set the switches appropriately in the general balance equation, apply the localization function, and solve the linear system.

Figure 19.6 shows the joint queue length/modulation state probabilities in a 19 processor batched infinite queue with 3 superimposed arrival streams and both individual and global breakdowns and repairs as in [1].

19.10 Conclusions

Early results of using these geometrically batched queues in networks appear in [10, 12]. In general, we find that models based on geometrically batched traffic give lower bounds on performance parameters. This agrees in general with conclusions drawn by Kouvatsos et al. (e.g. [8]), and we find that our provision of multiple independent traffic to queues allows a closer match in some circumstances. It should be noted that we do not currently model blocking. We are investigating the use of principles motivated by entropy maximization to enhance our parameter fitting procedure while avoiding unrealistic extrapolation.

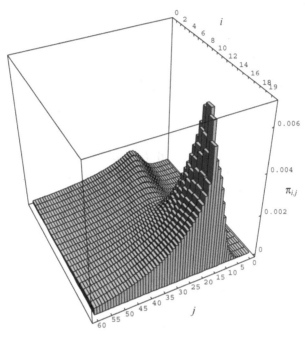

Figure 19.6 Queue length distribution for a 19 processor batched infinite queue with 3 superimposed arrival streams and both individual and global breakdowns and repairs as in [1].

Markov arrival processes, in which a job transaction may simultaneously accompany a change in modulation state, are straightforwardly implemented using off-diagonal rate matrices in a standard solution method for Markov modulated queues [16]. We also find that this can be incorporated into the formulation and solution regime we provide here for geometrically batched job transactions. The reason this is possible, is that the matrices of batch size distribution parameters are still diagonal, and hence susceptible to the same treatment as the unsynchronized case. The problem of formulating these localized balance equations is entirely separate to that of solving them. An example of their use can be found in [12], wherein queues with multiple arrival processes are formulated. Development of accompanying sojourn time calculations is on-going [6, 7].

Our methods facilitate routine implementation of new queueing paradigms based on geometrically batched processes. This is achieved by appropriate, simple modifications to the raw balance equations. Repeating regions, which map onto an eigensystem and can be solved using established methods, are revealed when the localized equations in a contiguous region are of the form $\sum_{i=-u_k}^{d_k} \mathbf{v}_{j+i} \cdot Q_{i+u_k} = \mathbf{0}$ for all levels j, $\epsilon^b_k + u_k \leq j \leq \epsilon^t_k - u_k$. Our approach facilitates the use of a large class of queues in networks with a highly expressive traffic definition. The formulation method is efficient, reliable and easily extensible. Efficient methods, based on either of the matrix geometric or spectral expansion approaches, provide solutions for individual queues that can be deployed in network models with application to internet and communications network modelling. The power of geometrically batched traffic has been amply demonstrated in work by Kouvatsos et al., and this latest contribution broadens the class of queue and traffic models amenable to practical application. Further detail of the analysis of the spectral expansion and matrix geometric solutions to these queues can be found in [18].

References

[1] I. Mitrani and R. Chakka, Spectral expansion solution for a class of Markov models: Application and comparison with the matrix-geometric method, *Performance Evaluation*, vol. 23, pp. 241–260, 1995.

[2] E. Gelenbe, Product form queueing networks with negative and positive customers, *Journal of Applied Probability*, vol. 28, pp. 656–663, 1991.

[3] D. A. Bini, G. Latouche and B. Meini, Solving matrix polynomial equations arising in queueing problems, *Linear Algebra and Its Applications*, vol. 340, pp. 225–244, 2002.

[4] R. Chakka and P. G. Harrison, A Markov modulated multi-server queue with negative customers – The MM CPP/GE/c/L G-queue. *Acta Informatica*, vol. 37, nos. 11–12, pp. 881–919, 2001.

[5] S. Wolfram, *The Mathematica Book*, 4th ed., Wolfram Media/Cambridge University Press, 1999.

[6] P. G. Harrison and H. Zatschler, Sojourn time distributions in modulated G-queues with batch processing, in *Proceedings of 1st International Conference on Quantitative Evaluation of Systems (QEST) 2004*, University of Twente, August 2004.

[7] P. G. Harrison, The MM CPP/GE/c G-Queue: Sojourn time distribution, *Queueing Systems*, vol. 41, pp. 271–298, January 2002.

[8] D. Kouvatsos and I. Awan, Entropy maximisation and open queueing networks with priorities and blocking, *Performance Evaluation*, vol. 51, pp. 191–227, 2003.

[9] E. Gelenbe (Ed.), *European Journal of Operational Research*, vol. 126, no. 2, October 2000.

[10] D. J. Thornley and H. Zatschler, Analysis and enhancement of network solutions using geometrically batched traffic, in *Proceedings of 19th Annual UK Performance Engineering Workshop*, Warwick, 2003.

[11] R. Chakka, T. V. Do and Z. Pandi, Generalised Markovian queues and application to performance analysis in telecommunications networks, in *Proceedings of the First International Working Conference on Performance Modelling and Evaluation of Heterogeneous Networks (HET-NETs 2003)*, 2003.

[12] P. G. Harrison, D. J. Thornley and H. Zatschler, Geometrically batched networks, in *Proceedings (ISCIS 17) Seventeenth International Symposium on Computer and Information Sciences*, University of Central Florida Orlando, Florida, October 28–30, 2002.

[13] M. Neuts, *Matrix-Geometric Solutions in Stochastic Models: An Algorithmic Approach*, Dover Publications, 1995.

[14] P. G. Harrison and N. M. Patel, *Performance Modelling of Communication Networks and Computer Architectures*, Addison-Wesley, 1993.

[15] D. A. Bini, G. Latouche and B. Meini, Solving matrix polynomial equation arising in queueing problems, *Linear Algebra and Its Applications*, vol. 340, pp. 225–244, 2002.

[16] D. J. Thornley, Synchronized negative customers in an unreliable server queue, in *Proceedings of the First International Working Conference on Performance Modelling and Evaluation of Heterogeneous Networks (HET-NETs 2003)*, 2003.

[17] D. J. Thornley, H. Zatschler and P. G. Harrison, An automated formulation of queues with multiple geometric batch processes, in *Proceedings of the First International Working Conference on Performance Modelling and Evaluation of Heterogeneous Networks (HET-NETs 2003)*, 2003.

[18] D. J. Thornley and H. Zatschler, Exploring accuracy and correctness in solution to matrix polynomial equations in queues, in *Proceedings of the Third International Conference on Quantitative Evaluation of Systems (QUEST 2006)*, August 2006.

[19] D. Kouvatsos, Entropy maximisation and queueing network models, *Annals of Operations Research*, vol. 48, pp. 63–126, 1994.

[20] P. G. Harrison and R. Chakka, The MMCPP/GE/c queue as a node model for ATM networks, in *Proceedings of the 12th UK Performance Engineering Workshop*, Edinburgh, 1996.

[21] I. Mitrani and R. Chakka, Spectral expansion solution for a class of Markov models Application and comparison with the matrix-geometric method, *Performance Evaluation*, vol. 23, no. 3, pp. 241–260, 1995.

[22] P. G. Harrison and R. Chakka, The Markov modulated CPP/GE/c/L queue with positive and negative customers, in *Proceedings of the 7th International Conference on Performance Modeling and Evaluation of ATM Networks*, Antwerp, 1999.

[23] R. Chakka and P. G. Harrison, The MMCPP/GE/c queue, *Queueing Systems*, vol. 38, no. 3, pp. 307–326, 2001.

[24] R. Chakka, T. V. Do and Z. Pándi, Exact solution for the MM-CPPk/GE/c/L G-queue and its application to the performance analysis of an optical packet switching multiplexer, in *Proceedings of the 10th International Conference on Analytical and Stochastic Modelling Techniques and Applications*, Nottingham, UK, June 2003.

[25] T. V. Do, R. Chakka and Z. Pándi, Novel analysis method for optical packet switching nodes, in *Proceedings of the ONDM 2003 Conference*, Budapest, Hungary, February 2003.

[26] D. Thornley and H. Zatschler, Analysis and enhancement of network solutions using geometrically batched traffic, in *Proceedings of the 19th UK Performance Engineering Workshop*, Warwick, 2003.

[27] T. V. Do, R. Chakka, O. Gemikonakli and D. Papp, Level dependent band-QBD processes: Steady state solution and applications, Working Paper, Second International Working Conference on Performance Modelling and Evaluation of Heterogeneous

Networks, Craiglands Hotel, Ilkley, West Yorkshire, UK, HETNETs'04, 2004.

[28] D. J. Thornley, Construction of novel continuous time Markovian queues for exact solution, in *Proceedings of the 24th UK Performance Engineering Workshop (UKPEW 2008)*, London, 2008.

[29] R. Chakka and T. V. Do, The MM Sigma k=1..K CPPk/GE/c/L G-queue with heterogeneous servers: Steady state solution and an application to performance evaluation, *Performance Evaluation*, vol. 64, no. 3, pp. 191–209, March 2007.

20

Collaborative Environment for Tool Sharing in the Framework of Euro-NGI Network of Excellence

A. E. Garcia and K. D. Hackbarth

Communication Engineering Department, University of Cantabria, 39005 Santander, Spain; e-mail: {agarcia, klaus}@tlmat.unican.es

Abstract

The objective of the Network of Excellence Euro-NGI is the design and engineering of the Next Generation Internet, NGI. This objective covers a large and complex set of issues, from technological aspects, such as the logical and physical layer design, QoS management, traffic engineering, network security and link restoration and so on, up to social and economical issues like quality of service from the users' point of view, network pricing and costing and service tariff issues. Furthermore all the subjects previously mentioned have to be studied in a large set of heterogeneous networks, xDSL, DWDM, UMTS, WLAN, all of them based on the IP protocol. In order to study all these issues, some of the different research labs participating in the Euro-NGI have already developed, and are currently developing, a large set of software tools oriented mainly to two work lines: tools for network planning and simulation; and tools for performance measurement, statistical analysis and test beds. A specific objective of the Network is to develop a sort of Macro-Tool which provides a homogeneous environment for the software tools developed by the research labs allowing for their interrelations. The development of such a kind of environment attaches several coordination and integration problems. Therefore the optimal solution is the integration of all tools under a common user access interface, one of the best options

D. D. Kouvatsos (ed.), Traffic and Performance Engineering for Heterogeneous Networks, 459–476.

being the web interface. The Telematic Engineering Group, G.I.T., of the University of Cantabria is developing a web portal named Euro-NGI Planning Tools Portal which implements this concept. Therefore this article exposes the specification, design and current state of the implementation of the portal, explaining several innovative aspects in the field of remote software tool execution, secure access and customized interfaces and sharing environments for simultaneous execution.

Keywords: Software sharing environment, remote execution, network planning tools.

20.1 Introduction

The objective of the Network of Excellence Euro-NGI is the design and engineering of the Next Generation Internet, NGI. This objective covers a large and complex set of issues, from technological aspects, as the logical and physical layer design, QoS management, traffic engineering, network security and link restoration and so on, up to social and economical issues like quality of service from the user's point of view, network pricing and costing and service tariff issues.

Furthermore, all the subjects previously mentioned have to be studied in a large set of heterogeneous networks, xDSL, DWDM, UMTS and WLAN, all of them based on the IP protocol.

In order to study all these issues, some of the different research labs participating in the Euro-NGI have already developed and are currently developing, a large set of software tools oriented mainly to two work lines, as Figure 20.1 shows:

- Tools for network planning and simulation
- Tools for performance measurement, statistical analysis and test beds

Following the previous model, we identify multiple interrelations among different work-packages, where Figure 20.2 shows the most important relationships which are directly related with the network planning process.

A specific objective of the Network is to develop a sort of Macro-Tool, which provides a homogeneous environment for the software tools developed by the research labs, allowing their interrelations. The development of such a kind of environment attaches several coordination and integration problems. Therefore the optimal solution is the integration of all tools under a common user access interface, one of the best options is the web interface.

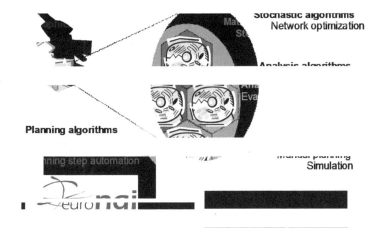

Figure 20.1 Software tools: Euro-NGI developing lines.

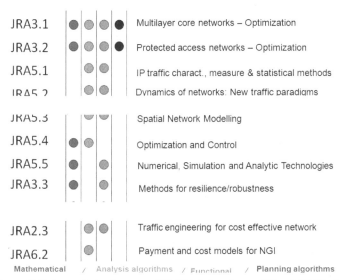

Figure 20.2 Interrelation between WP JRA 3.4 and rest of EuroNGI Workpackages.

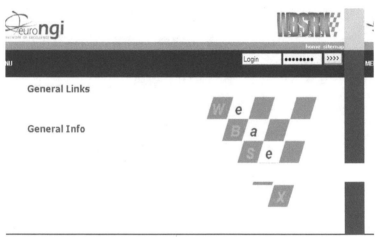

Figure 20.3 WeBaSeReX home page.

The Telematic Engineering Group, G.I.T., of the University of Cantabria developed a web portal named WeBaSeReX (standing for Web Based Service for Remote Execution – http://www.tlmat.unican.es/eurongi) showed in Figure 20.3, which implements this concept. Therefore, this article exposes the specification, design and current state of the implementation of the portal, explaining several innovative aspects in the field of remote software tool execution, secure access, customized interfaces, and sharing environments for simultaneous execution.

20.2 Portal Specification

The *Euro-NGI Planning Tools* web-based portal, or WeBaSeReX, is mainly used for information exchanges and communication between partners, mostly in the field of planning tools. Furthermore it provides the access to the corresponding tools developed or provided inside the Euro-NGI project.

These tools can be simulators, NS or Opnet modules, performance modelling tools, etc. Using WeBaSeReX, these tools can be executed before installing them on the local computer, as a demo of their main features. It can be also noted that WeBaSeReX allows single-user tools to be executed as multi-user ones.

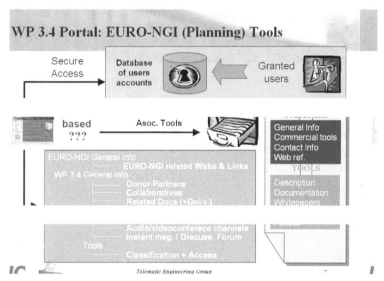

Figure 20.4 General structure of EURO-NGI Planning Tools Web.

20.2.1 Portal Sections

The access to this specific Portal is always performed from the general Euro-NGI Web site, which is already implemented. This fact assures a first control provided by this Web site. The main structure is shown in Figure 20.4.

From the contents point of view, the portal consists of the following sections:

- General information. This is a public section with special information about the Euro-NGI project, the specific work package that is developing the portal and the partners involved in it. Furthermore documents and related info are placed in this section.
- Communications. This section is accessible only for the specific work package members. It provides the following services:

  - Documental service. Web or FTP based, for making easy exchanges of documents between project members.
  - Audio-conference or video-conference channels for direct communication between project members.
  - Instant messaging, discussion forums, etc.

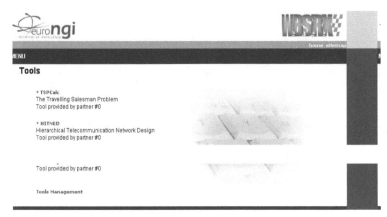

Figure 20.5 Tools section in the portal.

- Tools. This section stores the tools provided by project members. In a first step general info about shared tools (and other tools) and software packages appears. A second part is accessed only by a certain group of users with such privileges. There is a description of each tool, along with usage guide, the method used for sharing it, etc. (see Figure 20.5).

20.2.2 Portal Components

In the development of this Portal, several software aspects must be taken into account. The following platforms, solutions and technologies are proposed for this collaborative server:

- It is a web-based service, so a web server is required. Depending on the chosen platform and operative system, Apache or Microsoft IIS can be used.
- The portal has a web-based interface. It was designed using HTML and CSS (style sheets), but it can be improved adding other functionalities, such as XML content description, see [1], and PHP scripts, see [2] or Java applets [3] for further interactivity and security issues.
- Users must be identified for accessing some sections. User accounts are stored in the corresponding database. Open-source, multi-platform MySQL database server [4] is proposed for this task.

- Another database is used for storing information about partners and tools, so it can be easily queried and displayed. This information completes the general partners profiles defined by Euro-NGI.
- A secure access to contents is developed so only authenticated users from main Euro-NGI website can access private sections in the portal. A Java applet/servlet or PHP scripts might be used for adding the security functions to the HTML code.
- Portal includes a Search utility, taking advantage of the users-, partners-, and tools-database.
- Finally, the server implements a Management Control Panel for the portal setup: users, contents, available tools etc... can be managed (add, edit, delete) from this section.

20.2.3 Portal Users

Users in the *Euro-NGI Tools Portal* can be grouped by permissions level, ranging from users only able to access public sections, to users able to manage the Portal. Depending on the permissions established in the corresponding user account the user is granted to access different sections. There are three different types of users.

- Level 0: Users that are only allowed to access the public sections.
- Level 1: Users allowed to access the private section and to perform the corresponding operations with the planning tools. This level is subdivided in different groups to allow the customized access to the planning tools.
- Level 2: This level corresponds to the portal managers with total access and control of the whole portal.

Figure 20.6 shows the access scheme to the *Euro-NGI Portal*. More technical details about the access control are explained in Section 20.3.3.

20.3 Innovations

The *Euro-NGI Planning Tools Portal* has three main innovative lines which are enumerated below:

- Methods for tool sharing.
- Definition and implementation of sharing environments.
- Security and customization issues.

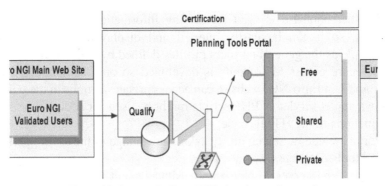

Figure 20.6 Users in Euro-NGI planning tools portal.

The underlying technology required to implement these three topics is already developed. It is commonly used, in a simplified form, for electronic commerce on Internet shops like e-bay® or Amazon®, and for electronic banking (the Bank BSCH.SA, www.gruposantander.es). However the application of this kind of shared environments in R&D activities is very limited. Therefore the Network of Excellence Euro-NGI constitutes the ideal framework to allow the exchange of knowledge through the shared use of the software tools. The next sections explain in more detail the three innovative lines previously introduced.

20.3.1 Methods for Tool Sharing

The fact of sharing a specific tool can be a conflictive issue by several reasons. First, the research labs may have signed privacy and/or commercial agreements with industrial companies, so they are not allowed to provide a freely use of the software tool. Second, there are some kind of competition between different research labs that work on the same topic, hence, any possible misunderstanding owning to software and knowledge sharing must be avoided. Third the different labs may work under several operative systems and with different program languages. Therefore it is a restrictive requirement to provide the corresponding interfaces to supply the integration of the different tools and algorithms.

Obviously, there is a broad spectrum of troubles. To overshot them the *Euro-NGI Portal* has to provide a broad spectrum of possibilities for tool and knowledge sharing. Figure 20.7 shows these chances which are explained below:

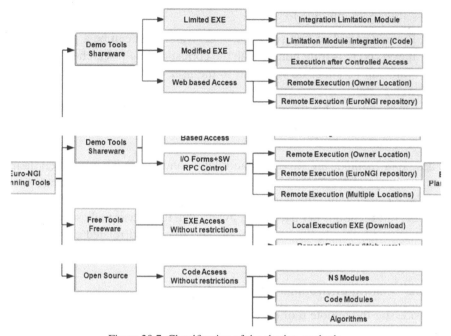

Figure 20.7 Classification of the sharing methods.

- Provide a demo or Shareware version of the tool: This is the optimal solution for the research labs which work jointly with some commercial company. They can provide either a limited version of the tool, as an example, with only a unique working scenario, or a time limited version which expires after a specific number of days or executions.
- Integration in the Web Portal, Web-ware. From the point of view of the authors this is the best option in order to interchange knowledge in the framework of the Euro-NGI. The integration of a specific tool into the web portal can be performed in two main ways:
  - Integrated software. In this option the input and output is performed using the corresponding web forms and the tool execution is performed in the machine hosting the portal.
  - Remote Execution. Again the data is introduced using the corresponding web forms. The software lies either in the Euro-NGI software tool

repository,[1] in a server of the research lab that provides the tool, or finally in multiple locations.

- Provide complete free software, Free-Ware: This can be a good option if everybody is truly honest with the corresponding researcher. The free-ware can be either downloaded to the user machine or executed in the web Portal.
- Open source. This means the complete access to the code of the tool or algorithm without any restriction, so everybody can use the code in their own software tools.

These four methods are strongly interrelated, and the division between them is not always clear. As an example, a shareware version of a software tool can be integrated into the portal with some restrictions about the maximum number of executions per user. It is obvious that the access to these tools have to be allowed only to the members of the project and related entities. The required security and customization aspects are exposed in Section 20.3.3.

20.3.2 Sharing Environments

Considering the remote tool execution, up to now we have assumed that the tools are only in a single and specific location. This place can be either a server in the same research lab that has developed the tool or the Euro-NGI repository. This means that in most cases only a single user can access and execute the corresponding tool. The other possibility is to install the application on several servers. In that case some mechanism is required to distribute the request of remote execution among the different servers. The set of the distribution mechanism, the servers where the application is installed and the methods for secure and Customization access conforms the concept of a *sharing environment.*

Most components of the *sharing environment* have been previously explained in the article. The new elements are the execution and communication servers. The communication server receives the connection request of the client. Then it establishes the secured and customized working environment with the client, using the information stored in the corresponding database. For this purpose it might use the *fingerprints* methodology explained in [5]. From this point the requirements about user-machine authentication and information confidentiality and integrity are fulfilled. After establishing this

---

[1] The Euro-NGI software tool repository is a single or set of application servers. It is a different entity of WEBASEREX, although they can be in the same machine.

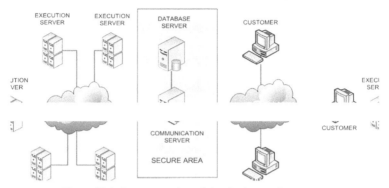

Figure 20.8 Representation of the sharing environments.

secure environment the client asks for the remote execution of a specific planning tool between the set of available tools. Then the execution server, using the information available on the database, establishes which one of the remote execution servers is going to provide the specific service to the client. For this purpose the database has to store information about the number, locations and specific facilities of each server. Furthermore the communication server has to feed the database with information about the usage level of each execution server to avoid possible congestions or service denials.

The other new element is the execution server. This equipment performs the service provision to the final client playing the same role as the Euro-NGI repository for the case of a single remote execution machine. However the execution server has an additional function. It has to inform about its state to the communication server, which stores this information into the database. That means that the communication server needs to know which execution servers are busy at a specific time moment to avoid a new client assignation to those servers. This type of execution server farm will allow the complete system to serve more users with higher quality of service. Note that because the communication server drives the different client request to different execution servers, the required bandwidth in each one of them is reduced.

The complete system with the different network elements is summarized in Figure 20.8. With this network configuration, the typical access process follows six main steps, as Figure 20.9 shows:

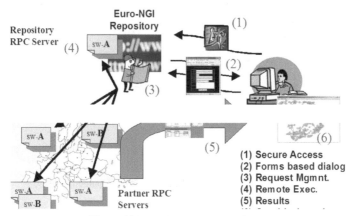

Figure 20.9 Basic remote execution process.

- Secure Access: This step grants to all EuroNGI users a predefined access profile with advanced security certification. Identification and validation are the main associated operations.
- Forms based dialog: Accordingly to the user profile, the access to the Portal contents follows a typical web dialog, using web forms to introduce the input parameters (individual values / files) to the stored and shared tools.
- Request Management: When a shared tool offers its execution on remote servers, the communication server connects to executable copies situated into remote locations (Execution servers). Communication server selects the best execution server (unloaded server) between the lists of candidates.
- Remote Execution: Assigned Execution server executes the program/algorithm using corresponding set of input parameters (from step 2).
- Results: Execution server sends the resulting files/values to the Communication Server which generates a response web form.
- Graphical results: Depending of results type, analysis could be made sending BMP/GIF generated by program, or using a graphical plug-in.

20.3.3 Security and Customization Issues

As it was commented before, there are three main entities (client, communication server including main Portal's database server and execution server). It is desirable to isolate the data warehouse most of the times from the rest of systems. As a consequence we consider two communication channels: client-Portal and communication server-execution server. The first channel represents the common access to the Portal. The second channel is an exclusive method for active sharing of programs and contents, including remote executions.

The first communication channel represents the interaction in the web browser between the customer and the Portal contents. The main Euro NGI Web server grants the access to the Tools Portal remotely.

The *Tools Portal* is a dedicated server integrated into the Euro-NGI main server. The unique entrance gate is the main server of Euro-NGI, www.eurongi.org, where the user is authenticated obtaining an active session. The session and the session identifier, ID, is maintained at www.eurongi.org, providing links to the dedicated server. When a user moves towards the dedicated server, the main server handovers the user to the dedicated server. This process is performed using the following procedure. The main server sends a HTTP Post message with the Euro-NGI user ID, the main server session ID, and unique hash for security to the dedicated server. The dedicated server handles the user and it creates a new local session. Then this local session is defined by two dedicated Java Applet and Java Server entities (client program and server program) establishing a private dialogue based on encrypted mechanisms (DES), see [6]. The main server hash is used as encryption key.

In the second channel the communication take place between Communication Server and execution servers. Starting from this point the communication between both servers is encrypted and the channel is secure. Although the validation key is transmitted by a non secure channel, the complete authentication is only obtained through the establishment of the encrypted communication. This channel uses SSH to guarantee a secure channel taking advantage of the facilities provided by Java. Note that it is totally transparent to the user.

The application of these two methods assures a complete confidentiality of the communicated data because the couple of safe channels are relatively immune to possible external attacks. Figure 20.10 shows the complexity of the procedure for the creation of secure channels.

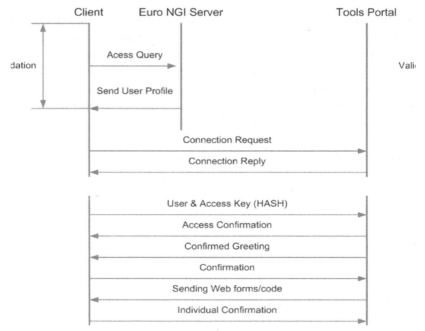

Figure 20.10 Secure channels establishment.

Once the user is validated, he proceeds by sending an encrypted greeting using the encryption key which is known by both server and client (session hash). If the greeting is decrypted correctly, authentication of the client is completed. Otherwise the user is expelled and the communication concludes. From this point the communication between the client and the server is encrypted and the channel is secure. In the following step, Input/Output web forms, programs or software are encrypted and sent, and the server confirms each one of them, as Figure 20.11 shows.

There are fundamentally mechanisms for user authentication in a protected data exchange environment [6]:

- Certificates: A certificate guarantees the end user authenticity of the communication. A secure case consists when both the client and the server posses a corresponding certificate. A trust authority certifies the identity of the possessor of the certificate. This method is useful in the case of client accesses to a server.

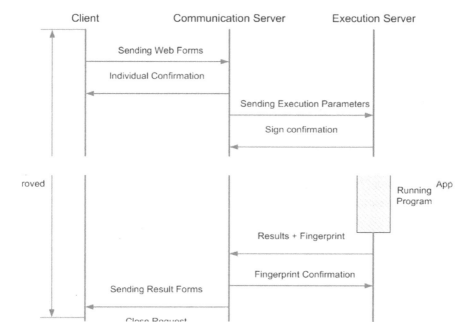

Figure 20.11 Secure execution.

- Digital signature: Digital signature assures the identity and integrity of the document. It encrypts the original message using a private key. The disadvantage of the digital signature is that it requires a considerable processing time with large files.
- Digital Fingerprints: Fingerprint is an alternative to the digital signature. The fingerprint encrypts an extract of the original message and hence a private key only encrypts the resulting "fingerprint". The receiver entity reads the signed message to check the authenticity of the document, by the generation of a summary with the public key comparing it with the summary that is included inside the message. The computational cost is relatively low.

While the user performs the execution or downloading, client and server do not exchange information. When the execution server concludes it sends the results. Each reception is confirmed and finally the *fingerprint* of the complete results is sent. By means of the exchange of keys the server checks the integrity and authenticity. If the confirmation is successful, the communication server proceeds by generating corresponding web pages with results of program execution.

PHP language offers some facilities for web sessions control. Standard PHP session management includes:

- Creation and initialization of sessions.
- Registration/un-registration of session variables.
- Session destruction.

With these functionalities, a basic session control can be achieved. But some problems appear when using it: users leaving the protected web page, and coming back later using the same browser window, won't need to authenticate again. This can be a security failure (e.g. using public terminals and leaving the browser window open at the end).

In order to solve these security problems, the Portal uses an improved session management and authentication methods. Their main features are:

- A session is created after authentication, and destroyed after leaving the web page. Further accesses, even in the same browser window, will require authenticating again. A PHP file is included in every protected page, checking if there is a valid and active session ID, and if the session variables (e.g. indicating if the user is authenticated, user name, and password) have the proper values.
- Passwords are never transmitted as plain text. The server generates a random number, and is send to the user. In the user terminal, using JavaScript code, this numeral is added to the password provided by the user. Then, the MD5 hash of it is calculated (again using JavaScript), and the result is sent back to the server. The server, finally, uses the generated number and PHP MD5 functions to check if the password is correct.

20.4 Conclusions and Future Work Lines

The article exposes the specification and first implementation of the web portal named Euro-NGI Planning Tool Portal. The objective of this research is to provide a homogeneous environment for the knowledge sharing between the different partners of the project. For this purpose multiple innovative techniques, in the field of Internet applications, have been used. The application of these methods to a cooperative research & development environment is also a very relevant issue. More details about the structure and implementation of WeBaSeReX appeared as publications in [7–9].

The most serious problem is the lack of security mainly when a public Internet access is used. This article presents a system and its implementation that allows carrying out remote exams with the security that nobody can break the confidentiality of the end point in the communication channel. This process is developed in using digital fingerprint in combination with encryption schemes implemented over common Internet programming languages. At the present time the first version of the portal is already implemented with the basic functionalities. Advance features like *fingerprints* with customized environments are currently under development and implementation.

The developed environment is applicable to diverse systems, for example, in susceptible communication channels of interception, as the air. Hence this proposed method can even extended to the new UMTS terminal (both Java-compatible), implementing it by means of the Java Micro Edition. As a consequence it might be able to integrate the tele-education into the mobile communication in a simple and secure way. Additionally, the software environment uses multi-platform tools (Java and XML) that allow exporting the application to any other environment.

Acknowledgements

The research exposed in this article has been developed under both the Network of Excellence Euro-NGI of the VI European Framework of the European Commission, IST-50/7613 and under the national research from the Spanish Ministry of Science and Technology TIC2003-05061.

References

[1] L. Argerich, K. Egervari, M. Anton, C. Lea, C. Lea, C. Killian, C. Hubbard and J. Fuller, *Professional PHP 4 XML*, Wrox, 2002.

[2] W. Choi, A. Kent, C. Lea, G. Prasad and C. Ullman, *Beginning PHP 4*, Wrox, 2000.

[3] W. Rockwell, *XML, XSLT, Java & JSP: A Case Study in Developing Web Application*, New Riders, 2001.

[4] M. Kofler, *MySQL*, APress, 2001.

[5] J. Álvarez, A. E. García, K.D. Hackbarth and R. Ortiz, Application of security protocols for tele-eduction environments: Virtual exams, in *Proceedings of IADAT – e2004, International Conference on Education*, July 7–9, Bilbao, Spain, 2004.

[6] S. McClure, J. Scambray and G. Kurtz, *Hacking Exposed: Network Security Secrets & Solutions*, McGraw-Hill Osborne Media, 1999.

[7] A. E. García, K. D. Hackbarth and R. Ortiz, Web-based service for remote execution: NGI network design application, in *Proceedings of 2005 NGI Conference*, 18–20 April 2005, IEEE Catalog Number 05EX998C, 2005.

[8] A. E. García, K. D. Hackbarth and R. Ortiz, WeBaSeReX: Next generation internet network design using web based remote execution environments, in *Proceedings of IEEE International Conference on Services Computing*, Orlando, USA, 11–15 July, pp. 200–209, 2005.

[9] A. E. García, K. D. Hackbarth and R. Ortiz, WeBaSeReX (Web based service for remote execution): Implementación de servicios de acceso seguro a aplicaciones compartidas, in *Proceedings TELECOM I+D 2005*, Madrid, 22–24 November 2005.

Author Index

Subject Index